Information at Sea

Johns Hopkins Studies in the History of Technology
Merritt Roe Smith, Series Editor

Information at Sea

Shipboard Command and Control in the U.S. Navy, from Mobile Bay to Okinawa

Timothy S. Wolters

The Johns Hopkins University Press
Baltimore

Printed in the United States of America on acid-free paper
9 8 7 6 5 4 3 2 1

The Johns Hopkins University Press
2715 North Charles Street
Baltimore, Maryland 21218-4363
www.press.jhu.edu

Library of Congress Cataloging-in-Publication Data
Wolters, Timothy S.
Information at sea : shipboard command and control in the U.S. Navy, from Mobile Bay to Okinawa / Timothy S. Wolters.
pages cm. — (Johns Hopkins studies in the history of technology)
Includes bibliographical references and index.
ISBN-13: 978-1-4214-1026-5 (hbk. : alk. paper)
ISBN-10: 1-4214-1026-5 (hbk. : alk. paper)
ISBN-13: 978-1-4214-1084-5 (electronic)
ISBN-10: 1-4214-1084-2 (electronic)
1. United States. Navy—Communication systems—History. 2. Command and control systems—United States—History. 3. Warships—United States—History. I. Title.
V283.W55 2013
359.3'3041097309041—dc23 2012048479

A catalog record for this book is available from the British Library.

The Johns Hopkins University Press uses environmentally friendly book materials, including recycled text paper that is composed of at least 30 percent post-consumer waste, whenever possible.

To Mom and Dad, my first teachers
and enduring inspirations to this day

Contents

Acknowledgments

This book has been a long yet rewarding journey, in no small part because of the remarkable scholars, teachers, friends, and family who have aided me along the way. In some ways the journey began years ago when my parents shared dinnertime stories about our family's history. Both of my grandfathers served in the U.S. Navy, one aboard a battleship in the 1930s, the other as a World War II draftee. Mom was born and raised in the historic navy town of Pensacola, and dad flew off carriers during the Vietnam War. No surprise that their son developed an interest in naval history!

The more proximate origins of this book lie in a conversation with Dave Mindell a little more than a decade ago. He pointed out that no one had ever really looked into the history of the CIC, and suggested that it might be a good topic for a book. Dave's suggestion echoed what Jon Sumida had often told me when I was a master's student at the University of Maryland. For far too long, he said, historians have neglected naval command and control. Into the archives I went, thinking my story would begin in the late 1930s with the U.S. Navy's development and adoption of radar. The more I worked in the archives, though, the farther back in time I had to go. The project evolved from one that focused on a specific artifact to one that explored an entire institution's efforts to develop and employ systems for managing information at sea.

Over the course of this project I have had the privilege of working with some exceptional mentors. Dave Mindell, Roe Smith, and David Rosenberg supervised my dissertation with just the right mix of criticism and encouragement. Roe, always the consummate professional, showed incredible patience as he worked with me toward publication in the Johns Hopkins Studies in the History of Technology series. Three others I also want to thank for their advice and assistance are Jon Sumida, Robert Friedel, and Len Rosenband. Jon and Robert were the teachers who really taught me to think like a historian, and Jon kindly shared some Royal Navy documents with me. Len devoted an unbelievable amount of time reading and critiquing my work, and the small note of thanks I offer here fails to do justice to the friendship and support he has provided.

Many others too have contributed in various ways to this book. Two anonymous referees gave insightful critiques of earlier drafts. John Walters read parts of the manuscript and was instrumental in tracking down government documents I never would have found without his help. David Boslaugh graciously sent me important material pertaining to the history of naval fighter direction. The late Steve Harwick helped me conduct research in the photographic division of what was then the Naval Historical Center, and I am saddened deeply that he passed away before this book came out. Fellow historians Denise Conover, Kate Epstein, Gerry Fitzgerald, Brendan Foley, Chris Havern, Ed Marolda, Randy Papadopoulos, Peter Shulman, John Sherwood, Tom Wildenberg, and Jonathan Winkler, all experts in closely related fields of inquiry, at one time or another listened patiently and offered advice as I worked through various ideas. Gwen Bingle, Kevin Borg, Vera Candiani, Larry Cohen, Maja Fjaestad, Maril Hazlett, Per Hogselius, Anders Houltz, Thomas Jehn, John Staudenmaier, Helen Watkins, Roz Williams, Matt Wisnioski, and Shana Worthen all made helpful suggestions on an early version of chapter 3 during a writing workshop sponsored by the Society for the History of Technology at the Woods Hole Oceanographic Institute. I also extend thanks to Jeff Barlow, Dave Lucsko, Mark Mandeles, Rob Martello, Larry McDonnell, Eden Medina, Jerry Miller, Ted Postol, Peter Swartz, Ed Wysocki, and Chen-Pang Yeang for offering thoughtful comments on various parts of this book.

While conducting research I had the chance to visit many archival repositories, where I met some very helpful archivists and librarians. I would especially like to thank Charles Johnson, Rebecca Livingston, Nate Patch, and Barry Zerby at the U.S. National Archives; Elizabeth Wilkinson at Purdue University's Archives and Special Collections, Craig Orr and Reuben Jackson at the National Museum of American History's Archives Center; Evelyn Cherpak at the Naval War College; Glenn Helm, Allen Knechtmann, and Davis Elliott at the Navy Department Library; Jennifer Bryan at the Naval Academy's Special Collections and Archives; and Ken Johnson, Kathy Lloyd, John Hodges, Curtis Utz, and John Walker at the U.S. Navy Operational Archives. I also thank Robert Hanshew of the Naval History and Heritage Command for helping me to locate many of the photographs reproduced in this book.

Research can be expensive, and I have benefited immensely from the generosity of many institutions. MIT extended me several teaching assistantships, the Dibner Institute provided two graduate student fellowships, the Naval History and Heritage Command awarded me the John D. Hayes dissertation fellowship, and the IEEE History Center granted me its Life Members' Fellowship in Electrical History. In 2003–4, I was honored to hold the Dewitt C. Ramsey Chair of Naval Aviation History at the Smithsonian's National Air and Space Museum. The museum gave me both funding and office space, but what I most enjoyed was the chance to collaborate with a great group of historians and museum specialists, including Dorothy Cochrane, Roger Conner, Tom Crouch, Phil Edwards, Peter Jakab, Jeremy Kinney, Roger Launius, Russ Lee, Michael Neufeld, Dom Pisano, Alex Spencer, and Bob van der Linden. Utah State University generously provided me a new faculty research grant that lasted from 2005 to 2008, and I would like to extend heartfelt thanks to my former colleagues from Logan, especially the history department junior faculty cohort: Alice Chapman, Chris Conte, Lawrence Culver, Victoria Grieve, Eric Kimball, Peter Mentzel, Colleen O'Neill, Jennifer Ritterhouse, Jamie Sanders, and Sue Shapiro. We would often discuss each other's work, and their influences are certainly present here.

Another group whose influence permeates this book are my navy shipmates. Working in the modern-day version of the institution one is studying certainly provides valuable perspective, but it also takes up considerable time. I have always found today's naval officers and sailors more than willing to discuss command and control issues, and during three reserve command tours I have been blessed with superior executive officers, senior enlisted leaders, and others who have made these tours both easier and more enjoyable.

Since relocating to Ames in 2010 I have received tremendous support and encouragement from my colleagues at Iowa State University. Department chairs Charlie Dobbs and Pam Riney-Kehrberg have been unwavering in their support, administrative assistant Jennifer Rivera has been incredibly helpful, and the College of Liberal Arts and Sciences generously awarded me a grant to conduct some final research on the Okinawa campaign. The Johns Hopkins University Press has been obliging during the entire publication process, with editor Bob Brugger showing steadfast patience throughout. Thanks also to Brian MacDonald, who copy edited the final manuscript.

Lastly I would like to thank my amazing wife, Karen, and our two wonderful children, Caroline and Jason. When we married, Karen probably never imagined she would help conduct research in the National Archives, proofread a never-ending stream of drafts, or listen to stories about coherers and strip ciphers over dinner. All this she did and more. No one could be a greater inspiration, and this book would never have seen the light of day if not for her love, support, and encouragement. Caroline and Jason in their own ways too have made this book possible, sometimes by giving daddy much needed quiet time but a great deal more so just by being themselves. For their sacrifices and enduring love, I am eternally grateful.

Ames, Iowa
March 2013

Information at Sea

Introduction

> It is the duty of historians and students to seek to know how active combat commanders think and reason.
>
> William F. Halsey Jr., 1950

American history is full of famous naval officers, from Revolutionary War hero John Paul Jones to *Apollo 11* mission commander Neil Armstrong. Why, then, investigate comparative nobodies like Edward Very, John Hudgins, Benjamin Miessner, Samuel Robison, Morris Smellow, and Caleb Laning? Even among leading historians, the work of these individuals remains virtually unknown. That anonymity is puzzling. Without their pioneering efforts to develop technologies and processes in information management, America's victory in World War II would have cost even more in blood and treasure than it ultimately did.[1]

For much of modern history, warships have been the most technologically complex creations of the nation-state. Perhaps nowhere today is this more evident than in Portsmouth, England, where a visitor can tour *Mary Rose*, *Victory*, and *Warrior*. The first of these vessels served in the navy of Henry VIII, the second achieved renown as Horatio Nelson's flagship at Trafalgar, and the third was the Royal Navy's earliest iron-hulled warship. To borrow a phrase from one historian of technology, these ships provide stirring visual evidence of the technological sublime.[2]

Americans, too, embrace their historic warships. In Boston, one can see sailing frigate *Constitution*; in Charleston, Confederate submarine *Hunley*; in Philadelphia, steel cruiser *Olympia*; and in San Diego, aircraft carrier *Midway*. Farther west, lying in the mud of Pearl Harbor, rests *Arizona*. The ill-fated battleship serves as a somber reminder that, for all their grandeur, warships are not built for tourists. From Salamis to the Cold War, naval affairs have influenced world history. At times, they even have decided the fates of nations.

Little wonder that historians since Herodotus have studied navies and naval operations.

At heart, navies consist of two interrelated and indispensable components: people and machines. Much has been written on both subjects. In studies of the former, one finds stories of the mighty and the oppressed, the skilled and the incompetent, the bold and the timid, the exotic and the mundane. Whether investigating admirals or seamen, this genre, at its best, provides a window into the human experience. Historical studies of naval machines generally emphasize the warship. For navies of the industrial era, innovations in warship design are pivotal: the introduction of the ironclad, the adoption of the all-big-gun battleship, the rise of the aircraft carrier, the development of the nuclear-powered submarine. Underlying this narrative is a theme of progress. Individuals who promoted the adoption of new warship types were progressive; those who supported older paradigms were unreasonably resistant to change.

For many historians, a textbook example of technological conservatism is the U.S. Navy's failure to recognize the full potential of carrier aviation before the Second World War.[3] According to this school of thought, American naval officers retained an irrational devotion to the battleship despite overwhelming evidence that carrier warfare was the wave of the future. The battleship was a sacred vessel, a warship type whose "reputation as terror of the high seas persisted long beyond the point that pure logic might have dictated." As one scholar colorfully proclaims, the supporters of naval aviation were like "orphan boys being raised by a committee of wealthy men each of whom had his own sons to consider first."[4]

Recent scholarship has exposed the oversimplifications that lie behind such views.[5] American naval officers inhabited a complex world of shifting political, financial, institutional, and operational environments, and their thinking about the interrelationships between warship types was both multifaceted and diverse. Yet even the best of this recent work maintains a focus on what one might label the brawn of navies: ships, aircraft, ordnance, and propulsion. Largely missing from the literature is an examination of the brains behind this brawn—that is, the people, equipment, procedures, and facilities used to coordinate naval operations between ships or between forces at sea and ashore.[6] Collectively, these are known as command and control systems.[7]

Perhaps inevitably, improvements in the brawn of navies significantly increased the difficulties of command and control. Literally and figuratively, aircraft and submarines added a new dimension to naval operations. Combined with nineteenth-century innovations in steam-driven screw propulsion, automobile torpedoes, and long-range gunnery, the adoption of the submarine and the airplane limited the effectiveness of existing methods of command and control. No longer could a commander make informed tactical decisions unaided by people, devices, and methods for processing information.

This book argues that the most difficult problem naval commanders faced from the latter part of the nineteenth century through the Second World War was how to coordinate geographically dispersed and dimensionally diverse seaborne forces. In so doing, it offers an alternative conceptual framework in which the focus of analysis shifts away from progressive changes in warship types to the tactical integration of organized groups of vessels and aircraft. Such an approach reveals that the apparent conservatism of American naval officers did not result not from adherence to some "battleship technological paradigm";[8] instead, it derived from inherent limitations in the fleet's command and control systems. To operate effectively, naval officers had to learn how to manage a sea of information.

Indeed, the development and employment of increasingly sophisticated shipboard command and control systems fundamentally altered the human experience of warfare at sea.[9] New devices and techniques changed the way information was acquired, displayed, and processed. In this new environment, successful leadership derived not only from an individual's audacity and innate mental aptitude but also from an ability to master the cognitive skills critical for processing vast quantities of information. In short, naval operations became an extraordinarily difficult thinking-man's game.

Even in the age of sail, of course, commanders struggled with the need to make decisions in the face of incomplete or misleading information. On one level, then, command at sea changed little with the industrialization of naval warfare. A commander still had to be able to lead people under demanding conditions in an unpredictable environment. On another level, though, the nature of command underwent a fundamental shift during the first half of the twentieth century. This shift profoundly altered the cognitive experience of command.

The distinction between mental processes carried out to bring one immediately closer to a desired end state and those performed to make clearer information that is obscure or difficult to process has attracted the attention of cognitive scientists and other scholars, who label the first as *pragmatic* action and the second as *epistemic* action.[10] From the Civil War to the end of World War II, the U.S. Navy devoted ever-greater attention to systems that could clarify information and simplify decision-making processes. These systems not only aided senior officers in carrying out epistemic cognitive actions; they also created an environment in which commanders had to rely increasingly on junior officers and enlisted personnel for the information necessary to make sound command decisions.

An important consequence of these developments was that shipboard decision making became distributed, with relatively junior personnel assuming unprecedented responsibility.[11] To be sure, ultimate authority continued to rest with a vessel's commanding officer or the officer in tactical command (OTC). Yet the extent to which a single individual could follow and dictate the course of events gradually declined. Successful exercise of command came to require the employment of information technologies that could distribute problem solving throughout the chain of command. The transition from an environment in which commanders could make informed tactical decisions with limited input from subordinates to one characterized by epistemic actions and distributed cognition is an unexplored but important aspect of American history. Shipboard command and control systems lie at the heart of this transition.

Regrettably, most historians have been content to place such systems inside a black box, neglecting their design, construction, and use. When noted at all, command and control activities appear as outcomes rather than processes. Samuel Eliot Morison's recounting of events at the start of the Battle of the Philippine Sea (19–20 June 1944) provides a typical example:

> As early as 0530 bogeys [unidentified aircraft] appeared on radar screens in the direction of Guam; a Hellcat from *Monterey* tallyhoed [visually sighted] two Judys [a type of Japanese aircraft] in that direction and destroyed one, and about half an hour later a Val was shot down by destroyers of the battle line. The first phase of the action was on.[12]

This simple, faceless description ignores the complexity of these developments. Morison never explains just how electronic images, appearing on small screens in dark rooms deep inside U.S. Navy ships, led to the destruction of fast-moving Japanese aircraft dozens of miles away. Neglect of such processes contributes to a general sense that American naval officers coordinated forces by "good sense and the seat of the pants" both before and during the Second World War.[13]

The reality was quite different: U.S. naval successes derived from decades of research and experimentation intended to improve the fleet's systems for managing information. During World War II, these efforts culminated in the Combat Information Center (CIC), an integrated human-machine system where officers and enlisted personnel used automated, semiautomated, and manual techniques to collect, organize, process, evaluate, and disseminate information. Significantly, midlevel personnel developed the CIC and many of the command and control systems that preceded it. These systems show that the lived experience of military innovation proved far less hierarchical than one might think.[14]

Operationally, combat information centers functioned not only as machines processing information for humans; they also consisted of humans processing information inside of powerful machines. American naval leaders understood that these and other shipboard command and control systems required a new breed of personnel, one in which rank and class distinctions weighed less than the ability to perform difficult intellectual tasks under pressure. How well the U.S. Navy trained people to do such tasks and, more critically, how well they did them mattered greatly. These individuals worked and lived in an environment where even small mistakes or brief delays could mean the difference between life and death, victory and defeat. This book explores these men and their predecessors, the choices they made, the worlds they inhabited, and the technologies they employed.

CHAPTER ONE

Flags, Flares, and Lights

A World before Wireless

> When it became evident that the difficulties between the United States and Spain . . . were in a fair way of settlement without recourse to the guns and torpedoes of our navy, it was decided to take advantage of the accumulation of war-ships at the Florida rendezvous, and have a general drill . . . The present month, accordingly, will be one of great importance in our naval history.
>
> Unnamed Correspondent, 1874

An International Crisis

Captain Joseph Fry must have longed for a better vessel. His ship, *Virginius*, had been built to serve as a Confederate blockade-runner. Now, eight years after the Confederacy had fallen, Fry was hoping *Virginius* had one more run left. His cargo consisted of contraband bound for Cuban rebels, but on this late October afternoon a Spanish corvette had discovered *Virginius* off the Cuban coast. Fry altered course while his crew jettisoned illicit stores overboard. For a time the situation looked salvageable, as hours passed and the corvette failed to close. Fry knew the coming darkness would offer him an opportunity for escape.

In the heat of the chase, *Virginius*'s commanding officer probably did not consider the irony of his situation. Born into a Florida family of moderate affluence, Joseph Fry accepted a midshipman's warrant in 1841 at the age of fifteen, fought in the Mexican-American War, and participated in Matthew Calbraith Perry's expedition to Japan. Yet Fry suffered from chronic seasickness, and eventually the Navy Department assigned him to permanent lighthouse duty in New Orleans. Had the Civil War not intervened, there he might have remained for the duration of his naval career. When war came, however, Fry resigned his commission and joined the Confederate navy. During four years of service, he commanded six vessels, five of which Union forces burned or sank. Known for his courage under fire, Fry offered a stark counterpoint to the old adage that fortune favors the brave.

After the war, Fry and his family fell on hard times. In 1873, with his resources exhausted and seven children to feed, he jumped at the opportunity to earn $150 per month commanding *Virginius*. The ship's agent was vague about the crew, cargo, and ports of call but emphasized that *Virginius* sailed under United States registry. Fry probably knew better than to ask too many questions.

Darkness finally came, but clear skies and a full moon did not auger well for Fry and his crew. *Virginius* began to slow, the hours at full speed having taken their toll. Leaks in the ship's rattled caulking weighed down *Virginius* and allowed the Spanish corvette to close within striking distance. After several near misses, disaster struck. A Spanish shell penetrated one of *Virginius*'s smokestacks. The chase was over. Fry surrendered his ship, and a boarding party raised Spain's national ensign in place of the American flag. The captured vessel was towed to Santiago de Cuba, where Spanish authorities imprisoned *Virginius*'s entire crew.

To the Spanish military commander in Santiago, General Don Juan Burriel, *Virginius* was an infuriating symbol of Cuban treachery and American duplicity. Almost immediately, he executed four known insurgents. But what would he do with Fry and the others, especially those who were British or American citizens? Given Burriel's reputation for ruthlessness and brutality, Fry already knew the answer. Over the heated protests of the senior American and British diplomats in Santiago, the former naval officer and thirty-six others were convicted in a sham trial. Executions were scheduled for 8 November, but Burriel expedited the process when he discovered that a British warship was steaming toward Santiago. On the morning of 7 November, Spanish authorities took the condemned men to receive last rites. Shortly thereafter, they were marched to a wall on the city's outskirts. Fry, the only prisoner not bound, went down the line saying farewell. Upon reaching the end, he took off his cap and turned around. The ensuing volley brought both death and an international crisis.[1]

News of the brutal slayings reached Washington on 12 November. President Ulysses S. Grant was morally outraged and immediately ordered the navy to assemble a fleet in Key West. Public sentiment for war was strong. The *New York Times* declared that the U.S. government had a duty "to use all the force necessary to prevent any recurrence of the injury and insult we have already suffered."[2] The British ambassador in Washington cabled home his opinion that the affair would drift into hostilities. And one politician stated boldly that, if international law furnished no precedence for action, then the United States should furnish one for international law.

In spite of the bellicose rhetoric, peace ultimately prevailed. Grant's secretary of state, Hamilton Fish, long had opposed U.S. intervention in Cuba, and he worked diligently to diffuse the situation. British support strengthened Fish's hand, and in late November the Spanish government agreed to American demands. Spain released the remaining prisoners, sent *Virginius* back to the United States, removed Burriel from his post, and eventually paid reparations to the victims' families. The Spanish government also sought to prosecute Burriel, but the wheels of justice moved slowly. The former despot died of natural causes while still awaiting trial for his crimes.[3]

Several days before his untimely death, Joseph Fry took a moment to write to President Grant. Deeply concerned about his family's finances, the former officer indicated he was due back pay from his pre–Civil War naval service and asked the president to ensure that his family received it. Fry closed his brief missive with a postscript echoing the sentiments of many naval officers and merchant mariners of the period. The country and its navy were miserably weak, he argued, "when a vessel can be captured on the high seas . . . and her captain, crew, and passengers shot without appeal to the protection of the United States."[4]

As if on cue, Fry's death set in motion a series of events that demonstrated the weaknesses and operational limitations of the American navy. At the end of 1873, there were 117 warships in the navy's inventory. Only 4 of these had been laid down and commissioned after the Civil War. The rest were an eclectic mix of sailing ships, gunboats, monitors, armed merchantmen, and steam-powered sloops and frigates. More than one-quarter of the navy's warships had entered service before 1860, and most of the vessels commissioned after that date had been built for coast defense and riverine warfare.[5]

In response to the *Virginius* crisis, Secretary of the Navy George M. Robeson directed available warships in the Atlantic Ocean and Mediterranean Sea to assemble at Key West, Florida. One by one the vessels arrived, but Hamilton Fish's diplomatic successes obviated the need for a fleet sortie. In mid-December, Spanish authorities released *Virginius* and freed all remaining prisoners. One U.S. Navy steam sloop transported the survivors to safety, while another took *Virginius* in tow with orders to proceed to New York.[6] En route, a storm struck, and the notorious vessel foundered off the North Carolina coast.[7]

The crisis having passed, Secretary Robeson concluded that the fleet concentration presented a golden opportunity to assess the navy's tactical proficiency. He sent one of the service's most tactically skilled officers, Foxhall A. Parker, to Key West. Arriving on 20 January 1874, Parker immediately assumed duties as chief of staff to the commander of the assembled fleet, Rear Admiral A. (Augustus) Ludlow Case.[8] Displeased with the fleet's material condition, Case commented that the vessels had taken too long to assemble and informed his superior that many of them required extensive repairs.[9] Nonetheless, he welcomed an opportunity to improve the fleet's operational effectiveness. Case was particularly excited about Parker's arrival, for his new chief of staff brought both valuable expertise and a new tactical signal book.[10]

Case's criticisms accurately reflected the concerns of most post–Civil War naval officers. Yet so too did his desire to improve the navy's ability to operate as a coordinated body. Naval officers wanted better warships, of course, but these required large appropriations at a time when most in Congress believed the country had little need for a strong navy.[11] Many of those who would be called upon to fight a maritime conflict disagreed, but the resources they desired were slow in coming. So naval personnel sought improvements where they could. And one of the things they sought to improve was shipboard command and control.

As If One Machine

During the age of sail, warships at sea communicated via two different methods. The first method, the dispatch vessel, was slow but permitted officers to exchange lengthy, handwritten messages. The second and theoretically more rapid method was naval signaling. Naval signals could be visual or audible, but many factors influenced their efficacy, especially during the smoke, noise, and chaos of battle. By the end of the eighteenth century, the British Royal Navy had developed one of the world's most effective systems of visual signaling, a system that had helped Britain achieve victory during the first major naval battle of the Napoleonic Wars, the so-called Glorious First of June (1 June 1794).[12] No surprise, then, that the fledging U.S. Navy's first signal book plagiarized heavily from the British. Published by a U.S. Navy captain in 1797, it adopted the Royal Navy's numerary signaling system, a type in which different flags represented the numbers zero through nine. According to this signal book, for example, hoisting flags for the number "224" sent out the grim query: "Are your dead buried and the ship well washed?"[13] The American navy's adoption of numeric codes that were familiar to the British worked reasonably well until the War of 1812, when for obvious reasons a new signal book had to be issued.[14] With minor revisions, it remained in force for over four decades.

Eventually, concerns about signaling effectiveness and security prompted the Navy Department to compile a new signal book. Completed in 1858, *Signals for the Use of the United States Navy* contained more than one thousand numeric codes for both flaghoist and night signaling. Some of these signals were for tactical maneuvers, such as "Form the first order of steaming" (flag signal 971, night signal 80), while others conveyed specific information, such as "Enemy appears to be coming out of port" (flag signal 551, no night signal), or "Overboard, a man has fallen" (flag signal 839, night signal 97).[15] A tactical manual published the following year praised *Signals for the Use of the United States Navy*, pointing out that even the best commander would struggle if subordinates had trouble understanding the orders signaled to them.[16] Unfortunately for the U.S. Navy, officers who resigned their commissions to fight for the Confederacy took with them knowledge of its contents. Faced with this breach of security, the service hastily issued a new signal book for coastal operations.[17] Although it adopted a uniform code for day and night signals (i.e., signal numbers had the same meaning whether transmitted at day or at night), most officers found the new book unwieldy. In response, the Bureau of Navigation directed naval personnel to use the army's semaphore system whenever possible.[18]

The Union army's semaphore system had been developed in the late 1850s. Called Myer's wigwag, the system was binary and required only a single flag. A flag wave to the sender's left signified "one," while a flag wave to the right signified "two."[19] To spell "tide," for example, a signalman would make the motions for "two" (t), "one" (i), "two-two-two" (d), and "one-two" (e).[20] The army used Myer's wigwag extensively during the Civil War, and it remained in service until the eve of the First World War.[21]

For the U.S. Navy, Myer's wigwag offered multiple advantages. It was easy to learn, easy to use, and improved interservice communications. For ship-to-ship communications, however, Myer's semaphore system had distinct disadvantages. As an alphabetic code, it was unwieldy for tactical maneuvers. Moreover, because Myer's system involved directional movement (left and right flag waves), messages could not be received simultaneously on both sides of a transmitting vessel. Finally, the rapid movements of Myer's wigwag could be confusing, especially to officers more accustomed to the protocols and procedures of numerary-based flaghoist signaling.

In practice, then, the service used a combination of Myer's wigwag, flaghoist, and night signaling during the Civil War. The first of these worked well during operations that involved only a few vessels, but commanders generally used one of the latter two methods for coordinating warship movements during large-scale operations like amphibious assaults on Confederate ports. Recognizing the limitations of Myer's wigwag and wanting to improve the navy's overall signaling capabilities, in May 1863 Secretary of the Navy Gideon Welles referred the issue to an interagency commission he recently had established to investigate "all subjects of a scientific character."[22] After deliberating for more than a month, the commission put forward its recommendations for improvement.[23] These included issuing a new signal book, adopting methods for "secret" (enciphered) signaling, and investigating further new communications techniques such as chronosemic signaling.[24]

Operational personnel echoed the board's sentiments. In March 1864 America's first admiral, David Glasgow Farragut, exasperatedly wrote to the Bureau of Navigation after a ship-to-ship communications failure resulted in one Union warship accidentally firing on another near Mobile Bay. Farragut beseeched the bureau chief "to do something to simplify the signals," writing that signaling had become "a matter of the greatest anxiety to the officers in this fleet." The rear admiral believed adoption of a simple encryption system would alleviate the need for frequent, handwritten changes to the signal book. He proposed adding or subtracting prearranged numbers to all numeric signals: "The officer making the signal adds 2 (or any number you please); the signal officer reads it . . . but subtracts 2 from it and looks in the signal book and finds (as intended)."[25] The chief of the Bureau of Navigation responded to Farragut's plea the following month, directing signal officers to employ a similar technique whenever they wanted to "mask" their signals.[26]

Already famous for his victory at the Battle of New Orleans (24–25 April 1862), Farragut would achieve further fame when he decided to "damn the torpedoes" at the entrance to Mobile Bay on 5 August 1864. Indeed, the Battle of Mobile Bay illustrates well the types of command and control challenges naval commanders could encounter during the Civil War.[27] As part of his battle plan Farragut decided to steam into Mobile Bay in two parallel, line-ahead columns. He wanted his four ironclads in a line nearest the Confederate fort guarding Mobile Bay's east entrance and his fourteen wooden vessels paired together in a second line just west of the ironclads (i.e., a column two across and seven deep). After discussing the matter with his commanding officers, Farragut chose to position his flagship, the wooden screw-sloop USS *Hartford*, second in line. Another screw-sloop, USS *Brooklyn*, would lead the column.

About half past five on the morning of 5 August 1864, *Hartford* raised four flags that ordered the fleet to get underway from anchorage. Within five minutes, every ship had responded, and Farragut's forces began steaming northward toward the bay. For a long while the admiral needed no other signals, but shortly past seven o'clock he noticed two army signal officers milling about on *Hartford*'s gun deck. These individuals had reported aboard several days earlier so that the navy would have some experienced experts in Myer's wigwag for communicating with the army once Farragut's forces reached Mobile. The admiral considered these individuals so valuable he previously had ordered them to remain below decks during the action. He quickly reiterated that order, and the two men made their way back down "to the stifling hold."[28]

Meanwhile, Farragut noticed that *Brooklyn* had fallen too far behind the column of ironclads, which were now to the northeast. In fact, *Brooklyn* appeared to have stopped. The admiral turned to his flag lieutenant and told him to signal *Brooklyn* to "go on." Signalmen quickly hoisted three flags representing "665," the signal code for that specific order. *Brooklyn*'s commanding officer copied the signal but opted to respond in Myer's wigwag. Unfortunately, no one above decks on *Hartford* was proficient in Myer's wigwag, so Farragut had to summon one of the army signal officers he recently had sent below. After a few tense minutes, the admiral received word that *Brooklyn* had stopped because the lead ironclad, USS *Tecumseh*, had veered too far west and was blocking the way up the channel. *Brooklyn*'s commanding officer inquired of Farragut, "What shall we do?"[29]

Farragut told his flag lieutenant to inform the army signal officer to order *Brooklyn* to "Go ahead."[30] About the same time, another problem had arisen. Both sides were furiously firing their guns, and a smoke pall had descended over the scene of battle. Farragut could no longer see what was going on around him. To improve his vision, he climbed up into *Hartford*'s rigging, where a member of the crew lashed him into place dozens of feet above the gun deck. Farragut's only command and control apparatus was his spyglass and a nearby speaking tube that allowed him to communicate with those below.

From this lofty perch Farragut saw disaster strike when *Tecumseh* hit a stationary torpedo a little after 7:30 a.m. The ironclad sank within minutes. This prompted *Brooklyn* to back down, putting the ship on a collision course with *Hartford*. Farragut ordered a Myer's wigwag message again sent to *Brooklyn*: "Tell the monitors to go ahead and then take your place."[31] *Brooklyn*'s commanding officer seems not to have comprehended the message, apparently believing that Farragut would not want him to go outside the eastern portion of the channel through a field of torpedoes. Exasperated, Farragut decided to lead by example. He ordered *Hartford*'s pilot to proceed through the torpedoes, by some accounts cursing the infernal machines for good measure.[32] At this point in the battle, Farragut needed no signals to tell the warships behind him what to do. They dutifully followed their commander through the torpedoes and into naval lore.

Farragut was certainly not the only prominent naval officer to take an interest in signaling. In the fall of 1863, Stephen B. Luce, then serving as the commanding officer of one of the navy's ironclads, wrote his superior to recommend that all young officers "be directed to practice [visual] signals constantly between the different vessels," arguing that such practice would be "of great practical value in the fleet."[33] Shortly after the war, David Dixon Porter voiced concerns about the service's signaling systems to Secretary of the Navy Gideon Welles. Porter called Myer's code "the best that has been invented" but argued that "it has its objections and . . . is not altogether suitable for squadron signalling." Porter asked Welles to support fully experiments then being conducted by Luce on magnesium lights, suggesting that such lights might even allow ships to "talk rapidly to each other, at a distance of ten or twelve miles apart."[34]

The navy's work with magnesium lights never led to a viable signaling system, but efforts to improve other areas of shipboard command and control were more successful. In 1867 the service finally adopted clear procedures for secret signals, and two years later the Bureau of Navigation issued an improved signal book.[35] Yet the most important change in the immediate postwar years was institutional. On 19 July 1869, Secretary of the Navy George M. Robeson formally approved the establishment of a signal office within the

Bureau of Navigation.[36] The new office sought to improve ship-to-ship communications through four interrelated and frequently interconnected means: interservice cooperation, standardization, training, and experimentation.

Fortunately for the navy, the army was amenable to pooling resources at a time when both services faced significant budgetary constraints. At the end of July, Robeson wrote his counterpart in the War Department to ask whether naval officers could attend the army's signal school in Arlington, Virginia.[37] After several midgrade officers tested the waters by attending a few weeks of the course, the army's chief signal officer responded affirmatively, offering "every convenience and facility" in the power of his office.[38] Shortly thereafter the navy sent a "full and regular class" of eight junior officers to the army's signal school.[39] By September 1870, more than two dozen naval officers had passed the school's full course of study.[40]

The navy's signal office also utilized the army's supply network to acquire new signaling equipment. Orders for items such as signal books, telescopes, and binoculars often went to the same companies that supplied the army, usually accompanied by explicit instructions to match the style, kind, quality, and cost of those previously furnished to the army's signal service.[41] The service usually acquired items of a more sensitive nature, such as code cards and cipher discs, directly from the army.[42]

Another effort to improve ship-to-ship communications focused on standardizing the signaling equipment carried on board American warships. Toward this end, the Bureau of Navigation dictated a standard allowance for each vessel. The signal office was responsible for providing manuals and "signal kits," the latter of which contained flags, flares, and torches.[43] Most vessels received only one kit (flagships received two), so the signal office monitored closely the contents of each kit.[44] The office also worked diligently to correct signal book errors and to distribute new signaling apparatus uniformly throughout the service.[45]

Standardized equipment loads were important because effective signaling required seamless compatibility between sender and recipient. Yet standardization was only part of the equation. Naval officers and signal quartermasters had to be proficient at using their signaling equipment, and that equipment had to perform as expected. The signal office thus implemented an ambitious training program designed to create and maintain proficiency. Simultaneously, it used this program to foster improvements in the service's signaling technologies and procedures.

The naval officers who had attended the army's signal school in 1869–70 were at the vanguard of this training program, and by the end of 1871 more than fifty officers had completed all or part of the signal school's course of instruction.[46] These young officers, a cadre skilled in the latest signaling techniques and procedures, reported back to their respective ships where they provided training to all personnel involved in signaling. The signal office required them to file quarterly reports detailing the names and ranks of all persons under instruction, the hours of practice with day and night signals, and the progress made by each individual in learning and executing those signals. The chief signal officer found these reports very useful and admonished those who failed to submit them.[47] Most commanding officers supported the training programs implemented by their signal officers, and the signal office itself satisfyingly noted that the emphasis on training resulted in numerous "suggestions and inventions relating to signals and signaling."[48]

Evaluating these suggestions and inventions was one of the signal office's most important responsibilities. Many suggestions came from officers who thought the navy's signal books could be improved. In early 1871, for example, one junior officer wrote the signal office identifying 138 errors in the signal book of 1869.[49] Others believed the signal book should be split and issued in two editions: one for general information, the other for tactical maneuvers. In both instances, the signal office acted quickly to implement these ideas. It issued corrections to the official signal book in October 1871 and promulgated the U.S. Navy's first tactical signal book just three months later.[50]

While suggestions for enhancing the navy's signal books usually came from officers, private inventors were more likely to propose new or improved signaling technologies. One of the most successful of these inventors was Martha J. Coston, a widow who began experimenting with signal flares in the late 1840s. Coston's early efforts were rather unsuccessful, but after a decade of work she ultimately perfected a tricolor system of red, white, and green flares.[51] Convinced that Coston's system held significant promise, the secretary of the navy convened a special board to evaluate it. After weeks of testing, the board pronounced Coston's night signals to be "decidedly superior" and wholeheartedly recommended their adoption.[52] During the Civil War, Coston's manufacturing company furnished more than one million signal flares to the U.S. Navy.[53]

After the war, Coston continued to develop her flares. On the whole, naval officers found them to be reliable, but there were still too many instances where Coston's flares simply failed to ignite.[54] This could cause problems for messages sent by numeric code. For example, to send the order "use shrapnel against the enemy" via the navy's signal book of 1869, a warship would light off a flare that burned red, then green (6), a flare that burned green, then white (8), another flare that burned red, then green (6), and a flare that burned only red (4). If the last flare failed to ignite, the recipient would instead read, "Can you spare me some bread?" (686). Errors were not always this obvious (or humorous), and in some instances they could significantly alter a commander's intent. For example, in the same scenario as above the order "act at your own discretion" (1463) would instead be read as an order to "continue the action" (146).[55]

The navy therefore showed great interest when Martha Coston completed work on a redesigned signal flare early in 1871. Coston modified her system by installing a friction ignition mechanism in place of the percussion cap that previously had been used to ignite the flare's fuse. She also modified the chemical mixture of the flares in an effort to achieve longer burn times.[56] Initially, the service appears to have been somewhat skeptical of these "alleged improvements."[57] Any such skepticism quickly disappeared. After the signal office conducted a series of experiments with Coston's new flares in April 1871, it determined that they were more than satisfactory. In a glowing recommendation to his superior, the chief signal officer commented that "for quickness, certainty, and efficiency, this new Night Signal combined with the modes of igniting it, submitted by Mrs. Coston, surpasses the present Coston Signal now in use in the Navy, and may be considered a great improvement."[58]

Of course, some suggestions and proposed inventions were unlikely to improve the navy's signaling capabilities. The signal office nevertheless exhibited due diligence evaluating all that it received. For example, in 1872 the chief signal officer tasked a member on his staff, Lieutenant George A. Norris, to evaluate a new system of chronosemic signals. The navy already employed this type of signaling in foggy weather, but the chief signal officer thought it might be modified for tactical signaling. Norris, an 1870 graduate of the army's signal school, thoroughly evaluated several alternatives but ultimately concluded that the amount of time required to send and receive chronosemic signals was too lengthy for tactical maneuvering.[59] On another occasion, Norris rejected a private inventor's proposed new signaling system with the terse commentary: "Mr. Ward's system is far inferior to the General Service Code now in use in the Navy, and he is evidently behind the times."[60]

Unfortunately for the navy, not all new signaling technologies and procedures adopted by the signal office achieved the desired results. At times, ideas that looked promising on paper or items that performed well on land were unsuccessful when implemented operationally. One example of this was the tactical signal book distributed by the signal office in early 1872. The logic behind its issuance was sound. The official signal book contained thousands of numeric codes, only a few hundred of which were for tactical maneuvers. By placing all tactical signals in a separate manual, the likelihood of errors was reduced. Regrettably, no copy of this tactical signal book is known to have survived, so details about it are sparse. Yet existing documents suggest that most officers found the new book inadequate, and within months of its issuance the Bureau of Navigation convened a special board to examine and report all necessary or advisable changes to the "U.S. Naval Tactics Signal Book now in use."[61]

The most distinguished expert to appear before the board was Captain Foxhall A. Parker. A Virginia native, Parker was born in 1821 and grew up in a naval family. He entered the navy as a midshipman in 1837 and served in a variety of billets over the next two decades. Unlike his younger brother William, who accepted a commission in the Confederate States Navy, Parker remained loyal to the Union when the Civil War erupted. During the war, Parker commanded a gunboat and wrote a book on squadron tactics for steam vessels.[62] In late 1864 he assumed command of the Potomac Flotilla, supporting the Wilderness campaign and the sieges of Petersburg and Richmond. He wrote and published several books immediately after the war, including the influential and widely respected *Fleet Tactics under Steam*.[63] In the early 1870s Parker commanded the flagship of the European Squadron and had just reported as the North Atlantic Squadron's chief of staff when the Navy Department recalled him to Washington to fix deficiencies in the service's newly issued tactical signal book.[64]

As directed, Parker corrected and revised the navy's tactical signal book over the summer of 1872. That fall he submitted his handiwork to the board, which was highly satisfied with the changes made.[65] In addition to correcting obvious errors, Parker simplified the signals for existing evolutions and added new ones for tactical maneuvers such as ramming and torpedo attacks. The board was so concerned about the inadequacies of the existing tactical signal book that it recommended immediate publication of those sections of Parker's manuscript that dealt with steam tactics, a recommendation that certainly calls into question portrayals of the postwar American navy as an institution filled with anti-steam reactionaries. In the board's own words: "[We do] not contemplate the continuance of the many Fleet Evolutions under sail or under sail and steam combined, for the reason that in the probability of an engagement, the Fleet would be put under steam."[66] The board later added a section to the naval tactics signal book for fleet evolutions under sail but stressed that this section was specifically for peacetime cruising.[67]

The board's members showed great respect for Parker's expertise, and they praised him for his intellect and "the many valuable and useful suggestions" he had made.[68] The chief of the Bureau of Navigation (who controlled the detailing of officers) took notice and gave Parker permanent orders to Washington, D.C. His new assignment, which became effective on 1 July 1873, was chief signal officer of the U.S. Navy.[69]

Foxhall Parker was only a few months into his tenure as the navy's chief signal officer when the Spanish seized *Virginius* and executed Joseph Fry and dozens of his crewmen. The navy's new tactical signal book was then in the final stages of publication, and Parker pushed to get it out to the fleet as quickly as possible.[70] Believing war might be imminent, he also initiated experiments to test the reliability of Coston's improved signal lights and the service's ciphers and codes.[71] In short, Parker sought to make ready the service's existing shipboard command and control systems for combat against the Spanish navy.

The service's new signal officer was not the first to grapple with issues of command and control in the era of steam warships. The amphibious assaults of the American Civil War were complex military operations, and communications problems were not uncommon. As combat veterans, Parker and his contemporaries experienced firsthand many of these problems. Nor were American naval officers alone in this regard. In 1866 a quantitatively inferior Austrian fleet defeated an Italian fleet at Lissa largely because the Italian commander moved to a new flagship just before the battle, and his subordinates kept looking to the wrong vessel for signals. The Italian commander also belatedly discovered, undoubtedly to his horror, that his new flagship was carrying an incomplete set of flags.[72]

Yet Parker was one of the most vocal proponents of the idea that screw-driven, steam-powered warships had changed the nature of command at sea. Influenced by the work of fellow officers both in the United States and abroad, he argued:

> Now that, through the agency of steam, war has become not less a science at sea than on land; when the ocean is a great chess-board, upon which the skillful looker-on sees many a move not apparent to the contestants, whose brains have become heated with the strife, the *role* of the admiral approximates to that of the general, and he should, like the latter, take post whence, without being an active participant in it, he may overlook the whole *sea of battle*, and signal to the fleet such formations as he shall find necessary. He should, in other words, be the *mind* of the fleet.[73]

Articulating such a vision was one thing. Executing it, as Parker would soon discover, was quite another.

Eventually, the United States government resolved the *Virginius* crisis through diplomatic means, eliminating any chance of the fleet engagement that had so occupied Parker's thoughts in the waning days of 1873. But the war scare presented Secretary of the Navy George Robeson with an exceptional opportunity to exercise the vessels assembled in Key West. Accordingly, Robeson directed the senior officer present, A. Ludlow Case, to conduct one month of fleet maneuvers. Case, previously in command of the European Squadron, arrived in Key West on 3 January 1874. Parker joined Case about two weeks later and served as his chief of staff throughout the exercises.[74]

Case commanded more than two dozen warships, but many of these were unsuited for coordinated fleet maneuvers. Several had steam plants in desperate need of repair, and nearly one-quarter of his warships were monitors, a type of ironclad with low freeboard, limited speed, and poor steerage.[75] Case and Parker thus exercised with only about half their available vessels, which they organized into three divisions of four ships each. A dispatch vessel accompanied the fleet's flagship, USS *Wabash*, which roamed between and around the other warships during the exercises.[76] *Wabash* had been flagship of the European Squadron and was one of the most powerful ships in the assembled fleet. That Case and Parker did not incorporate it into a division was significant and reflected their mutual understanding of a naval commander's appropriate role in the age of steam.

Like several of their contemporaries, Case and Parker believed steam power permitted a fleet to operate more easily "as if one machine."[77] The commander's role had changed, they reasoned, becoming even more critical than it had been in the age of sail. During that era, naval warfare moved more slowly and commanders led by example. Battles frequently turned into melees, with individual vessels fighting one another at close range. In such an environment, fleet maneuvers became impossible, and the best commanders trusted their subordinates to carry out any plan of attack or defense with initiative and valor. Naval legend Lord Horatio Nelson stood as the archetypal commander. "In case Signals can neither be seen or [*sic*] perfectly understood," he wrote famously, "no captain can do very wrong if he places his Ship alongside that of an Enemy."[78]

In the late 1850s, however, some naval tacticians began to reconsider this point of view. Several French officers argued that the center of battle was not necessarily the proper place for a fleet commander, and a few years later one of the Royal Navy's leading tactical thinkers argued that the "flag ship of the future" should be a swift, quick-turning vessel from which a commander "can best view the movements of his own and the enemy's fleet, and where his signals can be at once perceived from all quarters."[79] American naval officers incorporated similar concepts into their own tactical doctrine. The service's *Introduction to the Naval Signal Code*, first published in 1867, called on fleet commanders to summon the "moral courage" to lead their forces from a detached flagship rather than in Nelsonian fashion, and Parker himself, as indicated above, clearly ascribed to this new vision of naval command.[80]

Fourteen ships in all, Case's fleet got underway early on the morning of 4 February 1874. After a month of preparations, which included the distribution of Parker's new tactical signal book, Case was anxious to test his forces.[81] The initial results were not particularly encouraging. An evolution that could be done in forty minutes took an hour to complete, and even after several days of practice, vessels still had difficulty performing basic maneuvers at slow rates of steaming.[82] Putting a positive spin on his first week of fleet command, an obviously disappointed Case informed Secretary Robeson that the exercises had "gone off as well as could be expected" under the circumstances.[83]

The circumstances to which Case referred were an inexperienced officer corps and an aged assemblage of warships. The admiral could do little about the latter. Of the warships assigned to him, only one had been laid down and commissioned after the Civil War. Case's flagship *Wabash* dated from the 1850s, and the entire fleet carried guns inferior to those employed by major European navies. Perhaps worst of all, Case and Parker had to conduct all maneuvers at speeds of no greater than four and one-half knots, which was the top speed of the fleet's slowest vessel.[84] Recapping the entire situation, Admiral David Dixon Porter likened his service to "a foot-soldier armed with a pistol encountering a mounted man clad in armor and carrying a breech-loading rifle."[85]

A generation would pass before American warships matched those of European navies, in part because commerce raiding (*guerre de course*) remained the nation's principal maritime strategy until the 1890s. Yet Case and Parker's efforts to give their officers valuable experience in fleet operations reveal that a growing cadre of the navy's senior leadership saw such operations as a cornerstone of the American naval profession. As demonstrated by the *Virginius* crisis, that profession might be called upon to fight a war on short notice. No matter how meager the equipment on hand, naval leaders saw efforts to improve the service's operational capabilities as vital. From this perspective, the Key West maneuvers were not entirely lamentable, for officers acquired both precious experience and an appreciation of the practical difficulties surrounding ship-to-ship communications during fleet operations.

Neither Case nor Parker recorded his unvarnished opinion of the U.S. Navy's first week of fleet maneuvers, but surely each hoped for a better showing during the second week of exercises. After a long weekend at anchor in "Florida Bay," the fleet went to sea again on 11 February. Initially, events did not go smoothly. Case had to withdraw one warship from the exercises after a boiler failure, and Parker found the underway evolution rather disorderly.[86] By the end of the week, however, both officers noted considerable progress in the assembled ships' ability to function as a coordinated body. Parker observed "marked improvement" throughout the fleet, and Case went so far as to write that the fleet could now operate "as if one rudder guided all."[87]

This progress permitted Case and Parker to schedule more complex maneuvers during the third and fourth weeks of the exercises. By and large, the fleet executed these maneuvers with precision and skill. According to Parker's journals, there were few instances where an evolution was performed poorly or a signal had to be repeated.[88] Despite continuing materiel problems, which actually resulted in cancellation of two additional weeks of tactical maneuvers, Case and Parker were pleased with the way their ships had come together as a fighting force. Both men recognized that the navy was in desperate need of modern ships and ordnance, but they also foresaw a future that involved fleet operations. Case awaited the day when the country could field a fleet of large and powerful warships, and Parker exhorted his colleagues to adopt the motto: "*The ships of our Union and the union of our ships: may they be like our States, 'one and inseparable.'*"[89]

To a large extent, the fleet's ability to operate as a coordinated body derived from the tactical signal book Parker had produced the previous year. Parker approvingly wrote that the new book provided "the means of throwing our whole fleet, or any division or squadron of it, into groups offensive or defensive."[90] Case also praised Parker's new system of tactics, stating that they were "so simple in their form and execution, that with the general intelligence of the officers in handling ships, and the use of steam for their propulsion, they become almost machine work."[91] Of course, Case wrote such words after weeks of practice by his ships and their crews. To the regret of many naval officers, more than a decade would pass before major exercises became part of the U.S. Navy's operating practices.

Elevated Signaling

Foxhall Parker returned to the signal office in Washington with a new appreciation for the practical difficulties of ship-to-ship communications. He confronted two major challenges. The first was the issue of reliability. All too often, naval personnel transmitted signals incorrectly or misinterpreted them on the receiving end. A story recounted by Bradley Fiske in his memoirs illustrates well the types of communications problems that could result when personnel were not proficient. One night in either 1879 or 1880, Fiske witnessed the following when his ship, USS *Powhatan*, signaled a nearby vessel:

> It was not really necessary to signal, because the night was so calm that a man with a good voice could have shouted the message, and it would have been heard on board the other ship. But signaling is often done for purposes of practice, and so the message was signaled . . . The surprising answer came back, "Our commander is dead." So the *Powhatan* again signaled the same message as before. To this a long-drawn-out answer came back, "Our commander is ill." The *Powhatan* again repeated the original message, and the answer came back, "Our commander is absent." As more than two hours had already been consumed, a little dinghy was despatched with a note, explaining what the message was, and asking what the answer had intended to be. The dinghy returned in ten minutes with a note saying that the answer returned each time had been, "I do not understand."[92]

The second practical difficulty facing Parker was that of range. Even in good weather shipboard personnel could not accurately read flags at distances

Steam sloops USS *Franklin* and USS *Richmond* at anchor in Villefranche, France, c. 1871. As a result of a war scare with Spain, *Franklin* and other American warships assembled at Key West, Florida, in late 1873 and early 1874. There they tested Foxhall Parker's new signal book and conducted the first fleet maneuvers in U.S. history. *Naval History and Heritage Command photograph NH 61880.*

greater than about four nautical miles. The flags themselves generally could be seen, but their colors started to blend, making them difficult to interpret.[93] At night, the range at which personnel could send and receive signals doubled to roughly eight miles, although for small warships six miles was a more realistic distance.[94] Parker certainly grasped the potential value of any system that maximized the range at which vessels could communicate.

With these goals in mind—greater reliability and longer communication ranges—Parker continued to direct an aggressive testing program within the signal office. In an effort to minimize signaling errors, he scrutinized the quarterly reports from the fleet for helpful suggestions and made modifications to the signal book based on lessons learned during the Key West maneuvers. Hoping to increase the range at which signals could be sent and received, his office initiated a series of trials with signals projected from mortars and commenced experiments designed to ascertain the potential usefulness of a new technology, the electric searchlight.[95]

Yet the most significant signaling development of the period had its origins in an unsolicited letter sent to Parker by a twenty-six-year-old lieutenant then serving in the Bureau of Ordnance. In August 1874 Edward W. Very wrote the signal office to propose a new system of signaling. According to Very, his innovative system enhanced both the simplicity and the rapidity of ship-to-ship communications. It consisted of two parts. The first was a hand-held pistol, likened by Very to "a Colt revolver," that could shoot flares "to a height of not less than 300 ft." To accompany this device, he proposed a new signal code, one that used only two colors instead of the three required for Coston's signals.[96]

Very argued that his system offered many advantages over Coston's flares. For a sender, the pistol was easily loaded, gave better control over ignition, and could be fired from any open deck. For a recipient, the advantages were even more pronounced. To begin, the height of the signal extended the range at which it could be seen and minimized the likelihood spars or rigging would obstruct a recipient's view. Also, the entire signal was seen at once. This permitted a recipient to more easily distinguish between colors (of which there were now only two) and to read signals more quickly. Finally, Very claimed his system was simpler than Coston's, arguing that "the principal object of any improvement in signals should be to reduce the bad effects of carelessness and clumsiness in both maker and receiver to a minimum."[97]

What prompted Very to devise a new signaling system? A definitive answer remains elusive, but extant sources suggest he participated in the Key West exercises and witnessed firsthand the value of reliable shipboard command and control systems for fleet operations. A native of Maine, Very entered the Naval Academy during the Civil War, graduating in 1867. Little is known about his early years of commissioned service, although he probably served on several different warships before transferring to the European Squadron in the spring of 1871. Shortly thereafter Very was promoted to lieutenant, a rank he still held more than thirteen years later when a prominent ordnance manufacturer lured him from the service with a lucrative job offer.[98]

Parker almost certainly replied to Very, and in fact the navy's chief signal officer previously had expressed an interest in elevated signals.[99] Regrettably, this reply appears not to have survived, but existing records reveal that Parker's office tested several different methods for elevated signaling in the mid-1870s. For example, in June 1874 it conducted a series of experiments that demonstrated the limited utility of signal mortars. The fuses were temperamental, possibly because shells left the mortar so quickly, and the green signals "failed to ignite in a majority of cases."[100] The following year the signal office purchased rockets "for experimental purposes," and at one point Parker requested a book entitled *Pyrotechny and the Art of Making Fireworks at Little Cost*.[101] Parker's office also solicited bids for Roman candles, but neither these nor the experimental rockets worked particularly well for naval signaling.[102]

Recapping these experiments, in February 1876 Parker addressed the issue of nighttime signaling in a letter to his immediate superior. The navy's chief signal officer called the current system of Coston's signals "entirely reliable" and wrote that, when mistakes did arise, "the fault will invariably be found, not in the signal itself, but in those who are using it as a means of communication between ship and ship." Yet Parker also argued that a reliable system of elevated signals would be very beneficial, in part because Coston's flares illuminated a vessel enough to reveal its characteristics to an enemy. He indicated that tests with signal mortars and rockets had been unsuccessful and posited that Roman candles, despite their problems, offered the best hope for the future. Parker nonetheless warned his superior that a reliable signaling system based on Roman candles was still many months, if not years, away.[103]

Because there exists no correspondence between Parker and Very other than the latter's proposal of August 1874, one is hard pressed to ascertain exactly when the signal office commenced tests on Very's system. Parker's letter of February 1876 contains no mention of any such tests, but one year later his successor as chief signal officer, John C. Beaumont, wrote that his office had conducted "many experiments" with the system, concluded that it was "the best yet offered," and recommended the purchase of numerous signal pistols.[104] Beaumont later indicated that Parker had supervised some unsuccessful trials in Boston, so the signal office probably began testing Very's system in the spring or summer of 1876.[105]

What explains the delay between the signal office's receipt of Very's proposal and its initiation of testing? A definitive answer is unattainable; however, one can state with near certainty that Parker would have tested Very's system earlier if practicable. After the Key West maneuvers, his office aggressively sought a reliable system of elevated signaling, and it would not have rejected a priori a promising system like Very's. Because there is no evidence of personal animosity between Parker and Very, the most likely explanation is that the latter was simply not ready to subject his system to trials. Very's own proposal supports such a conclusion. In that proposal, the young naval officer stated he had performed "but a few experiments" and wrote that, although his ideas had been completely thought out, the system itself remained "in embryo."[106] Patent records provide further support for this explanation, as Very did not file a patent application for his new signal cartridges until February 1877.[107]

By the fall of 1877, tests of Very's signaling system were going well. The navy's chief signal officer personally supervised many of the trials, noting with satisfaction that only three flares out of seventy-nine tested, or slightly less than 4 percent, failed to ignite properly. More importantly, experiments indicated that the new signals could be seen at distances of up to ten nautical miles. Calling such results "extremely satisfactory," the chief signal officer concluded that "the 'Very Signal' meets the demands of the service more satisfactorily than any other system now known."[108]

These experiments led the Bureau of Navigation to authorize use of Very's signals in the fleet, but for the time being the service chose not to acquire them on a large scale.[109] At first glance, this decision appears to reinforce the idea that the navy was a conservative institution resistant to new technology. A closer look actually leads one to the opposite conclusion. In an acquisition strategy the service would pursue regularly in future decades, the signal office postponed widespread adoption of a communications technology in order to invest its limited resources in a research and development program for that technology. Far from being conservative, this strategy carried some degree of risk. The *Virginius* crisis had shown how quickly a maritime conflict might arise, but the navy's leadership purposefully delayed introduction of a new signaling system into the fleet. Why?

Four reasons stand out. First, the existing system employed by the fleet for nighttime signaling, Coston's flares, worked well. As previously discussed, in the early 1870s Martha Coston had replaced her percussion cap igniter with a friction ignition mechanism, an improvement that significantly reduced signaling duds. Indeed, in 1874 the chief signal officer approvingly noted that during repeated experiments with Coston's new flares, "not a single failure to ignite occurred."[110] Second, the navy faced severe budgetary constraints throughout the 1870s, and the service's leadership recognized that funds for the wholesale replacement of signal flares would be extremely hard to obtain. Whatever system the fleet acquired, it probably would remain in service for some time. Third, while armed conflict was always a possibility, the likelihood of war, at least in the short term, was low. Lastly, and perhaps most critically, the signal office believed it could improve upon Very's work. The 4 percent failure rate obtained by the signal office in the fall of 1877 resulted, in part, from optimal conditions: a clear night with recently manufactured flares operated by experienced personnel. The misfire rate under typical shipboard conditions would be higher, and during combat even a single signaling failure could have dire consequences. Simply put, the navy's leadership concluded that projected improvements to Very's system outweighed the risks incurred by pursuing a research and development program for that system.

Accordingly, the Bureau of Navigation tasked an officer to conduct further experiments with Very's signals. The individual chosen for the assignment was Lieutenant Edwin Longnecker, an 1865 graduate of the Naval Academy who had participated in the Key West maneuvers. The bureau probably chose Longnecker because several years earlier he had been stationed at the United States Torpedo Station in Newport, Rhode Island, where he would have gained experience working with chemical explosives.[111] Longnecker's orders tasked him to experiment with Very's signals and to improve, if practicable, their "altitude, certainty of ignition, security against breakage, [and] uniformity of burning." He started experimenting in early November, test firing nearly five hundred flares over the next twelve weeks. Of particular interest to Longnecker was the pressure at which signal cartridges should be packed. Through trial-and-error experimentation, he determined that a press setting of just less than four hundred pounds, combined with a fifteen-day period in which the cartridges were allowed to "dry and harden," produced the most reliable flares. By inserting elastic wads into Very's signal cartridges, Longnecker also increased the maximum altitude to which flares rose before reaching their apogees.[112]

Additional experiments conducted by the signal office in the spring of 1878 demonstrated the value of Longnecker's work. In late April and early May, two officers performed tests to determine the reliability of the new signal cartridges. Out of 150 flares fired, only 1 failed to ignite.[113] The chief signal officer reported these favorable results to the chief of the Bureau of Navigation, who replied that he was "very much gratified."[114] Once again, however, the navy chose to continue development rather than to pursue production and distribution. Lieutenant John H. Moore, one of the officers who had conducted the aforementioned tests, believed he could improve further Very's system by reducing the time interval between signals.[115] To send the hypothetical signal "258" under Very's original scheme, for example, an operator had to fire a cartridge containing one red flare (2); pause, then fire a cartridge containing one white and one red flare (5); pause again, then fire a cartridge containing three red flares (8).[116] These pauses were critical for ensuring a recipient did not mistake a late-igniting flare for a unique digit. Moore believed more rapid signaling was possible if all digits consisted of three flares, an arrangement that also reduced significantly the likelihood a signal would be misinterpreted if a flare failed to ignite. That summer Moore requested funds to develop the idea.[117] Predictably, he received full support from the signal office.[118]

Details about Moore's work remain elusive, but at the end of 1878 the chief signal officer reported that the "combination signals" suggested and developed by Moore were ready for testing.[119] In a series of experiments supervised by officers not attached to the signal office, a special board confirmed the superiority of Very's signals over other systems of nighttime signaling.[120] With this favorable report in hand, the signal office prepared for the final phase of its research and development program: trials at sea by fleet personnel.

Even before proceeding to sea trials, however, the signal office welcomed a new addition to its staff. Sometime during the first eight months of 1879, most likely in July or August, Lieutenant Edward W. Very reported for duty.[121] Very probably requested this assignment, although possibly a high-ranking officer in the Bureau of Navigation simply decided he was the best man for the job. Either way, Very was an appropriate choice. Even with his duties at the Bureau of Ordnance, Very had followed closely efforts to improve his signaling system.[122] Certainly there was no one more qualified to shepherd that system through the final stages of development.

Generally pleased with the signal office's improvements to his signal cartridges, Very initially spent a majority of his time developing a better pistol from which to fire those cartridges. In August he requested the resources to conduct experiments toward that end, and in September he submitted drawings and a prototype pistol to the chief of the Bureau of Navigation for review. Very claimed superiority for his new device in six areas: "Greater and *more regular* altitudes for the stars . . . Greater compactness . . . Cheaper manufacture . . . greater rapidity in loading . . . Decrease in liability to misfire [and] Decrease in weight."[123]

While officers in the bureau took time to examine this new signal pistol, Very kept busy. He thoroughly investigated the use of electric searchlights for signaling purposes, submitting a detailed report on the subject in January 1880.[124] Five years earlier the signal office began experimenting with these devices, but progress had been slow.[125] Existing dynamos were inefficient and temperamental, and the sea's harsh environment wreaked havoc on the electrical connections between dynamo and searchlight. True to form, Foxhall Parker became an enthusiastic proponent of the new technology. Yet he also realized that some time would pass before lights supplanted flares as the fleet's primary means of nighttime signaling.[126] Shipboard experiments conducted in 1879 confirmed Parker's analysis, demonstrating that unsolved technical problems still bedeviled the navy. Very put forward potential solutions to several of these problems and encouraged the signal office to implement "a specific programme of experiments," particularly with state-of-the-art European dynamos and searchlights.[127] Two decades later the navy would follow a similar approach when adopting wireless telegraphy.

Very's research on searchlights may have prompted him to consider more broadly the signaling practices and procedures of the U.S. Navy, for in January 1880 he also penned a remarkable letter in which he recommended changes to some of those practices and procedures. Very's goal in writing this missive was classically progressive: he sought to use institutional reforms to achieve desired efficiencies. His proposals certainly supported this end. Yet Very's letter is remarkable for a different reason; namely, it reveals a keen awareness of the need for more effective methods of shipboard information management.

In his letter, Very suggested some minor administrative reforms, such as consolidating accounts for signal stores and removing obsolete equipment from ships' signal kits, but his most significant proposals focused on changes in two areas: training and analysis. Regarding the former, he proposed abolishing the requirement for a special signal officer on each cruising vessel, arguing that the practice caused more harm than good. First, Very reasoned, it placed critical expertise in the hands of just a few individuals at a time when every naval officer "should be thoroughly acquainted" with signaling procedures. Second, the policy of detailing a special signal officer restricted training opportunities because a single individual was primarily responsible for all signaling instruction. Very argued that every division officer should be made "as much responsible for the progress of the men of his division in this particular [area] as in great guns or small arms." Finally, Very proposed that certain crewmen should receive extra instruction, writing: "It is always useful and sometimes of the greatest importance that quartermasters and lookouts should be thoroughly conversant with different signals."[128] In short, Very's training proposals sought to increase the number of shipboard personnel who could facilitate command and control processes.

A signal quartermaster raises a flag off the deck of USS *Portsmouth*, c. 1880s. Even with large flags such as this one, the maximum distance at which signals could be read with a spyglass was about four nautical miles. *Naval History and Heritage Command photograph NH 42683.*

Very organized a second set of proposals around a closely related issue: how best to evaluate the efficacy of the navy's signaling systems. As previously discussed, in 1869 the Bureau of Navigation began requiring every vessel in active service to submit a quarterly signal report providing the names and ranks of all persons under instruction, the amount of time they had practiced with day and night signals, and their progress in learning and executing those signals. Such reports theoretically provided the signal office with a measure of the fleet's signaling proficiency. On one level, this was true. Yet Very saw these "quarterly reports of instruction" as inadequate. He argued that they encouraged personnel to view signaling in isolation rather than as a means to an end. To solve this problem, Very proposed incorporating all signaling exercises into each ship's "routine drills."[129]

According to Very, if the service made signaling exercises a matter of routine, then separate reports on signaling instruction would no longer be necessary.

In place of these now "superfluous" reports, he proposed a new type of quarterly report, one that permitted the signal office to analyze systematically the fleet's actual communications capabilities. Very wanted each vessel "to keep a record of every signal made and received . . . and to make a Quarterly Report thereof to its flagship together with all useful information with regard to signalling." He explained his reasoning in some detail:

> The Naval Signal Book is an empirical work . . . It is a commonly known fact that there are many necessary signals which cannot possibly be transmitted by the General Code. There are many which are of constant use and which might well be modified so as to [be] economical in the number of flags displayed or in the manner of expression . . . It is only through reports on the actual use of these signals that the errors can be ascertained. It is the province of the Signal Office to examine and keep revised the signal book and this work can only be accomplished through the study of reports. I therefore recommend that in order that our signal system may be perfected, that each vessel of a squadron keep a record of all signals made by day or night.[130]

Once more, Very's proposals advocated institutional procedures that would improve shipboard command and control, this time by providing a more effective way for the signal office to collect and analyze signaling data.

Eventually the navy would adopt most of Very's suggested reforms, but in 1880 the signal office faced a more immediate concern. Specifically, it needed to draft and incorporate into the general signal book rules and procedures for using Very's flares. Very spent most of the year on this task, completing his work in late October.[131] After more than four years of research and development, the navy's new nighttime signaling system was finally ready for a comprehensive set of sea trials.[132]

Before commencing these trials, however, the chief of the Bureau of Navigation wanted to obtain the legal rights to Very's improved signal pistol. The chief may have remembered earlier disputes between the service and Martha Coston over her signal flares, and as a rule the bureau sought to avoid any type of patent litigation.[133] Very was willing to negotiate with the Navy Department, and in early July the two parties came to terms. The service agreed to pay the inventor a lump sum of $280 for "the working drawings and patterns" of his "improved night-signal pistol." In return, Very relinquished all future claims to his invention.[134]

Just two weeks after Very and the Navy Department finalized this agreement, the Bureau of Navigation issued revised signaling instructions to the fleet. The bureau distributed Very's system widely, and in January 1882 it directed the commanders of the North Atlantic, European, and Pacific squadrons to conduct comparative tests of Very's and Coston's signals.[135] The commanders of the North Atlantic and European squadrons assigned this task to their warships' commanding officers, while the Pacific Squadron convened a special board to evaluate the relative merits of the two signaling systems.[136] Overwhelmingly, the officers performing these trials reported that Very's signals were superior to Coston's. Succinctly reflecting the tenor of these reports, the individuals assigned to the Pacific Squadron's special board wrote that "the greater rapidity and decidedly greater visibility and distinctiveness of the Very's [signals] give them, taking everything into consideration as at present issued for service, an appreciable advantage over the Coston's."[137] In August, the signal office performed a series of experiments to ascertain the optimal type of signal cartridge for Very's pistol, but by then debate over the best available nighttime signaling system had ended.[138] On 21 October 1882 Secretary of the Navy William E. Chandler issued a general order officially adopting "the Very System of Night Signals . . . for use in the Navy."[139] The service finally had achieved Foxhall Parker's goal of making available to the fleet a reliable system of elevated signaling.[140]

The French Connection

Very's system extended the range at which naval signals could be sent and received to about twelve nautical miles, an impressive distance for signal flares.[141] Yet adoption of the new technology was no panacea. Ignition problems persisted, especially for flares long exposed to the ocean environment, and one of the biggest advantages of Very's system—360-degree visibility at a considerable distance—was actually a liability for any commander who wished to communicate as covertly as possible.[142] The navy therefore continued to pursue promising new communications systems, including electric light signals, messenger pigeons, and signal balloons. Although pigeons and balloons never achieved the success hoped for by their advocates, in 1891 the service adopted an electric light system to complement Very's signals.

Throughout the 1880s, most electric signaling systems were operationally impractical. While improved dynamos and better insulators partially solved the issues identified by Edward Very in his report on electric searchlights, other problems remained. Even the best dynamos required frequent maintenance, and none could be operated routinely twenty-four hours a day. In addition, the wiring of warships was both complex and expensive. As one expert described the matter: "The leading of wires on board a vessel is, of itself, a laborious undertaking . . . wires must be out of the way of an enemy's shot and shell as much as possible, they must be accessible, must be protected from chafe, salt water, or undue strains, and the various circuits must be conveniently located for those who are to use them."[143] Most critically, though, existing searchlights were relatively fragile and produced light that varied considerably from minute to minute.[144] Searchlight manufacturers had some success correcting the latter problem, but the former remained an acute issue well into the twentieth century. For obvious reasons, no commander wanted to rely on a signaling system that might be knocked out at the first exchange of gunfire.

No surprise, then, that the first electric light signaling system adopted by the service employed not searchlights but rather incandescent lamps suspended from a ship's mast. Impetus for acquiring such a system appears to have come from John Grimes Walker, who assumed command of the United States Squadron of Evolution on 1 October 1889.[145] Walker, who had been chief of the Bureau of Navigation for an unprecedented eight years, persuaded Secretary of the Navy Benjamin Franklin Tracy to place him in charge of this new squadron, the purpose of which was to test tactics and to promote the "New American Navy."[146] Almost immediately, Walker demonstrated a strong interest in ship-to-ship communications. Just seven weeks after taking command, he wrote Tracy to request a "double allowance" of night signals, informing the secretary that his current supply of Very's flares was "entirely inadequate for the amount of night signalling which I propose to do in this Squadron."[147]

The Squadron of Evolution sailed for Europe in early December, arriving in Lisbon, Portugal, shortly before Christmas. After a brief holiday stand down, Walker began to exercise his forces in earnest. As had been the case in the aftermath of the *Virginius* crisis, American warships initially had difficulty executing even simple maneuvers.[148] Repeated practice improved their ability to operate as a coordinated body but also revealed an unexpected problem with Very's signals.[149] In executing rapid nighttime maneuvers at high speeds, Walker discovered he could not convey tactical instructions at operationally adequate rates. Simply put, Very's system, which many officers had extolled for its rapidity less than a decade earlier, was gradually being rendered obsolete by improvements in warship propulsion. Compounding the matter, Walker noted that frequent maneuvers made the use of Very's signals quite expensive.[150]

To overcome these dilemmas, Walker suggested "a system of night signals which consists in general of a number of double electric lamps, which are permanently set up in some prominent position, along one of the back stays, or from the masthead to the end of the signal yard." According to Walker, the navies of England, Spain, France, Italy, and Germany either were experimenting with or already had adopted such signals. The admiral reported that he had examined each foreign system and concluded that "the one devised by Captain Ardois . . . is the best yet devised."[151] On the basis of Walker's recommendation, the Navy Department acquired Ardois lights for every vessel in the Squadron of Evolution.

Because the Parisian company that made the Ardois system required several months to manufacture its product, the squadron did not receive the new signal lights until January 1891.[152] Walker immediately tasked his flag lieutenant, Sidney A. Staunton, to evaluate the new system. An 1871 graduate of the Naval Academy, Staunton was one of the services' leading signal experts. He had served as flag lieutenant to the European Squadron commander for two years in the mid-1880s and had held the same post in the Squadron of Evolution from the day Walker assumed command in October 1889.[153] Staunton eagerly embraced the assignment, inspecting the Ardois lights on every ship and analyzing their effectiveness during maneuvers. In mid-May he reported to Walker that, although there existed a few minor defects, these were "trivial and of easy remedy." Staunton believed the new system had "no rival" at distances less than three miles and thought it greatly improved the squadron's ability to maneuver at night. He even reminded Walker of the lights' recent utility on "a dark rainy night" when "the Squadron [maneuvered] with all the ease and rapidity which usually attend the best circumstances in daylight."[154] The admiral forwarded Staunton's analysis to the Bureau of Equipment with a ringing endorsement, writing: "The practical working of this apparatus has fully confirmed the high opinion which I had formed of its capacities and usefulness during my cruise in the Mediterranean."[155] Walker also informed Secretary Tracy that the new system was invaluable at rendering nighttime communications "rapid and certain."[156]

How exactly did Ardois signals work? Fortuitously, a junior officer in the Squadron of Evolution named Albert P. Niblack had a strong interest in ship-to-ship communications and published a detailed description of the Ardois system. According to Niblack, it cost about eighteen hundred dollars per vessel and consisted of five evenly spaced electric lanterns on a cable strung between masthead and deck.[157] Each lantern held a white light and a red light, but an operator could illuminate only one color at a time in any given lantern. By turning all five lights to one color a ship could communicate via binary code (e.g., all white equals "zero," all red equals "one"); it also could employ a special keyboard to transmit letters, numbers, or directions. For example, to send the direction "west," an operator first would transmit the code for "compass." He accomplished this task by turning the pointer on his keyboard to a slot that illuminated (from top to bottom) one white light, three red lights, and another white light. The recipient would respond by turning his keyboard to the same slot, thereby displaying an identical signal. Seeing that the intended recipient had received the code for compass, the sender then would turn his keyboard pointer to the slot for west, which illuminated three red lights, a white light, and another red light. The receiver again would repeat the signal to confirm receipt.[158]

Niblack was not the only junior officer interested in naval signaling, but throughout the 1890s he was one of the more influential. Born in Vincennes, Indiana, on 25 July 1859, Niblack graduated from the Naval Academy in 1880. After he served two years of sea duty, the Bureau of Navigation detailed him to the Smithsonian Institution, where he received extensive scientific training in preparation for a coast survey assignment in Alaska.[159] Niblack returned to the Smithsonian after several years in the Pacific Northwest, and in the summer of 1889 he requested extended duty aboard USS *Chicago*, soon to become flagship of the Squadron of Evolution.[160] The Navy Department honored Niblack's request, and *Chicago*'s commanding officer assigned him to several important positions, including assistant in charge of the ship's electrical plant, signal officer, and special aide to the executive officer.[161]

These assignments certainly help explain Niblack's interest in ship-to-ship communications systems. That interest, along with Niblack's sharp intellect and strong work ethic, caught Walker's attention.[162] In March 1891 the admiral tasked Niblack and two other officers to investigate an improved signal flare developed by Martha Coston, and over the next year and a half Niblack published a series of articles proposing changes to the Ardois system. He also submitted a formal report to Walker, who convened a special board to consider his subordinate's proposals.[163] The service adopted many of Niblack's ideas, but not all of them. In particular, it rejected the young officer's proposal to cut the number of Ardois lights from five to four.

Niblack posited that a four-light system offered three advantages. First, it was more economical, reducing the cost of each system by nearly five hundred dollars. Second, many vessels with masts too short for a five-light configuration would be able to employ a smaller four-light system. Most significantly, however, Niblack argued that a four-light configuration was easier to use, in part because it decreased the number of signals an operator had to memorize by more than 50 percent.[164]

Opponents of the four-light system could hardly disagree with the first of these arguments, but by 1893 the service faced fewer fiscal constraints than at any time since the Civil War.[165] With respect to Niblack's second claim, not all officers agreed that smaller warships should receive Ardois lights. Indeed, fellow signal expert Sidney Staunton asserted that practical experience had shown just the opposite. Staunton noted that during his time in the Squadron of Evolution, "ships not supplied with the Ardois were furnished with the descriptions and instructions and were required to take in the Ardois signals with the remainder of the squadron, answering the call and the signal with [Very's flares] . . . no difficulty was experienced from the first."[166] Yet Niblack's third proposition, with its emphasis on simplicity, generated the most debate.

Niblack believed a reliable, slightly slower signaling system was superior to a speedier one more liable to errors, and his four-light proposal was actually part of a broader plan to simplify ship-to-ship communications. The innovative officer wanted to reduce the total number of naval signal codes from four down to two. For nighttime signals, Niblack sought to replace the standard three-flare Very code with a new four-flare code, which would allow all signals sent by either Very's flares or Ardois lights to employ the same four-element scheme. He also proposed readopting Myer's code, a system the service had abandoned after taking up Morse code in 1886.[167] Here Niblack was not alone, as many officers found Morse code inferior to its predecessor.[168]

The Navy Department implemented Niblack's recommendations for the four-element Very's signal and Myer's code, but it kept the five-light Ardois system for several more years.[169] A few officers, most notably Staunton, challenged Niblack's assumption that operators would find a four-light Ardois system easier to use. According to Staunton, "Any convenient and rational code is quickly learned by men who constantly use it. The [five-light] Ardois system has 62 elements or groups of lights. It is not strictly necessary to memorize then, as they are marked on the transmitter." Drawing again from his experiences in the Squadron of Evolution, Staunton claimed that signalmen "soon learned them [Ardois signals] by simple use and repetition as a child learns the alphabet, and would call them out when displayed without reference to the keyboard."[170]

In essence, Staunton and Niblack agreed in principle but differed on what was operationally practicable.[171] The former saw the five-light Ardois system as an efficient way to transmit time-critical tactical signals, whereas the latter viewed it as an overly complex system that might confuse naval personnel in time of battle. To a certain extent, both were correct. During the Spanish-American War (1898), skilled operators often communicated with Ardois lights and experienced few difficulties. On the other hand, the navy's rapid wartime expansion resulted in a shortage of these skilled operators. Moreover, because all major naval engagements of the war took place during daylight, the fleet never had to employ the Ardois system in the type of nighttime action that had so concerned Niblack.

Although the service did not embrace all of Niblack's proposals, for several more years it drew heavily on his expertise in naval signaling. After Niblack detached from the Squadron of Evolution, he spent the winter of 1893–94 revising the navy's general signal book for the Bureau of Navigation.[172] This work was critically important, for by the mid-1890s the service's various squadrons increasingly were conducting coordinated tactical maneuvers. In August 1894 North Atlantic Squadron commander Richard W. Meade selected the still relatively junior Niblack to serve as his flag lieutenant, later reporting that the young officer "assisted me greatly with additions to Signal Code and Fleet General Orders . . . I consider him eminently fit for any service."[173] Niblack retained a strong interest in naval communications and fleet tactics after this tour, but his career would ultimately take him down a different path. In the late 1890s Niblack's outstanding reputation and noted language skills garnered him coveted attaché duty in Europe, and he spent an atypical portion of the remainder of his career filling important shore billets for the navy.[174]

The work of Staunton, Niblack, and other young officers, as well as the considerable support they received from senior commanders like Walker and Meade, indicates that there was widespread institutional interest in acquiring better systems for tactical command and control. Such interest was not limited to officers, as enlisted personnel sometimes also put forward useful suggestions for improvement.[175] Yet squadron commanders were the ones who could most readily initiate and practice tactical maneuvers, something they did with evermore regularity during the 1890s, especially in areas of the world where international tensions ran high.[176] To a considerable extent such maneuvers also reflected the changing strategy of the U.S. Navy, which was shifting from a focus on commerce raiding to one on squadron or fleet warfare.

Illustrative of this evolving *mentalité* were the actions of Francis M. Bunce, Meade's successor as commander of the navy's North Atlantic Squadron. The Civil War veteran had less sea time than most of his contemporaries, and his appointment came as a surprise to many observers.[177] Yet, even with a limited background in fleet maneuvers, Bunce exercised his ships in tactical drills as often as possible, striving "to keep the force under my command ready at all times for any duty or emergency." The new commander-in-chief took a special interest in ship-to-ship signals, directing each vessel under his command to report "at the end of each cruise at sea for tactical drills, every suggestion for improvement of the Fleet Drill Book and the Signal Codes that may occur . . . from [both] personal experience and observation." Bunce also had the New York Navy Yard modify his flagship's after bridge and signal house so he would possess better shipboard command facilities.[178] In effect, Bunce's activities show that by the mid-1890s Edward Very's vision—of a world in which American naval officers viewed signaling not in isolation but rather as part of an integrated whole—had become standard operating practice.[179]

In the aftermath of the Civil War, a growing number of American naval officers envisaged a service whose future lay not in coast defense or *guerre de course* but rather in fleet operations. Rapid, reliable, and secure signaling systems were central to this vision, and many officers, both senior and junior, worked hard to develop new and improved means of ship-to-ship communications. To be sure, the U.S. Navy rarely operated its warships in assembled groups during the 1870s and 1880s. Yet this did not stop individuals like Foxhall A. Parker and Edward W. Very from spearheading the development of communications systems well suited for fleet operations.

By the time the New American Navy was large enough to exercise as a coordinated group, Parker and Very had passed from the scene. But the future anticipated by Parker in the 1870s had started to materialize, bringing new problems, such as faster warships, as well as new possibilities, particularly those enabled by shipboard electrification. Once again, older officers like John Grimes Walker and younger officers like Albert P. Niblack worked together to develop and improve the fleet's systems of tactical communication. While some ideas led nowhere, others significantly enhanced the fleet's ability to communicate.

The Spanish-American War tested these abilities, although admittedly that test could have been more difficult. Even so, when President William McKinley announced a blockade of Cuba on 21 April 1898, U.S. warships in Key West sortied. By 23 April, America's most powerful battleship, USS *Iowa*, and dozens of other vessels had taken up station off the Cuban coast. *Iowa*'s signal logs reveal that during the ship's two-day transit and first day on station, personnel on board transmitted or received eighty-six messages through four different means: flaghoist signals, Myer's wigwag, Very's flares, and Ardois lights.[180] All of these methods relied on visual reception, the dominant signaling paradigm of the nineteenth century. Yet a new communications technology was looming, one that eventually would provide naval personnel with access to tactically relevant, real-time information from beyond the horizon. That technology was naval radio, and it soon would add a new level of complexity to the science and art of warfare at sea.

CHAPTER TWO

Sparks and Arcs

The Navy Adopts Radio

> It is needless to point out how greatly [wireless] will facilitate fleet maneuvers and operations, particularly in the way of supplying and communicating information gained by scouts, and the greatly enlarged field of action in which a single commander may, at all times, be promptly supplied with all important information, and may at the same time control and distribute his fleet to the best advantage.
>
> Albert M. Beecher, 1902

There Was No Sun

Jack London, the famous American writer, was in the wrong place at a historic time. Early on the morning of Wednesday, 18 April 1906, a shift in the tectonic plates at the northern edge of the San Andreas Fault created a massive earthquake. The nearest metropolis, San Francisco, was hit hard. London, who lived some forty miles north of the city, quickly made his way south, arriving in San Francisco about twelve hours after the first tremors. The earthquake itself was a rather brief event, lasting no more than two and a half minutes. The aftermath lasted much longer.

The quake brought down many buildings, and hundreds caught in the destruction perished instantly. It also toppled hot stoves, electric utility poles, and fuel tanks. Gas lines everywhere severed. Within minutes, fires broke out all over the city. By the time London arrived that afternoon, San Francisco was a vast conflagration. Fortunes vanished, ashes fell like rain, and refugees filled the streets. Still, the fires burned. London wandered the city for hours, a night watchman armed with a pen instead of a gun. At last, the new day approached. London described the scene: "A sickly light was creeping over the face of things. Once only the sun broke through the smoke-pall . . . [it] was a rose color that pulsed and fluttered with lavender shades. Then it turned to mauve and yellow and dun. There was no sun. And so dawned the second day on stricken San Francisco."[1]

At the very moment the earth shook in northern California, the officers and crew of steel cruiser USS *Chicago* were busily preparing to get under way from San Diego. Unaware of the escalating disaster hundreds of miles to the north, *Chicago*'s commanding officer pulled his ship away from the pier shortly after sunrise. The trip was to be brief, as *Chicago*'s intended destination was Long Beach, California. A little before seven o'clock, the pilot debarked and the cruiser rounded San Diego harbor's final buoy. Steaming at nine knots, the warship settled into a typical underway routine. Around half past seven, a young officer named Stanford C. Hooper made his way to the bridge. In preparing to assume watch as officer of the deck, Hooper undoubtedly checked the navigational charts, the ship's logs, and the weather forecast. He then relieved the watch.[2]

Stanford Hooper was in many ways typical of a new generation of naval officer. Born in Colton, California, on 16 August 1884, Hooper's first encounter with modern communications technology came at an early age when his father taught him how to operate a telegraph. For several years that skill helped him land summer employment as a substitute for vacationing telegraph operators. Hooper first considered joining the U.S. Navy while in high school after his father took him on a tour of a warship anchored near San Diego. Not only was the teenager fascinated by the vessel's mechanical equipment; he also liked how the "attractive girls" interacted with the ship's officers. Hooper sought and eventually received an appointment to the U.S. Naval Academy in Annapolis, reporting for duty in the summer of 1901.[3]

Hooper entered a service that recently had combined its engineering and line officer communities. In 1899, passage of a sweeping personnel reform act codified what seemed obvious to many naval administrators: good officers needed to master the skills of both the engineer and the seaman.[4] This view of the naval profession appealed to technically oriented individuals like Hooper. Because he had attended only two years of high school, the young Californian struggled in some subjects. Yet Hooper's background in telegraphy helped him excel in courses on electricity and engineering. After four hard years, Hooper graduated near the middle of his class. The navy assigned him to *Chicago*, flagship of the Pacific Fleet.[5]

Hooper quickly qualified as officer of the deck, the watch he was standing on the morning of 18 April 1906. He had relieved the watch at eight o'clock sharp, noting that the sky was overcast and hazy. Neither Hooper nor his fellow crewmen had forewarning of what would happen next. At ten past ten, *Chicago*'s wireless operator received a disquieting message: a destructive earthquake had hit the city of San Francisco. Just seven minutes later, at the direction of the senior officer on board, Hooper ordered fires struck in all boilers and changed course to the north. *Chicago* was going to San Francisco.[6]

Even at full speed, the Pacific Fleet flagship was more than a day away from the stricken city. The ship's officers and crew nervously bided their time during the transit. On the evening of 19 April, the day that dawned without a sun, *Chicago* finally pulled into San Francisco. A little past seven-thirty, the cruiser anchored and began transferring ashore hoses and men to fight the many fires still raging throughout the city. The next morning, *Chicago*'s commanding officer sent ashore additional supplies, including brooms, tents, water beakers, cooking ovens, and leg irons, the last of these apparently for use on apprehended looters.[7]

Other U.S. Navy ships soon arrived to assist, and on 22 April *Chicago* moved to pier 21. At that time, *Chicago*'s commanding officer, Charles J. Badger, removed Hooper from the watch bill.[8] Badger learned that fires had destroyed nearly all of San Francisco's telegraph wires and that city officials were having great difficulty communicating with the rest of the nation. Yet the commander was in charge of two things that could help solve the problem. One was Hooper, the former telegrapher. The other was his spark gap wireless apparatus. So Badger assigned Hooper to a warehouse by the docks, placing the young officer in charge of all outgoing traffic. Hooper would relay pertinent traffic to *Chicago*'s wireless operator, who then transmitted it to a naval station still linked to the telegraph system.[9] In this way, *Chicago* helped connect San Francisco with the rest of the country while a hastily assembled team of repairmen began to fix the city's broken telegraph lines.

Badger's leadership and initiative during the great San Francisco earthquake garnered praise from his superior, who wrote that his subordinate's work "could not have been more efficiently and tactfully performed."[10] A few years later, both Badger and the talented Stanford C. Hooper would play key roles in the U.S. Navy's adoption of radio for tactical communications. But in 1906 much uncertainty still surrounded this new technology.

All too often scholars have minimized this uncertainty, criticizing the navy for failing to appreciate radio's advantages and for not embracing it sooner.

Inventor-entrepreneur Guglielmo Marconi demonstrated a system of wireless telegraphy to the service in 1899, yet the navy delayed several years before adopting the new technology. According to Susan J. Douglas, the leading historian on early naval radio in the United States, a contractual dispute between Marconi and the navy was at the heart of this delay. She argues that this dispute resulted from contrasting attitudes, conflicting cultures, and divergent approaches to innovation.[11] There is much truth to this interpretation; however, the reason the navy reasonably could delay its acquisition of radio equipment at century's turn was because the technology did not yet provide significantly improved capabilities. In other words, the service simultaneously recognized the potential of radio communications *and* its existing operational limitations. Douglas overlooks this vital point when she argues that the American sea service "was not the sort of organization in which technical sponsorship, especially of an invention that threatened autonomy and decentralization, was either desired or possible."[12] As the previous chapter has shown, the U.S. Navy was just that sort of organization.

During the latter nineteenth century, Foxhall Parker, Edward Very, Albert Niblack, and other American naval officers aggressively pursued technologies that would enhance the service's shipboard communications systems. Individuals like these did not disappear with the end of the century, nor did the navy as an institution suddenly abandon its efforts to improve shipboard command and control. *Chicago*'s activities in April 1906 present a microcosm of those efforts. Badger innovatively used the machine and human resources at his disposal to solve an unanticipated communications problem.

While true that a contract dispute with Marconi delayed slightly the navy's adoption of wireless telegraphy, this delay never seriously affected the service's war-fighting capabilities. In 1903 the North Atlantic Fleet employed wireless during its summer maneuvers, and by the end of that year the navy had purchased more than fifty sets of radio equipment. Yet the fleet did not earnestly begin to use wireless telegraphy for short-range communications until the early 1910s. Why? Because until then the technology did not provide superior capabilities to existing methods of ship-to-ship communications. To be sure, older technologies had their own limitations. When compared to contemporary radio technology, however, day and nighttime signaling systems were more reliable, especially under the anticipated conditions of battle. Articulating such sentiments, Bradley Fiske, an officer often recognized for his foresight and technical expertise, wrote that "for nine-tenths of our purposes the present system of signals are perfectly adequate," and he declared that although wireless telegraphy was "extremely convenient for all the ordinary purposes of fleet work . . . it possesses the inherent defect that, if used in the presence of the enemy, the enemy can interfere with its indications by sending out Hertzian [radio] waves themselves."[13]

Some scholars have relied on statements made by Fiske and others to highlight the navy's institutional conservatism toward radio communications. Historian Susan Douglas, for example, claims Fiske viewed naval radio as fundamentally useless.[14] Yet a principal concern for Fiske, as for many naval officers, was wartime signaling. These individuals did not oppose wireless per se; rather, they believed radio communications would be operationally ineffective under conditions of battle. According to Fiske, wireless telegraphy supplemented and improved existing methods of ship-to-ship communication, but only in the absence of an enemy.[15] Fiske not only recognized that an opponent could use intentional interference to disrupt wireless communications but also anticipated the emergence of efforts to collect intelligence from radio signals, asking rhetorically, "What admiral is going to fill the ether with Hertzian waves, and make a present to the enemy of the information that he is near?"[16]

In the long run, Fiske overestimated an enemy's ability to interfere with wireless communications (but not an enemy's ability to exploit communications intelligence). Yet he, and many of his fellow officers, had assessed accurately the operational environment as it existed at the turn of the century. As such, Fiske offers a point of departure for reconsidering the generally accepted interpretation, based largely on anecdotal evidence, that the U.S. Navy resisted the adoption of a promising new communications technology in the early 1900s because of individual and institutional conservatism.[17] A close examination of the navy's work with wireless during the first decade of the twentieth century exposes not a hidebound bureaucracy but rather a technically oriented institution operating under budgetary constraints.

Once the service adopted radio, personnel of many ranks worked hard to develop innovative techniques for employing the new technology. Significantly, wireless telegraphy became the first communications system to provide seaborne commanders with reliable access to tactically relevant, real-time information from beyond the horizon. Because such information increased the complexity of naval operations, officers and enlisted personnel began to develop devices, methods, and procedures for managing that complexity.

Sparks and Coherers

Institutional efforts to harness electromagnetism for ship-to-ship communications date to the late 1880s. In 1887 the resourceful Bradley Fiske converted his ship, USS *Atlanta*, into a giant electromagnet by wrapping wire around the steel hull. He did the same to another steel vessel and attempted to communicate between the two ships via induction telegraphy. An expert in electricity and electrical systems, Fiske undoubtedly was familiar with the induction telegraph developed by Granville Woods for railcar communications.[18] Would a similar system work at sea? The answer was no, but Fiske remained undeterred. In an ensuing experiment, he tried to signal through the water itself by passing current through copper plates immersed beneath *Atlanta* and a nearby pier. Fiske ultimately succeeded in communicating between the pier and his ship, a distance of about two hundred yards. With the assistance of another officer, he then tested his system on two moving vessels.[19] It failed to work, and Fiske's superior soon directed him to pursue another project, an intriguing one that sought to make a torpedo boat "disappear" through the careful positioning of large mirrors.[20]

At the same time Bradley Fiske was attempting to devise a practical system of electrical ship-to-ship communications, German physicist Heinrich Hertz was in his laboratory trying to develop a device that could produce rapid electrical oscillations. In late 1887 Hertz finally succeeded, soon thereafter publishing a series of papers validating the electromagnetic theories of James Clerk Maxwell.[21] Hertz's scientific apparatus was impractical for communications, and several years passed before inventors achieved progress toward that end. In early 1896 physicist Ernest Rutherford succeeded in transmitting and receiving "Hertzian waves" at a distance of about one-half mile, but the future Nobel Prize winner shifted his energies to other endeavors after a colleague estimated that initial capital expenditures for a wireless signaling system could exceed £100,000.[22]

Another individual interested in the practical applications of Hertzian waves was British naval officer Henry B. Jackson. The idea of using such waves for naval signaling first occurred to Jackson in the early 1890s. His official duties left him little time to investigate the idea, however, until he assumed command of HMS *Defiance*, one of the Royal Navy's torpedo schools, in January 1895. Jackson was one of England's foremost torpedo experts, having served multiple tours in billets that involved the maintenance, operation, and development of torpedoes.[23] Because many types of torpedoes were electrically detonated these tours gave Jackson strong theoretical and practical backgrounds in electricity and the use of electrical apparatus.[24]

Toward the end of 1895, Jackson read about a relatively new device for receiving the Hertzian waves created by the generation of an electric spark across an air gap. To receive these waves, Hertz had employed a "resonator," an instrument that generated a tiny spark when electromagnetic waves of sufficient intensity passed through its wires.[25] Yet Hertz's resonator, which had been a groundbreaking artifact in 1887, was obsolete within a decade. In the early 1890s, a French scientist discovered that the conductivity of metal filings in a glass tube increased markedly when exposed to electromagnetic waves. A British physicist turned this principle into a new type of receiving device, which he named the coherer (because the metal filings "cohered" when exposed to electromagnetic waves). The coherer was more sensitive than ordinary spark gap resonators, and it quickly became standard equipment in physics laboratories.[26]

Upon learning about the coherer, Jackson incorporated the new device into his research on the use of Hertzian waves for naval signaling. Very quickly Jackson and his staff determined that existing coherers were insufficiently sensitive for naval purposes, so the inventive officer spent several months developing one of his own design. In the summer of 1896, he succeeded in sending and receiving messages at distances of about one hundred yards.[27] Such was the state of *Defiance*'s experimental program on Hertzian waves when the Admiralty ordered Jackson to London for a meeting at the War Office. There, he met a man who claimed to have invented a system that could guide a self-propelled boat or torpedo via wireless commands. That inventor's name was Guglielmo Marconi.[28]

Claims of a working, wireless, remote-control system were certainly exaggerated given Marconi's technical capabilities in 1896, but Jackson recognized immediately how closely the inventor's work paralleled his own. Although he had concerns about the seaworthiness of Marconi's equipment, the future First Sea Lord volunteered to advise the young Italian "on the special difficulties he would have to guard against in fitting his apparatus on board ship."[29] For several years thereafter the two men remained in regular contact. In spite of this budding professional relationship, Marconi's efforts continued to revolve mainly around tests for the British post office.[30] Meanwhile, Jackson, now armed with substantial knowledge about Marconi's work, continued to experiment. In May 1897 he installed a transmitter on an old gunboat and signaled to *Defiance* from various locations around Plymouth Sound. Jackson reported he could communicate effectively at a distance of more than two nautical miles and expressed optimism that with additional funds he could extend this range 75 percent while also doubling the speed of transmission from five to ten words per minute.[31]

Largely for pecuniary reasons, Marconi severed his relationship with the British post office in July 1897 and established a limited liability corporation, the Wireless Telegraph and Signal Company. The new firm pursued a range of business opportunities, at various times seeking to sell its equipment to the War Office, the Lighthouse Authority, and giant insurance syndicate Lloyd's of London.[32] The company also sought a contract with the Royal Navy, but not until September 1898 did the service find Marconi's system of wireless telegraphy reliable enough to warrant negotiations. By then Jackson had detached from *Defiance* to assume duties as the British naval attaché to Paris. Nevertheless, the officer who relieved him frequently sought Jackson's opinion on technical matters related to naval signaling. In January 1899 Jackson encouraged the Admiralty to collaborate with the Wireless Telegraph and Signal Company, arguing that the joint experiences of *Defiance* and Marconi would "determine the best details of the necessary fittings for practical use in H. M. service, and that more efficiently and quickly than either would do alone."[33]

Not only did the Admiralty follow Jackson's advice; it also tasked him to acquire wireless equipment from Marconi for naval maneuvers scheduled for the summer of 1899.[34] Jackson complied and soon found himself transferred from the British Embassy in Paris to HMS *Juno*, one of three warships fitted with Marconi's apparatus. Marconi embarked on *Juno* with Jackson, and during the maneuvers all of the wireless-equipped ships succeeded in sending or receiving tactical information.[35] Just as importantly, Marconi's apparatus held up well under the harsh conditions of a seaborne environment. In his report about the exercise, Jackson argued in favor of acquiring the new technology, pointing especially to its value for scouting.[36] His superior concurred, recommending that " the proposals to fit certain ships with [wireless telegraphy] be carried into effect as soon as possible."[37]

Satisfied with the performance of Marconi's equipment, the Admiralty tried to reach a contractual agreement with the Wireless Telegraph and Signal Company. Yet British patent law as interpreted by the post office (the branch of government statutorily assigned to oversee the development of electric communications) prohibited the Admiralty from agreeing to Marconi's demand for the payment of royalties, and Marconi and his business partners were unwilling to sacrifice this form of remuneration.[38] In an effort to break this impasse, the Admiralty wrote the postmaster-general to argue that the extreme importance of wireless telegraphy for the navy in time of war justified considering it "altogether apart from that of the general adoption by H. M. Government for peace purposes."[39] The post office appears to have accepted this justification, and in May 1900 the Admiralty agreed to purchase thirty-two sets from Marconi's company for £100 apiece. It also agreed to pay annual royalties of the same amount (i.e., £3,200 per year) until Marconi's 1896 patent had expired.[40] By the end of the year, the Royal Navy possessed more than fifty wireless sets.[41] The world's most powerful navy had formally adopted naval radio.[42]

Unlike the Royal Navy, the U.S. Navy had performed no sustained tests with wireless telegraphy during the 1890s. As such, reports filed by the American naval attaché in London recounting the performance of Marconi's equipment during the Royal Navy's 1899 summer maneuvers generated considerable interest within the service.[43] One officer particularly intrigued by the attaché's reports was Rear Admiral Royal B. Bradford, chief of the Bureau of Equipment. While in command of a vessel attached to the Squadron of Evolution, Bradford had witnessed firsthand John Walker's adoption of the Ardois light system, and he was one of the navy's leading experts in electrical systems.[44] Little surprise, then, that Bradford arranged for several naval officers to observe Marconi's radio reporting of the America's Cup yacht races off the New Jersey coast during the fall of 1899.[45] Sensing a business opportunity, Marconi was willing to allow these officers to observe his wireless sets in operation, but he and his assistants refused to provide detailed technical information about their equipment, undoubtedly for fear of revealing proprietary secrets.[46]

Despite Marconi's penchant for secrecy, the observers, all of whom were electrical experts, provided thorough reports back to Bradford. One officer merely described the tests,[47] but the other three submitted reports articulating their opinions about the new technology and its potential usefulness for naval operations. All three found Marconi's system promising but imperfect. One officer wrote that "a great deal of guess-work" was required to interpret some of the messages; although the transmitter worked satisfactorily, he noted that there were too many instances "when the receiver would not respond to the signals sent."[48] Another expressed dismay at the conflicting technical explanations given by Marconi and his assistants, yet also praised the inventor for his "control of the art of Wireless Telegraphy."[49] The third officer appears to have captured the general impression of all four observers, writing: "Signor Marconi seems to get better practical results than other workers in the field, but there is reason to believe that the system is capable of development to a much greater extent in the not distant future."[50]

After the international yacht races were over, naval personnel helped Marconi install his wireless equipment on board two warships, USS *New York* and USS *Massachusetts*. A lighthouse in New Jersey also received a set of Marconi's apparatus, so that both ship-to-ship and ship-to-shore communications could be tested. On 23 October 1899, navy secretary John D. Long appointed a special board of three officers, two of whom had observed Marconi's radio reporting of the America's Cup, "for the purpose of investigating the Marconi system of wireless telegraphy."[51] Trials began a few days later and continued for one week, during which time the U.S. Navy transmitted its first official radio message. More tests were scheduled, but Marconi chose to cancel these tests. The inventor stated he was unable to demonstrate "the devices I use for preventing interference," claiming that available equipment was inadequate and that he was unwilling to demonstrate a technical capability not yet "completely patented and protected."[52] Possibly Marconi sensed the observers' skepticism as to whether he really had the means to prevent interference. From Marconi's perspective, shortening the trials and returning to England was a sound business decision. He already had a preliminary contract with the Royal Navy, and the War Office wanted to purchase his equipment for the Boer War.[53] Marconi might jeopardize these opportunities by staying in America, and the U.S. Navy could not be counted on to buy equipment that repeatedly failed what the service considered to be critical interference tests.[54] Nevertheless, Marconi offered to sell his system under terms similar to those he had reached with the Royal Navy. For twenty sets, Marconi wanted five hundred dollars per set, plus an annual royalty of the same amount for each set.[55]

From the U.S. Navy's perspective, the tests of Marconi's wireless system offered mixed results. On the one hand, they showcased the potential of radio communications. *New York* and *Massachusetts* communicated at ranges of up to thirty-six nautical miles, and operators successfully transmitted and received messages at average speeds of twelve words per minute.[56] In a remarkably prescient statement, several officers noted that, if the navy were to install Marconi equipment on its ships, "the best location of the instruments would be below, well protected, in easy communication with the Commanding Officer."[57]

Still, obvious problems existed, the most critical of which was interference. Unlike a modern radio, where multiple channels are available and tuning is automatic, all of Marconi's equipment radiated at approximately the same wavelength. In practice this meant that when two or more stations transmitted simultaneously, the sent messages could not be read. The naval officers assigned to comment on the trials, known collectively as the Marconi Board, described the problem of interference:

> When two transmitters are sending at the same time, all the receiving wires within range receive the impulses from transmitters, and the tapes, although unreadable, show unmistakably that such double sending is taking place. In every case, under a great number of varied conditions, the attempted interference was complete. Mr. Marconi, although he stated to the Board, before these attempts were made, that he could prevent interference, never explained how nor made any attempt to demonstrate that it could be done.[58]

Shortly after Marconi suspended testing, the Navy Department thanked Marconi's company for demonstrating its system of wireless telegraphy but expressed "hope that you will further perfect the system."[59] Secretary Long became more favorably inclined toward the new technology after reading the Marconi Board report and related correspondence from the chiefs of the Bureau of Equipment and the Bureau of Navigation, which recommended both further trials and establishment of a school to train naval personnel in wireless telegraphy.[60] Notwithstanding these recommendations, the secretary chose to postpone action. He noted that Congress already had appropriated funds to another government agency for wireless experiments and held that the navy should not duplicate those efforts.[61]

Royal Bradford, chief of the Bureau of Equipment, was frustrated by the secretary's decision. He argued that the experiments conducted by other U.S. government agencies had been unsuccessful and rather bluntly told the secretary he was "misinformed."[62] Bradford continued to argue for "the installation of the Marconi devices on board of several ships of the Navy," but he softened his stance after unexpectedly high operating expenses depleted his annual budget.[63] In February and then again in June, Congress had to pass two deficiency acts to prevent Bradford's bureau from overspending on routine equipment for naval vessels.[64]

While Bradford struggled with this budgetary shortfall, Marconi continued to investigate the interference problem, and in April 1900 the British government granted him a patent for a syntonic transmitter.[65] The heart of this so-called four sevens patent (patent no. 7,777) was an oscillation transformer, designated the jigger, which Marconi used to connect the transmitter's discharge circuit with its antenna. The inventor connected the coherer circuit and receiving antenna in a similar manner. Because the radiated energy in this four-circuit tuning system could be confined to a narrower range of frequencies, Marconi obtained two practical benefits: the ability to distinguish one set of signals from another, and a stronger response to feeble signals. In theory, at least, Marconi's four sevens patent provided the means both to prevent interference and to increase range.[66]

Like many inventions, Marconi's syntonic system did not work as well as intended. Keeping the transmitter and the receiver in syntony required frequent calibration, and skilled operators were a necessity. Worst of all, Marconi's apparatus remained extremely vulnerable to interference. The America's Cup races in 1901 highlighted this problem. As he had done two years earlier, Marconi agreed to report the contest, but during the races his shore station could not read any incoming traffic because a rival company intentionally interfered with the signals transmitted by observers at sea. The incident greatly embarrassed Marconi and reinforced American naval officers' concerns about the problem of interference.[67]

If Marconi's system could not reliably transmit and receive radio messages, then perhaps another wireless system could. This sentiment gained currency within the Navy Department after the Office of Naval Intelligence published a favorable report outlining Adolf Slaby's work for the German navy. This report indicated that Slaby had succeeded in transmitting and receiving messages between ships at distances comparable to those achieved by Marconi.[68] Slaby and others previously had claimed that bad syntony was a serious flaw in Marconi's system,[69] but there existed no hard evidence that any system of wireless telegraphy could reliably prevent interference. Faced with this reality, Bradford concluded that available systems were unsatisfactory for naval use and that American adoption of shipboard wireless telegraphy would have to wait until it had passed "beyond the experimental stage."[70]

Taken out of context, Bradford's conclusion suggests an institutional conservatism toward radio technology. Yet this was the same officer who less than two years earlier had written that wireless "promises to be very useful in the future for the naval service."[71] Bradford's initial enthusiasm had been tempered, first by budgetary constraints, then by reports that the shipboard wireless systems of foreign navies would likely be ineffective under conditions of battle. From Bradford's perspective, Marconi's insistence upon the payment of annual royalties must have appeared outlandish. The inventor's system failed to ensure ship-to-ship communications during wartime, and it was expensive. In addition, Bradford faced a difficult legal constraint: the law barred him from obligating funds beyond the current fiscal year.[72] Marconi's proposal, which essentially amounted to a leasing arrangement, was rather unusual for its time. Marconi no doubt believed royalties were justified by the development costs he already had incurred, but he and his associates seem to have overestimated the utility of their system while failing to comprehend fully the financial constraints faced by the institution to which they were trying to sell their product.

Seeing Marconi's terms as unreasonable and recognizing the existing operational limitations of naval radio, the Bureau of Equipment adopted a "wait-and-see" approach. It began investigating domestic patents and launched an aggressive intelligence gathering effort to determine the latest European developments in wireless telegraphy. This course of action made perfect sense for an institution constrained by both legal issues and an austere budget, because it meant that European nations would bear a sizable portion of the development costs of the new technology. The strategy was not without risk, of course, as a technical breakthrough might leave the U.S. Navy behind and put American forces in peril during a naval engagement. At the turn of the century, however, the likelihood of a major maritime conflict seemed remote. Spain had just been defeated, war with England was increasingly unlikely, relations with France and Russia remained friendly, and Japan was not yet seen as a significant threat.[73] War with Germany was possible, but relations with that country had improved after Samoa's partition in 1899.[74]

Exactly when Bradford adopted this acquisition strategy remains unclear, but by the summer of 1901 he was actively pursuing information on naval radio. In early August, the Bureau of Equipment sought out more than a dozen patents relating to wireless telegraphy, and the following month it requested that the American naval attachés in London, Paris, Saint Petersburg, Berlin,

Rome, and Vienna be directed to seek comprehensive information on wireless equipment and methods of operation.[75] In October, Bradford persuaded the Navy Department to detail an officer to visit and report on all domestic companies developing wireless telegraphy.[76] About the same time, Bradford asked retired naval officer Francis Barber, then living in France, to investigate and report on the existing state of wireless telegraphy in Europe.[77]

Barber was not necessarily the ideal person for such an assignment. Although the former officer was an electrical expert, had a good relationship with Bradford, and spoke both French and German, he tended to hold inventors in low esteem and often voiced opinions bordering on the xenophobic.[78] He also could be acerbic and petty. Barber sometimes clashed with the American naval attachés who also had been tasked to gather information on European wireless developments, and he especially disliked Marconi, at one point expressing hope that the inventor's company could be driven out of business.[79]

Barber's foibles notwithstanding, his reports spurred Bradford to reconsider the Bureau of Equipment's wait-and-see approach to naval radio. Barber argued that the secrecy surrounding European navies' efforts to adopt wireless telegraphy meant the United States would be slow to learn about technical improvements and that, despite radio's obvious defects, the navy should initiate experiments with the latest available equipment. Aware of the bureau's budgetary constraints, he suggested that Bradford encourage Congress to provide a special appropriation for the purchase of two complete sets of wireless apparatus from five different European companies, "and such other instruments as the Naval Attaches or myself may from time to time discover and recommend."[80]

Bradford took action about two weeks before Christmas, informing Secretary Long "the time has arrived when it is not only necessary to test the various systems in use, but to instruct Naval signalmen in the use of wireless telegraphy."[81] He asked Barber to determine the prices and product guarantees European wireless companies were willing to offer. In January, the bureau chief authorized the purchase of two sets of three different types: Slaby-Arco, Ducretet, and Rochefort.[82]

Bradford also requested that an officer and two enlisted naval electricians be sent abroad to study recent developments in European wireless telegraphy. Like the navy's chief signal officer three decades earlier, Bradford recognized the need for a cadre of skilled personnel who could disseminate technical knowledge throughout the service. Although three people marked a slow start toward this goal, the bureau chief expected these individuals to return as subject matter experts. They would then provide training for what the Bureau of Equipment anticipated would become a new rate of "wireless telegraphic operators."[83]

Indeed, Bradford's attention to the role that enlisted personnel would play in the diffusion of wireless technology echoed Edward Very's efforts to provide extra instruction to certain categories of enlisted men in order to increase the number of shipboard personnel who could aid in command and control processes. The Bureau of Equipment worked closely with the Bureau of Navigation, which maintained enlistment records, to select two skilled chief electricians for the European assignment. Of the two individuals chosen by the bureau, William C. Bean and James H. Bell, relatively little is known about Bell, but Bean clearly possessed appropriate skills for the assignment. He had served almost continuously in the dynamo rooms of American warships for nearly a decade and had developed an interest in wireless telegraphy after witnessing the original Marconi tests as chief electrician on board *New York*.[84]

The officer selected for the assignment was Lieutenant John M. Hudgins. A native of Virginia, Hudgins received an appointment to the Naval Academy at the age of eighteen, matriculating there in September 1890.[85] He performed well academically and in 1893 was selected for admission into the engineering corps. As such, during his senior year Hudgins studied steam engineering, higher mathematics, and "other subjects having an affinity with the profession of engineering."[86] Promoted to assistant engineer in July 1895, Hudgins served mainly on blockade duty during the Spanish-American War. He became a line officer after passage of the Naval Personnel Act of 1899, serving on several different ships through the fall of 1901, when he received orders to the Bureau of Equipment.[87] Undoubtedly aware of Hudgins's strong background in engineering, the bureau placed him in the office of the inspector of electrical appliances. When Bradford needed an officer to go abroad to investigate developments in wireless telegraphy, Hudgins was an obvious choice.[88]

After some minor delays, in April 1902 the Navy Department issued orders to Hudgins and the two naval electricians, instructing them to depart for Europe as soon as practicable.[89] Hudgins, Bean, and Bell arrived in Paris in early May, met with Barber, and immediately began investigating the wireless apparatus of Ducretet and Rochefort.[90] After several days in Paris, Hudgins traveled to southern France to witness tests of some French wireless equipment on board USS *Nashville*. He reported that *Nashville* and another vessel communicated via wireless at a distance of five miles but that the rolling and pitching of the ship made reception by coherer extremely challenging.[91] Hudgins rejoined his assistants in Paris upon the completion of testing, and in June the trio moved to Berlin to learn about the Slaby-Arco system and that of another potential supplier, Braun-Siemens-Halske.[92] Finally, the group traveled to England, where it spent eleven days examining the devices of Marconi and several other British firms.[93]

After returning to the United States, Hudgins submitted a formal report concerning his assignment. The report concluded that no existing system of wireless telegraphy worked in an entirely satisfactory manner:

> There was always either interference, lack of adjustment, or some fault either in sending or receiving apparatus which rendered accurate reception of a message difficult and doubtful. Important messages, cipher, proper names, or any sentence in which the context would not assist in deciphering, [had] to be repeated back and forth frequently for four or five times before the receiving station could make a copy without errors. Many of these errors, in most cases the majority of them, are directly attributable to the coherer, either in failure to cohere or decohere properly.

Hudgins argued that published reports of European radio equipment were willfully misleading, with "no experimenter publishing anything except his best results, sometimes out of hundreds of tests."[94]

Hudgins not only reported deficiencies but also sought to provide the Bureau of Equipment with practical information that could help it determine the best available wireless apparatus. For example, he noted that the Braun-Siemens-Halske equipment was less susceptible to inference than other systems and that German naval officers had been able to improve their Slaby-Arco sets through minor modifications. Providing the caveat that he had to draw his conclusions from limited data, Hudgins listed, in order, Marconi, Braun-Siemens-Halske, Slaby-Arco, Rochefort, and Ducretet as the sets most suited for adaptability to naval service. He was careful to point out that the apparent superiority of Marconi's system was "due principally to the greater skill and experience of the engineers in installing and operating the apparatus, and not in the apparatus itself."[95]

Shortly after Hudgins submitted his report, the Bureau of Equipment commenced tests on the European equipment it had purchased. The Navy Department appointed five officers to a special board, and Bradford tasked them to ascertain "the best existing system of wireless telegraph apparatus for use in the naval service."[96] These officers included Hudgins; Albert M. Beecher, who previously had been tasked to visit and report on American manufacturers of wireless apparatus; and Charles J. Badger, who would in two years' time assume command of *Chicago*.

Bradford gave this "Wireless Telegraph Board" specific instructions for determining the "best" system for naval use, informing its members that reliable, low-maintenance sets were superior to more powerful ones requiring frequent adjustment. Trials began on 18 August 1902.[97] Noticeably absent was Marconi's company, which despite Bradford's entreaties still refused to sell its equipment to the navy.[98] Initial tests ran through late November and involved apparatus provided by the firms Hudgins and his assistants had visited in France and Germany. The Slaby-Arco radios impressed the board the most, and in late winter the Bureau of Equipment placed a production order for twenty sets of Slaby-Arco equipment.[99]

Anxious to get wireless out to the navy's operating forces, Bradford wanted these radios installed on the warships of the North Atlantic Fleet as soon as possible, but manufacturing delays held up shipboard installations.[100] Informed of this setback, the bureau chief still pushed for delivery of as many sets as possible, "consistent with good workmanship."[101] The manufacturer obliged, shipping five sets in early July.[102] The Bureau of Equipment worked closely with the North Atlantic Fleet and the Bureau of Construction and Repair to install these systems, and by August three battleships, two cruisers, one gunboat, one auxiliary vessel, and a shore station in Maine had received wireless sets.[103]

Bradford cautioned the North Atlantic Fleet commander, Rear Admiral Albert S. Barker, not to expect too much from the new technology, informing him that the optimal arrangement for shipboard installations had not yet been determined and that few skilled operators were available. Nevertheless, the bureau chief encouraged his seagoing counterpart to make full use of the Slaby-Arco sets and suggested he take advantage of the extensive expertise of Hudgins, who had been assigned to support the North Atlantic Fleet.[104] Barker took Bradford's suggestions to heart. In early August, he supervised an exercise in which his forces attempted to stop an "enemy" from establishing a base on the New England coast. USS *Olympia*, one of the friendly fleet's wireless-equipped ships, sighted the hostile forces and transmitted this information to the commander-in-chief some thirty-five miles away. Friendly forces stopped the enemy fleet before it could reach the coast, and Barker enthusiastically acknowledged the usefulness of shipboard wireless telegraphy for seaborne commanders.[105]

Indeed, Barker's reaction to wireless telegraphy offers another counterpoint to historian Susan Douglas's portrayal of the U.S. Navy's adoption of radio as a process in which "officers on shore" had to "compel officers at sea to adopt the invention."[106] Barker, who had no previous experience with wireless, recognized immediately its potential significance for naval operations. As had a number of American naval officers before him, including seaborne commanders A. Ludlow Case and John Grimes Walker, Barker viewed the new communications technology not in isolation but rather as part of a larger system of command and control. Writing to his superior, Barker emphasized the importance of well-trained operators and recommended that the service reexamine the meaning of such standard terms as "within signal distance" and "Senior Officer Present."[107]

And just as Foxhall Parker had done after the service's maneuvers off of Key West in 1874, Barker's thoughts went beyond the particulars of a new communications method to include a broader consideration of command relationships in the naval operating environment. He thought the adoption of wireless would create a "natural desire for centralization" but argued that any such tendency had to be tempered. While Barker believed wireless ultimately would increase "the distance between ships in manoeuvres and in actual battle," he presciently noted that such dispersion would also magnify the importance of "the individual judgment of the Commanding and other officers."[108]

While Barker's forces were integrating radio into fleet operations, the Wireless Telegraph Board endeavored to complete its work. Testing in the fall of 1902 had led to the purchase of the Slaby-Arco radios used by Barker, but wireless companies continued to provide new apparatus to the navy. Just before Christmas the board received its first domestically manufactured radios, when a firm headed by American inventor Lee De Forest delivered two sets, and throughout the spring of 1903 the board tested improved equipment submitted by Ducretet, Rochefort, and Braun-Siemens-Halske. Two other wireless companies (one American, one British) also participated in these trials, but the board again found Slaby-Arco "the one best adapted to naval use." Specifically, it pointed to Slaby-Arco's superior performance in range, reliability, syntony, and "ease of manipulation by unskilled or poorly trained operators."[109]

The Wireless Telegraph Board issued its final report in late August 1903. Accepting the board's findings, the Bureau of Equipment ordered twenty-five additional Slaby-Arco radios, bringing the total number purchased to forty-seven.[110] These acquisitions marked only the beginning. The next year the chief of the Bureau of Equipment proposed equipping all American warships with wireless equipment, and by July 1905 nearly four dozen vessels were carrying the new technology.[111] By the end of the decade, the navy had spent nearly a million dollars installing radio apparatus at shore stations and on board naval vessels.[112]

After these initial purchases of Slaby-Arco apparatus, the navy gradually expanded the number of firms from which it bought wireless equipment. Although the manufacturers of the Slaby-Arco and Braun-Siemens-Halske systems removed one potential supplier when they combined to create Telefunken in 1903, opportunistic entrepreneurs established new wireless companies, many of them in the United States. Just weeks after the service placed its original production order for Slaby-Arco radios, Canadian American inventor Reginald Fessenden wrote the Navy Department to complain about its purchase of foreign-made apparatus. Fessenden claimed his system was superior to all others and that the navy owed it "an opportunity for test."[113] The department wrote back, chastising the inventor for failing to submit a proposal for the previous year's wireless trials but encouraging him to provide one for the next round of tests.[114]

After further delays and some rather disagreeable correspondence from Fessenden, the navy eventually received two sets of his apparatus for testing.[115] In the meantime, the service continued to purchase Slaby-Arco (Telefunken) radios.[116] As the quality of American wireless apparatus improved, however, the Bureau of Equipment shifted to an acquisition policy that favored domestic wireless firms.[117] In 1906 American companies supplied just over 40 percent of the navy's new radios, a figure that increased to 50 percent over the next two years.[118] By 1909, six of the service's eight preferred manufacturers were based primarily in the United States.[119]

Over time, choosing which wireless equipment to purchase became the easy part. Working with irascible inventors like Fessenden was difficult, of course, and at least two firms threatened to sue the navy for patent infringement.[120] Yet, as radios became standard equipment on board American warships, three complicated questions emerged. The first was how, and by whom, the new technology should be developed, tested, and operated. The second centered on where to install shipboard wireless apparatus. And the third involved determining what procedures and equipment were essential for maintaining ship-to-ship communications under hostile conditions.

Growing Pains

As the U.S. Navy grappled with these problems, half a world away Japanese and Russian fleets engaged one another at the Battle of Tsushima (27–28 May 1905). In an encounter described by one eminent naval historian as "perhaps the most decisive and complete naval victory in history," the Japanese annihilated their Russian counterparts.[121] During the peace that followed, members of the world's navies investigated numerous facets of the Russo-Japanese War, including radio's implications for warfare at sea. Most contemporary observers believed the Japanese used wireless more judiciously than the Russians did, especially at Tsushima, where Japanese warships used radio communications both to report the enemy's location and to provide essential tactical directions.[122]

Even before Tsushima, the Navy Department had worked to establish routine practices and procedures for wireless communications. In 1903 the Bureau of Equipment published formal instructions for the use of wireless apparatus; soon thereafter the service promulgated procedures for the transmission of official radio messages.[123] In the spring of 1904 the Naval War College incorporated a lecture on wireless telegraphy into its curriculum, and the following winter John Hudgins once again found himself detailed to the North Atlantic Fleet to participate in maneuvers designed, in part, to test "the defects, limitations and possible new uses of wireless telegraphy."[124]

News about the Battle of Tsushima nevertheless seems to have created a fresh sense of urgency within the service regarding the adoption of wireless. In July 1905 a search exercise run by the North Atlantic Fleet revealed that interference, both incidental and intentional, continued to be a major issue.[125] The head of the Bureau of Equipment accurately summarized matters when he wrote: "As receivers become more sensitive and the range of communication increases, the prevention of interference becomes more difficult and at the same time more important."[126] Of equal significance, that same exercise appears to have been the first in which one side succeeded in breaking the radio ciphers of the other.[127] The problem of compromised wireless communications, a concern raised by Bradley Fiske and others as early as 1902, had become a reality.[128]

The navy tested the operational capabilities of wireless again in January 1906, when warships of the newly named Atlantic Fleet (formerly the North Atlantic Fleet) conducted a search exercise during a scheduled transit from Virginia to the Caribbean.[129] The fleet commander-in-chief, Robley Evans, attempted to maintain radio contact between the scouting line and his main body but achieved only sporadic success. Reporting on the exercise, he wrote that ship-to-ship wireless communications still could not be used with certainty and argued that better training for operators and improved methods for interference prevention were essential for success.[130] The admiral's report was promulgated promptly throughout the fleet.[131]

Evans's observations accurately reflected the three main difficulties faced by the navy with respect to radio in the middle of the first decade of the twentieth century. To begin, commanders still could not depend on wireless for reliable ship-to-ship communications. To a large extent, this deficiency derived from the existing state of radio technology. Spark transmitters produced damped waves that could not be precisely tuned, and even though new types of detectors had begun to replace coherers, receivers remained temperamental and prone to failure under the harsh conditions of shipboard use. Exacerbating these technical limitations was a pool of operators generally lacking the skills required to obtain maximum performance out of the navy's radio equipment. Existing programs for the training of naval electricians offered limited instruction on the maintenance and repair of wireless sets, and officers did not receive formal training in radio communications. Furthermore, the U.S. Navy had only just begun to embrace the new methods of industrial research, organization, and management then being adopted by many American firms. In practice, this meant the service possessed limited institutional mechanisms for addressing the engineering difficulties associated with a complex technology such as radio.

Nowhere were these weaknesses more evident than in the navy's efforts to adopt wireless telephony. As early as 1905 the Bureau of Equipment had sought to acquire a reliable radiotelephone system. The bureau chief believed such a system could be employed at relatively short distances to improve tactical signaling, but he recognized that existing apparatus remained "as yet in a decidedly experimental stage."[132] That situation appeared to change in July 1907 when Lee De Forest, using his newly developed audion, demonstrated a working system of wireless telephony on Lake Erie.[133] According to one of De Forest's employees, Atlantic Fleet commander-in-chief Robley Evans soon thereafter visited De Forest to witness a demonstration of the inventor's new radiotelephone.[134] Assuming this story is true, Evans likely encouraged the Bureau of Equipment to acquire some of De Forest's wireless telephones for testing. In late August, the bureau requisitioned sets from De Forest's company for trials on two battleships.[135]

Perhaps because one of the navy's most skilled wireless operators, Chief William Bean, was stationed on one of the test vessels, trials with the De Forest radiotelephones went smoothly.[136] As a result, in November 1907 the Navy Department authorized installation of De Forest's wireless telephones on more than two dozen warships, including the Atlantic Fleet's sixteen battleships.[137] Unfortunately for all parties involved, the upcoming world cruise of Theodore Roosevelt's Great White Fleet—scheduled to depart the United States in mid-December—placed an arduous deadline on De Forest's firm.[138] De Forest later complained the navy had given him "an almost impossibly short time in which to manufacture and install this apparatus," forcing his employees to work "feverishly, day and night, without respite, almost without sleep, to meet the tough requirements."[139] This accelerated schedule took its toll, and the day before the fleet was to depart the Bureau of Equipment still had not accepted the new radiotelephones. Lieutenant George Sweet, assistant chief of the bureau's radio division, reportedly told one of De Forest's employees, "Those damn sets aren't working . . . I certainly cannot pass the sets for payment."[140] After a prolonged discussion regarding the future of wireless telephony (one that apparently took place over a few beers), Sweet changed his mind.[141] The Great White Fleet thus set sail with radiotelephones officially sanctioned by the Bureau of Equipment.

Sweet's about face benefited De Forest in the short run, but the long-term ramifications were serious for both the inventor and the navy. For various reasons, ranging from operator unfamiliarity with De Forest's arc transmitters to an inadequate supply of spare parts, the sets proved unreliable.[142] The telephones had been installed on ships' bridges for tactical maneuvers, but they could not be used for this purpose because, as one operator later recalled, "it was never possible to get all the sets working at one time."[143] Another critical problem was that De Forest's radiotelephones could not be operated simultaneously with older shipboard radio equipment. According to one observer, the big spark sets "simply swamped the small telephone sets," an issue that proved "a major detriment [from] the first day at sea."[144]

In spite of these problems, the Great White Fleet did not entirely abandon the De Forest wireless telephone. For example, although the audion tubes in De Forest's receivers burned out rapidly, operators employed innovative methods to keep them in working order. In another instance, after a casualty on one ship put the radiotelephone antenna out of commission, shipboard personnel shifted to the vessel's main antenna and discovered that this alteration improved both range and reliability. And while the fleet commander-in-chief chose not to employ De Forest's telephones for tactical maneuvers, he nevertheless authorized several warships to experiment with the new equipment.[145] Meanwhile, back in Washington, the Bureau of Equipment worked with De Forest's firm in an effort to fix known problems before the fleet departed the West Coast for New Zealand and Australia.[146]

Such improvements notwithstanding, one is hard pressed to view the Great White Fleet's experience with wireless telephones as anything other than a failure. The Bureau of Equipment certainly saw matters that way, unjustifiably placing all of the blame on De Forest's company.[147] Still, the bureau sought feedback from the fleet before ordering removal of De Forest's radiotelephones. In July 1909 the head of the bureau's radio division wrote Atlantic Fleet commander-in-chief Seaton Schroeder to inquire whether or not wireless telephones should be retained in service.[148] Schroeder informed the Bureau of Equipment that, while ten of his ships possessed complete sets, many of those sets had "parts missing or cannot, for other reasons, be made to operate satisfactorily." Yet Schroeder's frustration with the De Forest radiotelephone did not temper his enthusiasm for wireless telephony in general. He told the bureau chief that "much improvement has been made since the telephones in question were installed in the fleet" and requested that an "improved type be installed . . . for observation and future report." Without more reliable radiotelephones, Schroeder argued, any discussion about wireless telephony's usefulness to naval forces "would be merely academic."[149]

Lee De Forest believed he could fix the radiotelephone apparatus that had sailed with the Great White Fleet, but the Bureau of Equipment never gave him the chance to do so.[150] By 1909, the year Roosevelt's fleet returned home, the navy no longer had to rely solely on civilian companies for solutions to technical problems related to radio. Possibly inspired by George Sweet's regrettable experience certifying De Forest's radiotelephones, during the spring and summer of 1908 the Bureau of Equipment worked with another government agency, the National Bureau of Standards, to establish an internal organization with scientific and engineering expertise in wireless technology. This work led to the creation of the United States Naval Radio Telegraphic Laboratory, which opened in September under the direction of civilian physicist Louis W. Austin. Unsurprisingly, the first work undertaken by the new laboratory was an investigation of some European wireless telephones. After performing a series of tests, the lab concluded that the sets "required more skilled attention than would be easily available on shipboard, and . . . would take up too much space to be used as an auxiliary in the ordinary wireless room."[151]

Before the founding of the Naval Radio Telegraphic Laboratory, the U.S. Navy had utilized specially convened boards or reports from knowledgeable officers, such as Albert Niblack and John Hudgins, to acquire detailed information about new shipboard communications systems. The Naval Radio Telegraphic Laboratory differed from these in that it also tried to provide solutions to technical problems previously viewed as the purview of civilian manufacturers. Ironically, in one instance the investigation of an engineering problem actually led to the establishment of an important scientific formula.[152] For the most part, however, the new laboratory focused on quantitative analysis and its practical application to naval radio. In particular, Austin and his staff sought more reliable radiotelephone equipment, pursued new ways to increase receiver sensitivity, analyzed the critical relationship between antenna resistance and wavelength, performed a series of successful experiments in long-distance radiotelegraphy, and explored possible means of preventing atmospheric interference.[153] From 1909 to 1914, the Naval Radio Telegraphic Laboratory published more than twenty technical papers, and its work noticeably aided the navy in acquiring better wireless equipment for the fleet.[154]

Another important organizational change that enhanced the service's ability to integrate wireless technology into fleet operations took place at the highest levels of the U.S. government. In late 1909, navy secretary George von Lengerke Meyer implemented a fundamental restructuring of the Navy Department. He believed the service had grown too big for effective control under its existing organizational form, viewing his most critical problem as "the lack of a branch dealing directly with the military use of the fleet and the lack of responsible expert advisers to aid the Secretary."[155] Accordingly, Meyer divided the Navy Department into four divisions—operations, personnel, material, and inspections—and placed the bureaus under their jurisdictions. A senior line officer headed each division.

With respect to naval radio, the most crucial aspect of Meyer's reorganization took place the following year, when Congress allowed him to abolish the Bureau of Equipment and distribute its responsibilities among the other bureaus. The Bureau of Steam Engineering, which was responsible for the design, construction, maintenance, and repair of shipboard machinery, acquired the Bureau of Equipment's radio division. Although personnel remained in their billets and retained their responsibilities, the division head reported to a new chief. This arrangement prevented loss of continuity while simultaneously creating a more suitable institutional home for wireless because of the Bureau of Steam Engineering's larger budget and stronger technical tradition. In combination, Meyer's restructuring and the founding of the Naval Radio Telegraphic Laboratory fostered an environment in which radio could more easily be acquired and evaluated.

Unfortunately, corresponding institutional improvements did not readily take place in the realm of training. Unlike the cohort of naval officers who had been able to attend the army's signal school in the early 1870s, officers interested in naval radio had no formal course of instruction available to them. As such, the only avenue for obtaining expertise in wireless was through practical experience. In some instances that experience came unexpectedly, as happened to Stanford Hooper following the San Francisco earthquake of 1906. In other cases, young naval officers purposefully sought out opportunities to work with the new technology. George Sweet, for example, learned of John Hudgins's pioneering exploits and wrote the Bureau of Equipment to request orders that would allow him to acquire expertise in wireless telegraphy. The bureau chief arranged for this, in the process telling his counterpart at the Bureau of Navigation about the need for officers knowledgeable in wireless and pointing to the benefits derived by assigning interested volunteers.[156] Still, the service would not make graduate education in radio available to officers until the early 1910s.[157]

For enlisted personnel, schools on each coast provided instruction in electrical science, the handling of electric machinery, and wireless telegraphy, but here as well the expectation existed that personnel would acquire needed skills through on-the-job experience.[158] Not only could such an approach be dangerous, as one unfortunate sailor discovered after inadequate electrical safety precautions left him lying unconscious on the deck of his ship; it also left the service critically short of skilled wireless operators.[159] John Hudgins felt strongly enough about the matter that in 1904 he personally wrote the secretary of the navy: "In my opinion, we are not getting one-half the service possible out of the apparatus in use, owing to the lack of skilled operators, and I . . . recommend that a systematic training of operators be begun under official supervision, either afloat or ashore, as deemed most expedient."[160] The chief of the Bureau of Equipment shared Hudgins's sentiments and sought to lengthen the existing course of instruction. The Bureau of Navigation acquiesced, extending it to a full five months.[161] The Bureau of Equipment also issued an instruction manual on wireless telegraphy for naval electricians, one that discussed both the theoretical underpinnings and the practical applications of marine radio. The principal author of this book was Lieutenant Commander Samuel S. Robison, an energetic officer who headed the Bureau of Equipment's radio division from 1904 to 1906. *Robison's Manual*, as the publication came to be known, went through eight revised editions and served as the navy's standard textbook on radio for the better part of two decades.[162] Later in his career Robison would draw upon his expertise in communications to implement significant reforms, but in 1906 those reforms still lay in the future.

To a certain extent, the service was hesitant to invest in training programs because wireless technology evolved so rapidly during the first decade of the twentieth century. Although the U.S. Navy would not even begin to test the first revolutionary advance in transmitter technology, the arc, until 1913, spark gap technology hardly remained static during this period. Inventors devoted numerous hours and considerable sums of money in pursuit of a spark gap transmitter that could produce undamped continuous waves, a big step toward overcoming the interference problem. Both Guglielmo Marconi and Reginald Fessenden developed devices that "quenched" sparks (i.e., the devices minimized damping), and other entrepreneurs attempted to do the same. According to George Clark, a civilian radio engineer hired by the Navy Department in 1908, the service purchased its first truly efficient spark gap transmitters from Telefunken in 1910. For several years Telefunken remained the navy's largest overall supplier of wireless apparatus, but the service also acquired quality quenched spark gap transmitters from other firms, including those led by Marconi, Fessenden, and De Forest.[163]

Many of these same firms also pioneered improvements in receiver technology. The coherer was an ineffective device on any moving vessel, but on a warship it was particularly problematic. As one frustrated naval officer informed his superior in 1905, "It is impossible to receive messages [by coherer] while the guns are firing."[164] Marconi recognized this difficulty and had begun to replace coherers with magnetic detectors as early as 1902, but the strained relationship between Marconi and the U.S. Navy precluded the latter from implementing this improvement.[165] The service instead replaced the coherer with the electrolytic detector, a device patented by Fessenden in May 1903.[166] Fessenden later incorporated this new detector into a balanced-circuit receiver designed to prevent interference, but tests carried out by the navy in 1905–6 demonstrated that the inventor's claims of a revolutionary new "interference preventer" were exaggerated greatly.[167] Over the next several years, however, Fessenden made significant improvements to his receivers. In 1910 the service again tested the inventor's equipment, this time achieving favorable results. As reported by an officer who participated in these trials, "[Fessenden's] interference preventer has been used to great advantage on numerous occasions in cutting out interference from other stations and atmospheric [interference] is greatly reduced by its use . . . Very sharp tuning can be obtained."[168]

During the years Fessenden was improving his receivers, another American firm offered a new product to the Bureau of Equipment: the crystal detector. Developed by inventor Greenleaf Whittier Pickard, this device took the oscillations of an incoming radio signal and rectified them through a crystal semiconductor.[169] Pickard's detector possessed some distinct advantages, including improved receiver sensitivity and potentially higher reception speeds, but it also required skilled operators to adjust the geometry and contact pressure between wire and crystal. Pickard established a company and incorporated his new invention into a receiver, the IP-76, which the Naval Radio Telegraphic Laboratory tested and recommended for purchase around 1909.[170]

Operators initially found the performance of the IP-76 somewhat disappointing, but in another example of the key role junior personnel often played in the development of naval command and control systems, an enlisted electrician named Benjamin Miessner made significant improvements to the metal "cat's whisker" that came into contact with the receiver's crystal semiconductor. Miessner was from a small town in Indiana and had enlisted in May 1908, two months shy of his eighteenth birthday.[171] As had been the case for Stanford Hooper, the navy's technical orientation appealed to Miessner, who joined the service to attend electrical school in New York, where he hoped "to get a thorough knowledge of the trade."[172] Upon graduation, the young electrician received orders to the Washington Navy Yard as a wireless operator. Frustrated by the difficulty of keeping Pickard's detectors in tune, Miessner experimented with them until he discovered that a curved metal contact, in combination with a strong spring, greatly enhanced receiver sensitivity.[173] The Navy Department embraced Miessner's invention, and for the next several years the IP-76 was the service's standard receiver on board warships and at shore stations.[174]

Like its magnetic and electrolytic counterparts, the crystal detector enabled wireless operators to listen for dots and dashes while they transcribed messages by hand, instead of receiving them in printed form on a paper tape. The shift from paper tapes to transcribed messages had significant implications. Data rates increased once coherers disappeared from shipboard radio rooms, and accuracy improved because operators could better discriminate

Wireless operators in front of their apparatus on board USS *Baltimore*, c. 1905. As shipboard command and control systems became more complex, relatively junior personnel like these enlisted operators played an increasingly critical role in naval operations. *Naval History and Heritage Command photograph NH 101374.*

between signals and random noise. These benefits were partially offset, however, by the need to rely on the ears of a single individual. Although a paper tape could be double-checked, an audible signal once received was gone forever, save for the scribbles of an operator who might be inexperienced, indifferent, incompetent, or distracted at the time of message receipt. For naval commanders, any viable system of tactical ship-to-ship communications had to minimize the likelihood such transgressions would occur.

The shift to effective quenched spark gap transmitters and more sensitive receivers was critically important because it minimized interference and made tactical maneuvers by radio a realistic possibility. In 1906 Robley Evans had struggled to maintain contact between the ships of his fleet during a search exercise, and the navy's ability to address the problems faced by its operating forces afloat was limited by institutional and technical constraints. The service also suffered a grim setback in its efforts to integrate wireless into fleet operations when a tragic shipboard accident claimed the life of Lieutenant John Hudgins. Hudgins was the navy's leading expert in wireless telegraphy, and his premature death, which occurred just days before the infamous San Francisco earthquake, was a painful loss for both Evans and the service.[175]

Hudgins's untimely death notwithstanding, the U.S. Navy pressed forward with efforts to obtain suitable equipment and develop effective procedures for ship-to-ship communications. New officers, similar to Stanford Hooper in their backgrounds and interests, followed in Hudgins's footsteps. One of these individuals was Daniel W. Wurtsbaugh, who sailed with the Great White Fleet as a lieutenant and assumed duties as the fleet's signal officer in April 1908.[176] Wurtsbaugh wanted to determine if wireless telegraphy could be reliably employed for tactical signaling and sought answers to two specific questions. First, he wanted to know if smaller antennas could be placed in locations relatively safe from enemy gunfire while still remaining effective at transmitting and receiving. Second, he sought to determine if operators could reliably receive messages during the loud discharges of a warship's main battery. Wurtsbaugh's proposed solution to the latter was to move the wireless room "below the protective deck and behind the thickest armor," establishing direct communication with the commanding officer through call bells and voice pipes. According to Wurtsbaugh, such an arrangement would permit the vessels of any fleet to be maneuvered "as readily and expeditiously by wireless as by flag signals."[177]

An early American submarine passes in front of USS *Kearsarge* during a review of the North Atlantic Fleet in 1905. One year later a turret explosion on *Kearsarge* would claim the life of the navy's foremost expert in wireless telegraphy, Lieutenant John Hudgins, but the service would press forward, conducting its first experiments with submarine radio in 1909. *Naval History and Heritage Command photograph NH 91217.*

Wurtsbaugh's ideas were not entirely new, but he was one of the first to recognize that radio technology had reached the point where tactical signaling by wireless was practicable. In order to maximize range and effectiveness, naval shipyards had placed early wireless antennas as high as possible, typically more than 120 feet above the waterline.[178] This left them highly vulnerable to enemy gunfire and made the employment of wireless for tactical signaling imprudent during battle. As early as June 1903, the Bureau of Equipment proposed installing wireless rooms "inside the armor protection" but noted that the long distances between highly elevated antennas and their receivers could create unacceptable signal losses.[179] Radio installations ordered by the Navy Department later that summer allowed some latitude on the matter, and the New York Navy Yard placed the wireless room on John Hudgins's *Kearsarge* below the protective deck.[180] Hudgins experimented with this arrangement during the spring of 1904, reporting that signal losses were "much less than expected."[181] Tests on board *Kentucky* the following spring confirmed Hudgins's observation, but the inferior performance of wireless receivers located lower in the ship prompted *Kentucky*'s navigator to recommend that radio rooms remain close to the main deck until further improvements could be made. The Bureau of Equipment appears to have accepted this reasoning, and shipyards continued to install wireless rooms on or near the main decks of American warships.[182]

Wurtsbaugh's investigations led to a renewed debate over the optimal location for shipboard wireless equipment. By means of an outrigger the innovative officer placed twenty-foot, single-wire antennas on the unengaged sides of two warships, making these aerials "reasonably safe from damage in action." The shorter aerials reduced wireless signaling distance to just five miles, but Wurtsbaugh improved this range fourfold by replacing single-wire antennas with double-wire aerials.[183] Persuaded that reliable tactical communications could be achieved at adequate ranges, the Atlantic Fleet commander-in-chief informed the Navy Department that the time had come to pursue aggressively the adoption of tactical wireless apparatus. He believed a "battle wireless room" should be at the heart of any such system and proposed installing such facilities in a central location below each vessel's protective deck. This battle wireless room would then be linked to both bridge and conning tower via voice tubes and telephones.[184] The navy's General Board encouraged further experimentation, and within weeks the Bureau of Equipment arranged for installation of a protected radio room aboard USS *Michigan*.[185]

Wurtsbaugh detached from the Atlantic Fleet in March 1909, but other young officers picked up where he left off.[186] Lieutenant George Sweet, the same individual who reluctantly had agreed to certify De Forest's radiotelephones on the eve of the Great White Fleet's departure from Hampton Roads, was one of those to do so. In the spring of 1910, Sweet assumed duties as flag lieutenant of the Atlantic Fleet's Fifth Division. He was well qualified for the job, having just spent two years in the Bureau of Equipment's radio division, and his immediate superior, Rear Admiral Sidney A. Staunton, was the same officer who in the 1890s had debated Albert Niblack over the best methods for transmitting time-critical tactical information via the means then available.[187]

Throughout the summer of 1910, Sweet and Staunton worked to determine experimentally the ideal shipboard arrangements for a battle antenna. Their tests revealed that the optimal configuration was deployment of a single, insulated, flexible wire suspended from a height of about eighty feet.

According to Sweet, this arrangement had two major advantages. First, it allowed the antenna to be replaced quickly if damaged in battle. Second, it minimized interference problems.[188] This was due partly to shorter aerials but also stemmed from Sweet's decision to use new quenched gap transmitters he had helped develop while at the Bureau of Equipment's radio division.[189] Like Wurtsbaugh, Sweet believed transmitters and receivers should be located in a protected area, but he suggested putting these devices in the conning tower so as to permit face-to-face communications between a warship's commanding officer and his wireless operator. Apparently, Sweet did not share Wurtsbaugh's faith in the reliability and effectiveness of shipboard interior communications systems.[190]

Sweet's work with wireless was cut short the following year by diphtheria, an illness that eventually led to his removal from sea duty.[191] Once again, however, aspiring naval officers stepped in to fill the void. This time Lieutenant Commander Thomas T. Craven took the lead. An 1896 graduate of the U.S. Naval Academy, Craven seems to have developed an interest in wireless while serving as a flag lieutenant during the world cruise of the Great White Fleet. In late 1911, he received orders to a key billet in George von Lengerke Meyer's reorganized Navy Department: director of target practice and engineering competitions.[192] Responsible for overseeing these important areas, Craven viewed his responsibilities in broad terms. When he and the head of the Bureau of Equipment's radio division began to discuss the potential development of battle wireless, Craven wanted to learn more.[193] How could target practice—and, by extension, the fleet's overall fighting capabilities—be improved by tactical wireless communications?

Craven soon learned of a young officer who might have a good answer to this question, none other than Stanford Hooper, who by then was on shore duty at the Naval Academy. Following his valuable work in the aftermath of the San Francisco earthquake, Hooper detached from *Chicago* and served on two other warships before receiving orders to instructor duty at Annapolis. There Hooper taught electricity, chemistry, and physics, subjects that allowed him to investigate wireless more thoroughly than he had been able to do during his years of sea duty.[194] The eager lieutenant set up several radios at the Naval Academy for instructional purposes, but most of his research took place at the U.S. Naval Radio Telegraphic Laboratory. Hooper had befriended laboratory director Louis W. Austin, who permitted him to use laboratory facilities on weekends.[195] Every other Saturday, Hooper would hop on a train and head over to Austin's lab, where he would study the data from recent experiments. With undoubted hyperbole, Hooper later claimed these visits were "the way I got my entire education in wireless."[196]

Austin's professional relationship with Hooper was similar to one he had with another young officer, Ensign Charles H. Maddox. Originally from Canada, Maddox graduated from the U.S. Naval Academy in 1909. He then served on two battleships before transferring to a torpedo boat stationed at Annapolis.[197] During the summer of 1911, his boat and another vessel conducted a series of experiments designed to investigate wireless telegraphy's suitability for tactical signaling.[198] Maddox must have impressed his superiors because shortly thereafter he received orders to observe and report upon the Atlantic Fleet's radio operations during autumn battle practice. The ensign did as assigned, filing a report that contained recommendations for improving the fleet's wireless organization, operating methods, and procedures. Among his suggestions, Maddox proposed the creation of a "fleet radio officer" billet on the commander-in-chief's staff. Maddox believed such a job was essential for bringing "order and discipline" to "the Fleet's air-waves."[199] A few years later the young officer would reap the benefits of this suggestion, but Maddox's immediate future was decidedly academic. In early 1912 the navy sent him to graduate school at Harvard, where he studied under some of the nation's leading experts in wireless telegraphy.[200]

Maddox departed for Harvard about the same time Thomas Craven assumed his duties as the director of target practice and engineering competitions. In preparing for the Atlantic Fleet's spring 1912 battle practice, Craven certainly would have read Maddox's report from the previous fall, but with the author in graduate school Craven needed someone else to help him evaluate the fleet's employment of wireless. That someone else was Stanford Hooper. Several weeks before the spring battle practice, Craven directed the Naval Academy instructor to draft instructions governing the use of wireless signals in battle. This assignment Hooper tackled with enthusiasm, later recalling:

> In due time I presented CDR Craven with a 50-page manuscript of the instruction he asked for. He started to read it and said, "Heavens, young man, this is as thick as the whole target practice instruction book; can't you boil it down to four or five pages[?] Let me do it. I can put it in just as we need it." So it turned out that he dictated in that forenoon about four pages containing all I had said in 50 and making it much clearer.[201]

The new instructions required all flag maneuvering signals made during target practice to be duplicated by wireless. They also gave every ship a distinct two-letter call sign and assigned a specific frequency to each division within the fleet.[202] For the first time, wireless telegraphy was going to aid an American fleet maneuvering at close quarters.

Years later Hooper would describe these maneuvers in very negative terms, and others have accepted his description of events. George Clark claimed that the spring 1912 battle practice showed the navy "did not take radio seriously," and historian Susan Douglas writes: "It was 1912, and radio was still being treated as an afterthought."[203] Hooper's actual report of the practice reveals a more nuanced story. The first day was indeed exasperating for Hooper, as many of the fleet's wireless operators "had not read the instructions carefully" and "did not understand what they were to do." The fleet commander-in-chief intervened the very next day, however, sending a message ordering personnel to review and comply closely with Craven's instructions. On the third day, Hooper observed marked improvement, noting that only two or three ships failed to follow instructions properly. Yet any opportunity for further progress ended abruptly the next day, when the fleet flagship's transmitter broke and could not be repaired at sea.[204] Although the fleet's employment of radio during this battle practice can hardly be seen as an unqualified success, difficulties arose because of a steep learning curve and a material casualty, not widespread institutional ambivalence toward wireless technology.

Like Maddox before him, Hooper sought ways to improve the fleet's wireless organization, operating methods, and procedures. For reasons ranging from inadequate shore-based schooling to insufficient shipboard practice time, Hooper found the quality of many radio operators lacking. He also faulted highly skilled operators who liked to "show off" by transmitting too fast, a practice that further marginalized weaker operators. Hooper's proposed solution to these issues matched one of Maddox's recommendations—namely, assignment of a commissioned wireless expert to the staff of the fleet commander-in-chief.[205] Favoring the idea, Craven consulted with the Bureau of Steam Engineering's radio division and the Bureau of Navigation (which was responsible for officer assignments), and in August 1912 Hooper found himself reporting to the Atlantic Fleet as the navy's first fleet radio officer.[206] Two years later, the service would tap Maddox to become the inaugural fleet radio officer in the Pacific.[207]

The creation of this specialized billet was an important institutional development, for it indicated that American naval leadership viewed efficient radio communications as an essential part of tactical proficiency. Hooper was an excellent choice for the billet, and he worked diligently to correct deficiencies. Soon after taking over as the Atlantic Fleet radio officer, he convinced his commander-in-chief to issue orders requiring every warship's commanding officer to appoint a radio officer. And although the Pacific Fleet would keep wireless under the jurisdiction of the fleet signal officer until Maddox arrived in 1914, that fleet's commander-in-chief issued similar orders in September 1912. Indeed, the Pacific Fleet commander-in-chief's wording left no doubt about the importance he ascribed to the job of shipboard radio officer:

> The officer so detailed will perform no other duty than that of radio officer . . . He will ordinarily be on duty at least six hours daily in the radio office, the hours to be definitely assigned by the commanding officer, during which time, when he becomes proficient, he will act as operator. In addition he will be held responsible for the efficiency of the radio force and outfit, and for the observance of the fleet regulations relating thereto; and to this end he will exercise constant surveillance over the operators and inspection of the radiogram record books.[208]

The assignment of dedicated radio officers had beneficial consequences, both because it distributed expertise throughout the fleet and because it created a cadre of young officers knowledgeable about the capabilities and limitations of radio communications. Many of these same officers would later rise to command battleships, cruisers, destroyers, carriers, and submarines.

Another of Hooper's recommendations addressed the still unresolved issue of where best to locate shipboard wireless equipment. Better receivers meant that the signal losses created by longer cable runs ceased to be a major issue, but there remained a substantial shipboard command and control dilemma. During maneuvers, commanding officers stood on the bridge or, in combat situations, moved to the nearby conning tower, an armored cylinder with slit windows that provided a panoramic field of view.[209] A radio on the main deck near the bridge permitted easy communication between a commanding officer and his wireless operator, sometimes directly, other times via messenger or voice tube. As the fleet integrated wireless into tactical operations, however, naval officers increasingly wanted to ensure that shipboard radio apparatus was well protected. One possible solution, as suggested by George

Sweet in 1910, was to move such apparatus into the conning tower. Yet, because this space was already cramped, few officers liked the idea. Another potential solution, one proposed by the Bureau of Equipment in 1903 and recommended again by Daniel Wurtsbaugh more than five years later, was to move the main wireless room beneath the protective deck. This arrangement introduced its own set of difficulties. Face-to-face communications between a commanding officer and a radio operator were precluded, voice tubes often worked poorly over lengthy distances, and messengers took too long to get between the radio room and the bridge, especially during combat operations when hatches were closed in order to facilitate damage control.

Hooper, the former telegrapher, had a clever solution to this dilemma. He proposed installing a remote key on the bridge. Under Hooper's arrangement, a receiver would remain on the bridge, but all transmitters would be placed safely below decks.[210] An officer wishing to give an order to or request information from another vessel (or vessels) could simply turn to his wireless operator and provide the relevant instructions. The operator would then key in the message, which would activate a transmitter below. Incoming messages could be received right on the bridge, and over time many naval officers learned to decipher the dots and dashes of Morse code as well as the operators themselves. The Bureau of Steam Engineering embraced Hooper's recommendation and made "installing sets in protected locations" its top priority for 1913.[211]

That year also witnessed other notable improvements in the fleet's employment of wireless. Hooper never felt his first commander-in-chief paid enough attention to radio, but things changed when a different officer assumed command of the Atlantic Fleet in January 1913.[212] Hooper's new boss was his old skipper from *Chicago*, Charles J. Badger, who knew Hooper well and was more aggressive than his predecessor in adopting reforms intended to better the fleet's system of wireless telegraphy. Hooper's highest priorities were enhancing circuit discipline and improving operator proficiency, two objectives fully supported by Badger. To minimize the transmission of unofficial messages, a rather common problem in the fleet, Hooper learned the unique sound of every warship's transmitter, and Badger authorized him to discipline offenders in the commander-in-chief's name. At one point a rogue operator attempted to disguise transmissions by attaching a tiny buzzer to his transmitter, but Hooper broke the ruse by sending out the following message: "The Flagship direction indicator and distance measurer show that you are operating with non-regulation transmitting apparatus and violating Fleet Regulations. Report at once to the Commander-in-Chief." No such device existed, of course, but the number of unofficial messages purportedly dropped after this event.[213]

Establishing circuit discipline was an easier task than increasing operator proficiency. Shore-based schools still were not producing enough operators, and many of the best took more lucrative commercial jobs at the end of their enlistments.[214] Hooper introduced a system to grade every radio operator in the fleet on his ability to send and receive messages and worked hard to teach subordinates how to read messages through static. While static remained a real problem, Hooper's goal was to prevent it from being used as an excuse by poor (or lazy) operators. As one contemporary sarcastically described the problem, "So many times the log would show 'unreadable; static' when it should have read 'Not able to read signal because I am not a good operator.' "[215]

Badger wholeheartedly backed Hooper's reforms, issuing detailed specifications to standardize wireless arrangements for the ships under his command. He recognized that radio had become indispensable to fleet operations and sought to create shipboard communications systems that were both robust and redundant. Badger specified that every vessel had to have a main radio room of at least two hundred square feet, along with two auxiliary stations, one in the conning tower and the other in a protected location "well abaft of the main sta[tion]." These stations were to be connected with one another and the bridge via an interior communications system consisting of numerous soundproof voice tubes. Personnel at any station needed to be able to operate transmitters in the main radio room via remote key, and Badger insisted on installing solenoids that would permit the bridge to transfer operations from the main to the auxiliary radio room with the flip of a switch.[216] The commander-in-chief also promoted the development of emergency battle radios that could be operated via hand generators and portable antennas.[217] Ten years earlier, Badger's contemporary Bradley Fiske had fretted about radio's ability to be heard through enemy interference. With that problem nearly solved, Badger wanted to ensure that radio would not be silenced by battle damage.

The shore establishment also contributed significantly to the fleet's ability to employ wireless effectively under hostile conditions. In the spring of 1913, Badger specified that all sending sets should be capable of shifting frequencies in two-hundred-meter increments within ten seconds, even though the equipment to accomplish this was still under development.[218] Such tuners, at the time referred to as "wave changers," were the brainchildren of two Navy Department civilian employees who independently began to develop these devices around 1911. At the heart of the wave changer was a switch that permitted operators to select transmitting frequencies without physically having to adjust their antenna leads. Fleetwide frequency changes that used to take minutes could henceforth be done in seconds.[219] The wave changer was a key advance because it ensured a fleet could communicate via wireless in the presence of an enemy intent on interfering with those communications. If the enemy tried to interfere with ship-to-ship radio, prearranged procedures would permit vessels to shift quickly to a new frequency. An enemy eventually might be able to interfere with the new frequency as well, but this would take time and the friendly fleet could simply change frequencies again. Any hostile force would always be one step behind.[220]

The Bureau of Steam Engineering tested a prototype wave changer in 1912 and began providing production models to the fleet in early 1914.[221] By the end of March, roughly half the Atlantic Fleet's battleships had received one. Hooper was ecstatic with the new devices and exhorted Badger to make them standard equipment on every ship in the fleet as quickly as possible. Only Hooper's own words can do justice to his enthusiasm:

> The problem of radio interference . . . which has heretofore appeared unsurmountable, has been satisfactorily solved by the introduction of wavelength shifters on board the battleships. Although but about one half of the battleships have yet been fitted with these tune shifters the efficient use of the few already installed has so increased the efficiency in radio . . . [I] recommended that the work of installing tune shifters on destroyers be carried on as rapidly as possible . . . The tune shifters are not only a great advantage in saving interference in time of peace but will be a tremendous asset to the radio of the Fleet in time of War.[222]

Hooper detached a few months after penning these words, but the navy's efforts to integrate wireless into fleet operations continued apace. On the eve of World War I, Naval Radio Telegraphic Laboratory director Louis Austin visited Berlin and London to obtain information about the latest developments in European wireless,[223] and in early 1915 the Atlantic Fleet conducted "lots of maneuvers, war games and tactics in which the radio took a very busy part."[224] During those maneuvers, participating warships employed improved wave changers and receivers, the combination of which eliminated most interference problems. As one knowledgeable participant reported, "The new type wave changer is all to the merry . . . The [new receiver] made a hit on the Flagship . . . On the drill grounds (here) where interference always gave us trouble things run smooth."[225]

With respect to radio, the excitement voiced by American naval personnel in the early 1910s contrasts starkly with the cautious assessments of the men tasked to investigate wireless technology at century's turn. Yet the former were no more technologically progressive than the latter. In 1900 the Royal Navy adopted an expensive and unproven technology because Britain was at war and the government was willing to fund its purchase. The United States, by contrast, faced no major threats, and the organization responsible for acquiring naval radio, the Bureau of Equipment, was cash-starved. The U.S. Navy thus adopted a wait-and-see approach, allowing European governments and, to a lesser extent, American firms to fund most of the initial research-and-development costs for radio technology. Yet when the service's emissaries reported that the secrecy surrounding European efforts to develop wireless was threatening to leave the U.S. Navy too far behind, the service acted with alacrity. It convened a special board to determine the best system for naval use and placed its first production orders shortly thereafter.

The Bureau of Equipment provided this apparatus to the fleet, prompting commanders to consider wireless telegraphy's role in naval operations. Radio's operational limitations gave pause for concern, but they did not temper personnel's willingness to test out the new technology. Beginning with John Hudgins, junior officers worked in concert with enlisted technical experts like William Bean, pioneering inventors like Greenleaf Pickard, newly created shore establishments like the Naval Radio Telegraphic Laboratory, and enterprising commanders like Charles Badger to integrate radio into the fleet's system of command and control.[226] Too often a lion's share of the credit for this success has gone to Stanford Hooper, and while there is no denying he played a key role in the navy's adoption of radio, his place of primacy is unfounded. Hooper's reputation benefited from the shortened careers of two key predecessors, John Hudgins and George Sweet, as well as his late-in-life decision to record his memoirs. While these recordings are an invaluable historical resource, they frequently overstate Hooper's role in the early development of naval radio.

Indeed, that development should be seen neither as the triumph of visionary young officers over encrusted old-timers nor as a series of unimpeded technical advances. While radio technology improved over time, it did so in fits and starts. The Great White Fleet's experience with radiotelephones, for example, threw a spotlight on some of the technical and institutional issues that the service needed to address. And although junior officers played a critical role in the navy's adoption of radio, they did so with the support and encouragement of numerous senior officers.

Over the long run, the most significant role of junior personnel is found not in their individual contributions but rather in their new responsibilities. Radio was the first technology to provide commanders with real-time information from beyond the horizon; it also gave them an ability to coordinate more rapidly the movement of forces. In other words, the battle space expanded while the time available to make decisions shrank. Commanders now had to process not only what they saw but also what they heard. Yet what they could hear had to be transformed from dots and dashes, perhaps encoded, into words and sentences. This required new arrangements for shipboard command and control, some of which were physical, such as the installation of voice tubes to connect a ship's bridge to its main radio room, while others were organizational. Senior officers continued to maintain overall responsibility for the actions of their ships, squadrons, or fleets, but operational responsibilities began to devolve. In an environment of expanded space and compressed time, the likelihood increased that a subordinate might have to make an operational decision before consulting a superior. Senior officers thus had a vested interest in ensuring that junior personnel possessed the skill sets needed to make prudent decisions.

One anecdote in particular illustrates how the personnel responsible for operating radio could both create and solve problems for seaborne commanders. On an otherwise routine day in 1914, the Atlantic Fleet's battleships were proceeding in line up a narrow channel into a Caribbean port. Charles Badger's flagship was in the lead when his senior enlisted radio operator decided to simulate some tactical signals for training purposes. Sitting on the bridge and believing his key was dead, the operator practiced sending a signal that ordered all vessels to turn right on the next command. But the bridge key was actually live, and Badger's ships copied the message. Before the signal to execute went out, however, Fleet Radio Officer Hooper intervened. He was down in the main radio room, had heard the unexpected message go out, and without consulting anyone immediately transmitted: "Annul the last signal." No doubt the perplexed conning officers on Badger's battleships were relieved to hear that message.[227] While this vignette offers a mere glimpse into radio's peacetime tactical role, it shows clearly how the new technology could affect command relationships and prerogatives. Clarifying such matters was difficult even under optimal conditions, but American naval officers were about to learn, albeit vicariously at first, the challenges faced by commanders attempting to coordinate geographically dispersed seaborne forces during war.

CHAPTER THREE

War and Peace

Coordinating Naval Forces

> A leader in planning a campaign or a battle thinks in terms of the limitations imposed by the ranges and arcs of train of his guns, of the speed of his ships, and the capacity of his bunkers. He must learn to think in terms of the limitations of his communications system too. These limitations are definite and any plans that overtax the system are destined to failure. He must know exactly what he can and cannot do.
>
> Harold W. Boynton, 1926

Like Hell Let Loose

On the afternoon of 31 May 1916, Admiral John Jellicoe stood impatiently on his flagship, HMS *Iron Duke*, awaiting information. Under his command were more than 150 vessels, among them dozens of the world's most powerful warships. At around half past two, the admiral received an enemy contact report transmitted via wireless telegraph by HMS *Galatea*. He immediately ordered his ships to generate steam for full speed and to prepare for action. Over the next two hours, Jellicoe received additional wireless reports from several of his subordinate commanders, many of whom reported that they were engaging the enemy. Until roughly half past four the admiral believed he was facing only a portion of the German High Seas Fleet, as shore-based personnel previously had sent him an intelligence message placing the German fleet's flagship in port. Abruptly, at 4:38 p.m., a British cruiser reported the enemy battle fleet in sight. Jellicoe received two more contact reports over the next ten minutes. Then nothing. For nearly an hour, the commander-in-chief of the British Grand Fleet received no new information.

On board *Iron Duke*, Jellicoe's staff kept a plot of the tactical picture, a bird's-eye view of the ocean's surface with pencil marks denoting own and opposing forces. But just how accurate was that tactical picture? Describing events after the war, Jellicoe recounted his frustrations:

> The first accurate information regarding the position of affairs . . . was timed 5.40 P.M., but received by me considerably later . . . At about 5.50 P.M. I received a wireless signal . . . reporting having sighted ships in action . . . There was, however, no clue as to the identity of these ships . . . The information so far received had not even been sufficient to justify me in altering the bearing of the guides . . . At this stage there was still great uncertainty as to the position of the enemy's Battle Fleet.[1]

As six o'clock arrived and darkness approached, *Iron Duke* came into contact with HMS *Lion*, flagship of the British battle cruiser force. At first, Jellicoe was confused. According to the plot, *Lion* should have been twelve miles straight ahead. Yet there *Lion* was, on *Iron Duke*'s starboard quarter, less than six miles away. Jellicoe's flagship signaled by searchlight to inquire: "Where is the enemy battle fleet?" After receiving an inadequate response, he repeated the inquiry, allegedly commenting to his staff: "I wish someone would tell me who is firing and what they are firing at." Not until 6:14 p.m. did Jellicoe receive an answer from *Lion* that enabled him to act: "Have sighted enemy's Battle Fleet bearing south-south-west."[2]

The frustration created by problematic communications during this naval battle, known to posterity as Jutland, was not limited to the fleet's highest-ranking officer. Aboard *Galatea*, the ship's leading telegraphist was also having a devil of a time. As he later described events:

> I shall never forget what it was like with those earphones on. Pandemonium . . . I had to report naturally every signal received. The Germans' wireless transmitters were by TELEFUNKEN, which all had a very high frequency note. It sounded as though all the Germans were transmitting at the same time. Also were all our own ships transmitting enemy reports dozens of times. At the same time the IRON DUKE was making MANOEUVERING messages every minute to the fleet. It was like hell let loose.[3]

These anecdotes offer but two examples of the command and control difficulties encountered by British naval personnel during the Battle of Jutland (31 May–1 June 1916). Of the methods then available for ship-to-ship communications, only wireless telegraphy permitted reliable communications beyond the horizon, and neither British nor German personnel possessed particularly effective systems for managing the information obtained through radio. In the case of the Royal Navy, inaccurate or incomplete information severely handicapped Jellicoe on multiple occasions.[4] One noted historian argues persuasively that the Grand Fleet's communications problems derived from a system of command that stressed obedience over initiative,[5] but technical limitations also contributed to the fleet's command and control troubles. Most British warships still used spark transmitters, and there was considerable mutual interference during the battle. One of the fleet's wireless operators later stated that sixty-five vessels had transmitters tuned to roughly the same frequency.[6] The sheer volume of signals greatly complicated matters as well. One officer estimated that during daylight action the warships of the Grand Fleet attempted to send, on average, one signal every sixty-seven seconds.[7] Last but not least, the command and control facilities available to Jellicoe and his subordinates were inadequate. Jellicoe's plotting organization consisted of only two staff officers, and they had at their disposal primarily paper, pencil, and protractor to keep track of the information reported by dozens of warships.[8] Simply put, Jellicoe and his subordinates lacked the means to attain a common tactical picture during the battle.

The inability of British naval officers to arrive at a common tactical picture, combined with the pervasiveness of a command system that discouraged individual initiative, created a "catch-22." On the one hand, lower-ranking commanders and individual commanding officers hesitated to take independent action for fear their superiors knew something they did not or had intentions of which they were unaware. On the other hand, senior commanders often possessed information no more tactically relevant than that held by their subordinates. In some instances, naval commanders failed to receive critical information; in others, battle damage inhibited ship-to-ship communications, such as when shells destroyed the wireless facilities of the British and German battle cruiser force flagships. The extent to which senior officers anticipated these problems remains a matter of debate, but there is no doubt that the "lessons" of Jutland influenced an entire generation of naval leaders in both Europe and the United States.[9]

The command and control difficulties faced by British and German naval commanders at Jutland far exceeded those previously experienced by the personnel of any other navy. Yet even after the United States entered the European conflict of which Jutland was a part, American naval officers gained limited experience coordinating the movements of large groups of geographically dispersed warships. Not until the 1920s would the U.S. Navy begin to evaluate systematically the fleet's ability to operate as a coordinated group, when the service implemented annual problem-solving exercises designed to provide "practical experience" for "maneuvers on a large scale."[10] During many of these exercises, American commanders found themselves in circumstances that undoubtedly would have been familiar to those present at Jutland. Confronted with difficult problems related to the command and control of naval forces, these individuals pursued practical solutions in a restrictive budgetary environment. In so doing, they demonstrated a keen awareness of the importance of information in the practice of warfare at sea.

Well before the Battle of Jutland, Secretary George von Lengerke Meyer had restructured the Navy Department in an effort to enhance the service's ability to administer certain managerial, operational, and technical matters. Many officers viewed Meyer's new system as a step on the path to a naval general staff, but some politicians—including Meyer's successor, Josephus Daniels—opposed the idea for fear it would dilute civilian authority. The outbreak of war in Europe gave senior naval officers and their congressional allies adequate leverage to overcome such opposition, and in March 1915 Congress established the position of chief of naval operations (CNO) to oversee fleet operations and prepare plans for war.[11] Although the CNO theoretically held centralized authority, in practice fleet commanders and bureau chiefs retained considerable autonomy. Nonetheless, the CNO's legislatively mandated role in war planning gave him tremendous influence. Over time, the office of the chief of naval operations (OPNAV) grew, and civilian secretaries increasingly relied on OPNAV for advice on technical matters, including those related to command and control.

Like most navy secretaries before him, Josephus Daniels came to his post as a result of political connections and not because of prior expertise in naval affairs. A native of North Carolina, Daniels was a successful journalist who backed numerous progressive causes. His support of Woodrow Wilson during the presidential campaign of 1912 led to a cabinet appointment, and with Wilson's inauguration Daniels became secretary of the navy. The new secretary possessed a fundamentalist zeal that could infuriate subordinates, such as when he prohibited alcoholic liquors on any ship or station, and according to one well-known naval historian he was narrow-minded, provincial, and indecisive.[12] Yet Daniels was an unabashed proponent of new technology. In his first complete annual report, Daniels invoked Alfred Tennyson and Jules Verne to promote aircraft and submarines, respectively, and he devoted four full pages to a discussion of naval radio.[13] Daniels also sought to reform the navy's personnel policies, expanding educational opportunities for enlisted men and refusing to promote officers with insufficient time at sea. Although he sometimes clashed with senior naval officers, Daniels generally fostered an institutional environment conducive to improving fleet operations.

In accord with his progressive beliefs, Daniels abhorred the patronage system and strove to bring efficiency and order to the Navy Department. He promoted policies for enlisting only "young men of the highest type," empowered commanding officers to discharge "undesirable" personnel,[14] and placed fleet requirements above all else, informing the president: "Every dollar expended ashore is wasted unless it promotes the efficiency of the fleet."[15] Daniels believed one way to accomplish this was to encourage the development of new technologies. With war raging in Europe, in mid-1915 he wrote inventor Thomas Edison to request assistance in creating a "department of invention and development" for the navy. Daniels thought naval officers, particularly those at sea, were ideally situated "to note where improvements are needed and to devise ways in which those improvements can be made," but he bemoaned the fact there was no central place where such ideas could be "worked out and perfected."[16]

Edison responded favorably to this request, and Daniels established what came to be known as the Naval Consulting Board. Although staffed by some of the country's most accomplished engineers, including Elmer Sperry, Franklin Sprague, and Willis Whitney, the new board was not particularly effective. Historian William McBride argues this ineffectiveness stemmed from senior naval officers' rejection of an organization that potentially challenged the service's technical competence, pointing out that the bureaus refused even to assign liaison officers to committees created by the board.[17] Yet McBride goes a little too far when he claims that until American entry into World War I only the Bureau of Ordnance possessed "any significant naval technical research facility."[18] In 1915 and again in 1916, the Bureau of Steam Engineering expanded considerably its research capabilities with respect to naval radio. These efforts would benefit U.S. naval operating forces during combat operations in 1917–18.

European events leading to the outbreak of war in August 1914 need not be recounted here. In the United States, Woodrow Wilson responded to the crisis by pronouncing American neutrality, and leaders on both sides of the Atlantic predicted a short war. When this forecast proved faulty, belligerents and neutrals adapted accordingly. Britain, fearful of attriting its fleet by operating too close to German minefields, coast artillery, and torpedo flotillas, maintained a distant blockade of the Central Powers. Germany sought to counter this blockade with a policy of unrestricted submarine warfare, announced to the world in February 1915. Unrestricted submarine warfare troubled the idealistic Wilson, whose displeasure turned to disgust when a U-boat sank the passenger liner *Lusitania* in early May. American naval officers sensed war was near, and many of them lobbied aggressively for naval expansion.[19] Yet Wilson averted war in 1915, and naval expansion would not earnestly begin until Congress passed a major shipbuilding bill in August 1916.[20] Rather than merely bide time, some U.S. Navy personnel chose instead to follow the precedent set by their forebears in the decades after the Civil War. New warships might be slow to arrive, but personnel could enhance "the efficiency of the fleet" by once again improving shipboard command and control.

Hepburn and Hooper

The previous chapter concluded with the story of Fleet Radio Officer Stanford C. Hooper potentially preventing a maritime disaster when he countermanded an order mistakenly sent by an inattentive radio operator. The Atlantic Fleet commander-in-chief, Charles J. Badger, either knew nothing of the incident or simply chose to ignore it, but there was no overlooking Hooper's superior operational knowledge of naval radio. As such, when hostilities erupted overseas, the Navy Department detached Hooper from the Atlantic Fleet, ordering him to Europe for duty as a radio observer. Commander Samuel S. Robison, former head of the Bureau of Equipment's radio division and one of the service's most knowledgeable radio experts, handpicked Hooper for the assignment. Sailing from New York as soon as he reasonably could, Hooper arrived in England late in the summer of 1914. With him he brought a portable receiver so that he could "listen in and see what was going on in the air over there."[21]

The thirty-year-old lieutenant reported directly to the American naval attaché in London but had wide latitude in ascertaining as much as he could about the radio operations of British and German warships. Hooper spent much of his time in England, where the British government gave him access to some of its wireless production facilities, and traveled to France, Belgium, and Holland, where he stayed in hotels that allowed him to hang a thin-wire antenna out his window. He reported excellent circuit discipline in the

British and German fleets but noted that vessels at sea frequently experienced interference when shore stations communicated with one another. He also spoke highly of British receivers, which included a special circuit for eliminating interference.[22]

Hooper completed his assignment and returned to the United States in early 1915. The Navy Department assigned him to a special board tasked with formulating an appropriate administrative plan for the operational and material aspects of naval radio. Hooper's recent experiences, first with the Atlantic Fleet and then as an observer in Europe, gave him a unique and valuable perspective on operational matters, so other board members gave him primary responsibility for crafting the board's final report. That report contained various recommendations, including one intended to prevent the U.S. Navy from experiencing the same interference issues Hooper had witnessed in Europe. Specifically, the board recommended the establishment of overlapping but distinct frequency bands for fleets and naval shore stations.[23] Wave changers installed on American warships beginning in early 1914 had been instrumental in limiting interference during ship-to-ship communications, but shore stations could not easily operate according to a fleet frequency plan. And although occasional interference was annoying in peacetime, it could be catastrophic in wartime. As such, the board recommended, and the service adopted, a 100–200 kilohertz frequency band for shore stations and a 150–750 kilohertz frequency band for fleet communications. Hooper later characterized this arrangement as "very satisfactory."[24]

While the establishment of separate frequency bands for fleets and shore stations further minimized the interference problem, another set of recommendations reflected changes of much broader historical significance. Specifically, the board proposed changing the U.S. Navy's management and training of enlisted radio operators. Viewed as a whole, the board's proposals in this area reveal that American naval officers were beginning to grasp how access to real-time information from beyond the horizon was changing the nature of command at sea. Class distinctions remained strong, of course, but the enlisted radio operator increasingly was seen as a critical member of the shipboard decision-making team. The board thus sought to provide these individuals with opportunities that would have been nearly unthinkable to earlier generations of naval personnel.

The two major problems with respect to enlisted radio operators, as Hooper and the other board members saw them, were quantity and quality. The quantitative problem was easier to state than to solve; simply put, the navy did not have enough enlisted radio operators. The job required considerable skill, and attrition was high at every level. Young men seeking to become radio operators first had to complete a demanding five-month course of training. Then they had to demonstrate their ability to "handle the key" in the fleet, where some individuals unexpectedly froze under the pressure. Finally, many talented operators served only a few years, leaving the service at the end of their enlistments to accept more lucrative commercial employment.

To remedy the operator shortage, the board suggested a two-pronged approach. At the entry level, it recommended reserving 4 percent of all enlistments for radiomen, calculating that the number of personnel entering electrical school had to net an average of twenty-two radio graduates per month. To minimize classroom attrition, the board proposed that every battleship and large cruiser "transfer two men, semi-annually, to the electrical class to assist in keeping the required number [of radiomen] in training." Yet the board also recognized that more recruits and fewer dropouts solved only half of the problem. Skilled operators had to be persuaded to reenlist, and the best way to do this was through inducements. The board therefore recommended guaranteed shore rotations and better promotion opportunities for these operators.[25]

Quality was another issue addressed by the board. Many recruits could not read or write well, and poor spelling was endemic. Bad handwriting could also cause problems, a fact that later led one board member to write the following on a requisition for typewriters: "Need them badly. One of the main reasons errors made by operators in writing code[d] messages."[26] The board suggested that previous knowledge of electricity and electrical systems be eliminated as a prerequisite for those interested in radio work, noting that personnel with the talent to be successful operators were harder to come by than electricians.

Neither of these recommendations did much to blur the divide between enlisted and officer, but the board put forward two other proposals that did. First, it recommended the establishment of "a post graduate course for Chief Radiomen." While board members had in mind something different from the type of postgraduate education pursued by officers, their proposal offers a remarkable reflection on how they perceived the intellectual capabilities of senior radiomen. Second, the board recommended that "all radio men above the rate of 3d class be instructed in Spanish and either French or German

. . . in order that they may acquire sufficient knowledge of these languages to qualify them to easily copy messages intercepted in these languages." From an operational perspective this idea made perfect sense, but it marked a sharp break from tradition. As gentlemen, officers long had been expected to hold proficiency in one or more foreign languages. To recommend that a group of enlisted personnel should also possess language skills highlights how class distinctions were beginning to fade as officers sought to empower those who would be aiding seaborne commanders in the operational environment.[27]

Although the board's foreign language proposal never took hold, the navy adopted many of its other recommendations. Most critically, the service instituted a "preliminary course of instruction for those men showing an aptitude for radio."[28] In this course, students learned only how to send and receive messages; those who achieved proficiency were then sent to the regular electrical course for radio operators. Drawing from the board's other proposals, the Navy Department also increased promotion opportunities for senior radiomen, expanded shipboard complements for radio officers, and compiled a list of civilian operators who were willing to join the navy in time of war.[29] In aggregate, these changes improved overall operational readiness.

Stanford Hooper, meanwhile, transitioned from Secretary Daniels's board on radio organization to his next assignment. Reminiscent of the previous moves Foxhall Parker and Edward Very had made from the fleet to the Signal Office, Hooper became head of the Bureau of Steam Engineering's radio division. He relieved another very capable officer, Arthur J. Hepburn, who had led the radio division for nearly three years. A Pennsylvania native, Hepburn was an 1897 graduate of the Naval Academy and, like Hooper, taught electricity and physics at the Naval Academy before being detailed from the Atlantic Fleet to the radio division.[30]

From 1912 to 1917, Hepburn and Hooper in turn led the Bureau of Steam Engineering's radio division through a period of fundamental change in wireless technology. The spark gap era started coming to a close in early 1913, when the navy investigated a new type of arc transmitter in experiments conducted between USS *Salem* and a shore station in Arlington, Virginia. The arc transmitter long had been an appealing theoretical concept because it generated nearly continuous waves, a type of waveform that enabled voice communications and precise tuning. Yet earlier trials with arc equipment had not been particularly successful, and the maker of the arc transmitters used in the *Salem* trials had to persuade the Bureau of Steam Engineering to give them a try. At the same time, the navy also was testing a "synchronous rotary spark gap" system designed by Reginald Fessenden. To the surprise of many, the arc outperformed Fessenden's more powerful spark gap apparatus.[31] Enthusiasm for these results was tempered, however, by the ineffectiveness of the receiving device used to read continuous wave signals. Fortuitously, during the same trials Fessenden's spark gap system employed a new and very successful reception method known as heterodyning.[32] Hepburn wondered if heterodyne techniques could be used to receive continuous wave signals. Expert radio aide George Clark, who personally observed the trials, assured him the answer was yes. Hepburn later recalled that the *Salem* tests "led more or less accidentally to what I consider the most important decision that I had to make during my whole tour," which was to move the navy from spark to arc transmitters.[33] He informed his superior of this decision in the spring of 1913, and the service bought nearly $530,000 worth of arc transmitters in 1914 and 1915.[34]

Hepburn shepherded through other important changes, both administrative and technical, while leading the radio division. He advocated the creation of fleet radio officer billets, pushed to get shipboard wireless apparatus moved to protected locations as quickly as possible, and vigorously promoted the development of wave changers. Hepburn also believed the radio division was spending too much time on operational matters, so he endorsed the establishment of the Naval Radio Service to issue regulations, oversee personnel, and coordinate commercial wireless activities.[35] One of Hepburn's biggest concerns was the lack of coordination between the various navy yards responsible for installing and servicing radio apparatus. To fix this problem he worked with the New York Navy Yard's civilian radio aide to standardize the service's numbering system for technical drawings and then mandated automatic forwarding of all drawings from each yard to every other yard.[36]

For Hepburn, standardized drawings were just one step in creating a geographically decentralized yet coherent research and design program for naval radio.[37] Hepburn believed commercial firms could not solve many of the unique problems encountered under combat conditions and felt the service needed to expand its own corps of civilian wireless experts, which in 1914 consisted primarily of the U.S. Naval Radio Telegraphic Laboratory and two radio engineers. The successful development of the wave changer by these two engineers, combined with the outbreak of World War I, gave Hepburn and Hooper the ammunition they needed to lobby successfully for a major expansion of the shore radio establishment. By mid-1915 the Navy Department had hired six new civilian radio experts, assigning each to a different navy yard.[38] The new personnel received funds to set up laboratories, and the radio division designated different yards as the principal facilities for certain types of apparatus. For example, the New York Navy Yard concentrated on wave changers, the Washington Navy Yard on receivers, and the Mare Island Navy Yard (Vallejo, California) on the motor-generators that powered most radio equipment.[39]

Another factor contributing to Hepburn's push for an expansion of the navy's shore radio establishment was his dissatisfaction with the quality of American-made receivers. The basic design of such receivers was sound, but they broke more frequently than those manufactured by the German firm Telefunken. The Naval Radio Telegraphic Lab investigated the problem and suggested some solutions, but it could not convince any domestic manufacturer to modify its production processes.[40] This response is not surprising, as nearly every commercial firm had an ambivalent relationship with the Navy Department. On the one hand, the service bought a lot of radio equipment and was a valuable potential customer. On the other, the navy was notorious for skirting patent law. Hepburn was no exception to this tendency, as revealed in a story he recounted years later:

> This reminds me of a little interesting session I had with Colonel [John] Firth who . . . was the head of the Wireless Specialty Company from who[m] we purchased a great many of our receiving sets . . . we didn't have much money and I finally told the Colonel, "Colonel, you have got to bring down the price of those crystal detectors we are buying from you . . . we have to have them and we haven't got the money to pay you $75.00. Now what I am going to do is pirate those sets and make them ourselves. We have got to have them. You would have a good claim against the government for infringement of your patent and you can go to law about it and in the course of two or three years you can undoubtedly get some compensation." . . . He came back in a few days and told me I had him over a barrel, that it was a dirty trick and a few other good natured remarks of the sort, but thinking it over he thought he could let us have those detectors for $15.00 a set.[41]

Hepburn apparently felt unencumbered by legal restraints when he thought American lives might be at stake.

As war clouds gathered over Europe during the summer of 1914, the radio division grew increasingly concerned about the receiver issue. Telefunken likely would be cut off as a supplier, and American manufacturers had shown little willingness to accede to the navy's demands for more rugged receivers. Adding fuel to the fire was the poor performance of shipboard receivers during the U.S. occupation of Veracruz (21–22 April 1914).[42] Feeling a sense of urgency, Hepburn adopted a dual strategy to address the problem. His first course of action was to have the Bureau of Steam Engineering take over the design of existing receivers. This began in 1914 and accelerated with the growth of the navy's shore radio establishment the following year. The lab at the Washington Navy Yard, which came to be known as the Radio Test Shop, eventually produced three very dependable receivers (designated types A, B, and C) widely used by American warships during the First World War and for several years thereafter.[43]

Hepburn's second approach to the receiver problem was to promote and encourage commercial firms pursuing a replacement for the crystal detector, especially Lee De Forest's Radio Telephone and Telegraph Company. De Forest had patented a two-element rectifying receiver in 1906 and a three-element one in 1908, but only in the summer of 1912 did he create a successful amplifier.[44] The following year the Bureau of Steam Engineering tested De Forest's improved audion, which could both detect *and* amplify incoming wireless signals. The results so impressed Hepburn that he immediately placed an order with De Forest's company. In early 1914 Hepburn sent some of these to Radio Officer Hooper, who gave them a thorough trial. Hooper reported back enthusiastically, asking the radio division to provide the new technology to every ship in the Atlantic Fleet. Hepburn demurred, as the average life-span of each audion bulb was only fifty hours, and keeping the entire force supplied would have been prohibitively expensive. The fleet initially received just six audions, while Hepburn exhorted De Forest to furnish the navy with longer-lasting bulbs.[45]

For more than a year the Radio Telephone and Telegraph Company remained the navy's principal supplier of amplifiers, but De Forest could never produce the longer-lasting bulbs desired by the radio division.[46] Unlike other vacuum-tube amplifiers, which contained a relatively high vacuum, De Forest's audion was a low-vacuum tube that functioned through the ionization of the gases contained therein.[47] Consistent manufacture of such a device was difficult, making the average life-span of different batches of audion bulbs highly variable. Experts at the Radio Test Shop investigated the problem and by 1915 had developed their own vacuum-tube amplifier unit. This important development gave the service the flexibility to buy vacuum tubes manufactured by other companies, including General Electric and Western Electric.[48]

When Stanford Hooper became head of the radio division in the spring of 1915, he took over an organization active on many fronts. Hepburn's previous work had led to an excellent system of coordination between the navy yards and the radio division. New laboratory facilities were in the works, and the Bureau of Steam Engineering was busy hiring additional radio experts. Nonetheless, Hooper would have to deal with numerous problems of his own. The navy's transition from sparks to arcs had just begun, and two other means of transmitting continuous waves, alternators and vacuum tube oscillators, were arriving on the scene. As for receivers, the service still needed longer-lasting vacuum tubes, even more so after it incorporated these devices into heterodyned reception systems.[49] Perhaps most critically, however, to a much greater extent than his predecessor Hooper would have to address the problem of how to keep the navy's growing air and subsurface forces connected with the rest of the fleet. And in two years time he would be dealing with all of these issues while leading the radio division into war.

Looking back on Arthur Hepburn's tenure as head of the Bureau of Steam Engineering's radio division, one of the navy's civilian radio experts remarked he was "a great believer in doing what the Fleet wanted. Not that others . . . did not similarly co-operate with Fleet requirements, but that this [cooperation] was especially evident during Hepburn's administration."[50] Hooper had the same approach and perhaps for that reason made few changes in the course his predecessor had charted. The new division chief moved forward with the navy's transition to the arc, overseeing installations at shore stations and on larger warships, and invested heavily in heterodyned receiving systems.[51] At the same time, he authorized modifications to older receivers so that spark gap signals would remain an effective means of fleet communications.[52]

Hooper pursued improvements in other areas as well. One important change, which clearly derived from his operational experiences, was in the tuning of shipboard radio apparatus. Over time, transmitters of all types would experience drift in their radiated frequencies, making reception difficult. Transmitters therefore needed to be calibrated periodically by measuring their outputs with a device called a wave meter, which enabled operators to implement appropriate adjustments. When Hooper was the fleet radio officer, U.S. warships did not carry wave meters, so these adjustments had to be

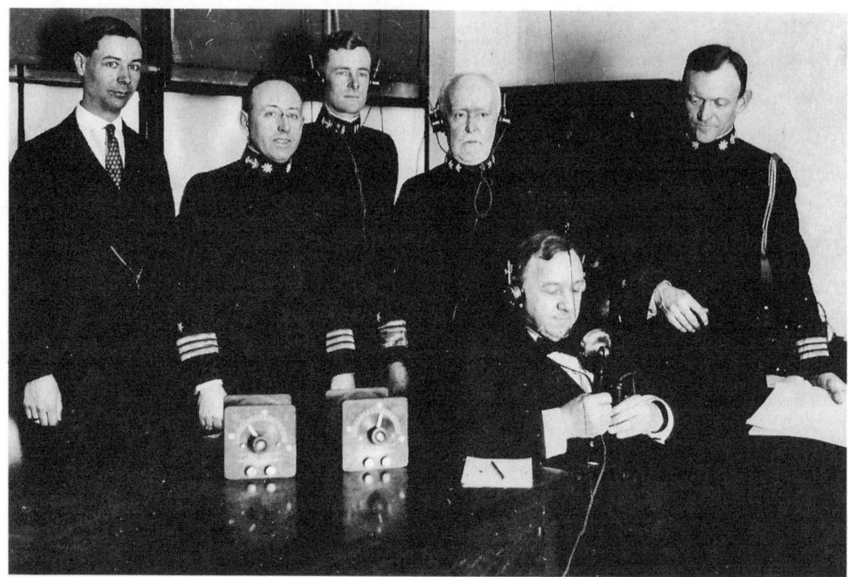

Commander Stanford C. Hooper (*second from left*) stands smiling in the background, as navy secretary Josephus Daniels (*seated*) sends greetings to President Woodrow Wilson, then at sea on his way to the Paris Peace Conference (December 1918). Hooper thrice served as head of the Bureau of Engineering's radio division and was a key player in the development of naval radio. *Naval History and Heritage Command photograph NH 57302.*

performed in a navy yard. Hooper viewed this situation as highly inefficient and sought to provide "a wave-meter for every ship." The high cost of such equipment prevented Hooper from immediately achieving his goal, but he did arrange for the radio division to purchase and distribute to the fleet as many wave meters as possible.[53]

Hooper was particularly proactive in trying to get radio on board naval aircraft, something he had encouraged since his days as an instructor at the Naval Academy. The service purchased its first airplanes while Hooper was teaching there, and it began experimenting with aircraft radio in 1912. The individual who conducted these experiments was Charles Maddox, the same officer who had observed and reported upon the Atlantic Fleet's radio operations during the fall battle practice of 1911. In the summer of 1912, Maddox had just returned to Washington after completing his first semester of graduate studies in radio at Harvard. The bureau knew aviation pioneer John Rodgers was testing out a new Wright biplane in Annapolis, so it sent Maddox over to investigate the possibilities of aircraft radio.[54]

Maddox did more than merely investigate. After Rodgers laid down strict rules regarding weight and size, Maddox assembled and installed a quenched spark gap transmitter and crystal receiver on Rodgers's plane. A leather belt attached to the engine's flywheel provided the requisite electrical power. Maddox then went aloft with Rodgers, later describing his unconventional arrangement in vivid terms: "The receiver was a small portable type . . . suspended by a strap around my neck to minimize vibrations . . . Lt. Rodgers acted as pilot, while I sat in the spare pilot's seat next to the engine. Strapped on my right knee was the sending key, and in one hand I held a screw driver, in the other a monkey wrench."[55] Unsurprisingly, reception was poor; only very strong signals could be heard above engine noises and plane vibrations. Efforts to transmit while airborne achieved greater success, and in July Maddox sent a message to a ship six thousand yards away. After a few equipment modifications, he succeeded in transmitting signals to vessels fifteen miles distant.[56]

Despite Maddox's pioneering work, for which he later received an official commendation, the Bureau of Steam Engineering did not further explore aircraft radio until the spring of 1915.[57] At that time, the radio division purchased four spark transmitters powered by electric batteries. The intended use of this apparatus was to spot the fall of shot from shipboard guns; however, trials revealed that they possessed a maximum effective range of only five miles when airborne.[58] Because the extreme range for target practice was then approaching nine nautical miles,[59] the bureau considered these transmitters inadequate. Such was the state of affairs with respect to aircraft radio when Hooper relieved Hepburn.

Fortuitously, the radio division was next door to the Office of Naval Aeronautics. One day Hooper and the individual in charge of that office began discussing the potential of aircraft radio, perhaps following an August 1915 test in which an aviator successfully relayed the fall of army mortar fire with Very signal flares.[60] The two officers agreed "something ought to be done," so Hooper asked Louis W. Austin of the Naval Radio Telegraphic Lab to calculate the attainable range of an airborne transmitter weighing not more than one hundred pounds.[61] Austin submitted his report in the spring of 1916, positing that with a fifty-foot trailing-wire antenna ranges of up to thirty miles could be achieved.[62] Spurred by this report, Hooper convinced the chief of the Bureau of Steam Engineering to assign a new officer to his division, one whose primary responsibility would be aircraft radio. This officer drafted detailed specifications for competitive bids, which the radio division put out that summer. Thirteen firms submitted bids; Hooper ultimately selected four to provide test models. All four models passed trials, and in December 1916 the navy ordered fifteen sets each from three different manufacturers.[63]

Even before awarding these contracts, Hooper had determined that there should be a single research facility for naval aircraft radio. The logical home for such a place was the recently established Naval Aeronautic Station in Pensacola, Florida, and in early 1916 Hooper convinced the bureau to allocate funds for the construction of laboratory facilities there. Hooper planned to send one of his officers down south but also wanted another civilian radio aide to help run what would become known as the Naval Aircraft Radio Laboratory.[64] At just that time Benjamin Miessner, the naval electrician who had invented the cat's whisker, showed up in the radio division seeking employment. Miessner had left the service upon expiration of his enlistment and had worked in industry for several years before matriculating at Purdue University in the fall of 1913. Although just one year away from a degree in electrical engineering, Miessner recently had decided to get married and needed a job. Hooper arranged for Miessner's hire, and the Navy Department sent him to Pensacola to lead research efforts at the laboratory being established there.[65]

The lab's earliest efforts involved testing airborne transmitters, but as Maddox had learned four years earlier, a bigger difficulty was the airborne reception of radio signals. In addition to acoustic disturbances created by engine noise and rushing wind, receivers were susceptible to electrical interference generated by vibrating equipment and the small sparks repeatedly emitted by airplane engines. The Naval Aircraft Radio Lab minimized the acoustic problem by crafting a flannel-lined leather helmet with soft rubber ear cups.[66] Research in this area also led to the development of an "aviation telephone" with noise-canceling headsets so pilots and their radio operators could easily communicate with one another while airborne.[67]

Eliminating electrical interference was a more difficult task. The most effective solution, fully shielding the engine's ignition system, was unacceptable to pilots. One radio expert later recalled that "the operating people were very fussy about their ignition . . . The greatest bug-bear of flight was engine trouble."[68] Also unacceptable to naval aviators was a bureau suggestion that airplanes could perhaps "glide, with engines stopped, for three or four minutes, when it is desired to obtain direction by radio."[69] As such, the Naval Aircraft Radio Laboratory pursued other ideas, two of which led to incremental improvements. First, working in conjunction with a civilian manufacturer, the lab developed shock-absorbing mounts. While these mounts did nothing to prevent engine interference, they minimized the internal electrical interference created by vibrating vacuum tubes. Second, the lab worked to design receivers less susceptible to the interference generated by an engine's spark plugs.[70] These efforts initially bore limited fruit, however, and even as late as June 1917 no U.S. Navy aircraft had been fitted with permanent radio apparatus.[71]

Like the airplane, the submarine presented the service with unique communications challenges. The navy purchased its first submarine, USS *Holland*, in April 1900. A few months later the service ordered six more boats, all of which entered commissioned service in 1903. These early American submarines were primarily for harbor defense, and there seems to have been little dialogue about how to communicate with such craft. Possibly this was because many officers still considered the submarine an experimental technology; in fact, the service would not commission a truly seagoing submarine until mid-1908.[72] The Navy Department began to investigate submarine wireless in earnest the following year, when the radio division oversaw the installation of experimental apparatus on several submarines.

Early trials conducted on surfaced submarines revealed that the small size and cramped spaces of existing boats were problematic but not insurmountable.[73] The big question, though, was whether a *submerged* submarine could receive radio signals. In 1910 the radio division arranged a series of experiments to explore the matter, informing Secretary Meyer of its desire "at this time to settle definitely the question of the feasibility of communicating with submarines entirely submerged."[74] In one group of tests, two submerged subs trailed floating-wire antennas behind them. They could receive radio signals at an operationally insignificant maximum distance of just six hundred yards.[75] In another experiment, a submarine submerged to different depths while moored to a pier. Signals could be heard to a submerged depth of fifteen feet but no deeper. The radio division concluded that subs would not be able to receive messages at ordinary operating depths and shifted its focus to acquiring better communications equipment for surfaced submarines. Permanent radio installations became standard with the boats that began to enter commissioned service in 1912.[76]

Stanford Hooper's interest in submarine radio was never as pronounced as his interest in aircraft radio, but his preferences did not keep him from investigating promising new ideas. One of those ideas was the floating-buoy antenna, a concept the service initially dismissed as impractical when proposed by a civilian inventor in 1912.[77] To receive messages, any such device would have to be deployed regularly, which would severely limit a submarine's mobility. Yet what if a sub needed to send an important message while submerged? The navy investigated this question in 1915 when Hooper had the Philadelphia Navy Yard explore the potential of a floating-buoy transmitter, one that could be deployed and retrieved from a submerged submarine. Tests carried out that November used a relatively small antenna mounted on a buoy that was to be "carried in a 'nest' in the submarine [and] so arranged that upon being released, it will float to the surface with its antenna."[78] Although personnel tried different arrangements, none produced signals strong enough to be considered practicable, and the radio division moved on to other issues.[79]

One of those issues was wireless telephony. As previously discussed the navy's first attempt to adopt voice radio came during the cruise of the Great White Fleet, which set sail with radiotelephones developed by Lee De Forest. For a variety of reasons, ranging from hasty installation to an inadequate supply of spare parts, this effort failed. Responsibility for that failure should be apportioned more or less equally between the navy and De Forest, but the Bureau of Equipment blamed De Forest and for several years thereafter remained skeptical of wireless telephony. By the mid-1910s, however, the continuous wave transmitters De Forest had used in his early radiotelephones no longer were novel, and the service again addressed the issue of wireless telephony. A civilian company, Western Electric, took the lead, and in September 1915 the navy's wireless station in Arlington, Virginia, exchanged voice messages with the naval radio station at Mare Island, California.[80]

Although the navy had helped pioneer wireless telephony, Western Electric's successful transcontinental demonstration convinced Hooper the service had fallen behind. He worked with Western Electric to acquire experimental sets for ship-to-ship tests, which the company delivered in February 1916. Trials revealed that these radiotelephones had a range of up to thirty miles, but like De Forest's earlier equipment, they operated within the same frequency range as existing apparatus. The bureau asked Western Electric to build sets that would not interfere with current shipboard radios, and in January 1917 the Atlantic Fleet took delivery of three state-of-the-art radiotelephones. Hooper tracked closely the trials of these radios, and reports from the fleet must have surpassed his most optimistic expectations. The Western Electric apparatus was the first naval radio to use vacuum tubes for both transmission and reception, and it could operate on three different channels. Furthermore, each channel was triple multiplexed, potentially permitting nine simultaneous ship-to-ship conversations.[81] In March the radio division acquired fifteen of the new devices with plans to test them thoroughly.[82] Those plans soon changed. Spurred by the German government's decision to resume unrestricted submarine warfare, in early April the United States declared war on Germany.

Reminders of War

The scores of books and articles written on the Battle of Jutland contrast sharply with the relative few on American naval operations during the First World War.[83] This fact stems largely from the ways in which the U.S. Navy contributed to the Allied war effort. While the service eventually provided a battleship division to support the British Grand Fleet, these vessels did not arrive in European waters until late 1917 and never engaged in fleet action.[84] Initially, a majority of American warships remained in the Western Hemisphere, where they safeguarded the nation's Atlantic seaboard and Caribbean possessions and, more critically, hedged against possible Allied defeat. Chief of Naval Operations William S. Benson philosophically opposed separating parts of the fleet, but over time he acquiesced to sending more vessels to European waters.[85] Most of these were destroyers and patrol craft employed in anti-submarine warfare operations. Congressional hearings later indicated that the Navy Department was not adequately prepared for the German submarine menace; when asked if there was "a sound, complete, and well defined plan" for that particular threat in the spring of 1917, Benson replied, "No, Mr. Chairman, there was not."[86] Given the general lack of American naval preparedness, one should not ascribe undue foresight to the Bureau of Steam Engineering's radio division. Yet the Western Electric radiotelephones that became available in 1917 were ideal for the convoy escort duty to which increasing numbers of ships and personnel were assigned.

The first American warships to depart for Britain were six destroyers under the command of Joseph K. Taussig. They pulled into Queenstown, Ireland, in early May; dozens more eventually would arrive. At first, Taussig's ships supported their British counterparts in conducting anti-submarine warfare patrols; however, as German U-boat attacks reached unprecedented levels of success, British and American leaders gradually shifted to a convoy system where merchant ships sailed in organized groups escorted by one or more warships.[87] For U.S. naval forces, both duties presented distinct command and control challenges.

William S. Sims, who headed U.S. naval forces operating in European waters during World War I, reportedly likened submarine hunting to searching for a needle in a haystack.[88] Unlike in World War II, when sonar and direction-finding equipment aided operating forces, in 1917–18 these technologies were in their infancy and most U-boat detections were visual.[89] Marine life, flotsam, and tired eyes led to numerous false sightings; and, whether real or imagined, the spotting of a hull or periscope required patrol craft to coordinate rapidly with one another. A problem for American warships arriving overseas, however, was that in most cases they were integrated into existing components of the Royal Navy. For example, the Atlantic Fleet's ninth battleship division became the sixth battle squadron of the Grand Fleet, and Taussig's destroyers operated as part of an integrated force under the overall command of a British flag officer.[90] Such integration created a fundamental command and control problem because neither American nor British shipboard radio systems had been built with combined operations in mind.

As junior partner, the U.S. Navy had to adapt, which in practice meant putting aside American radios and adopting British equipment and techniques. In some respects, the apparatus on board U.S. warships was superior to that carried by vessels of the Royal Navy. American transmitters emitted a higher-pitched note that operators could more easily copy through static, and American receivers were more sensitive, meaning that U.S. radiomen could hear weaker signals than their British counterparts. Generally speaking, these attributes were ideal for scouts that might be dozens of miles away from a fleet commander. British radios, on the other hand, possessed sharper tuning, which allowed ships operating in close proximity to transmit without fear of swamping receivers on nearby vessels. At the heart of the British system was a special circuit arrangement called the acceptor-rejector. It generated much interest among American naval personnel.[91]

The arrangement was so named because it consisted of two distinct parts. The acceptor enhanced the efficiency of a ship's main antenna, whereas the rejector was a shunt trap circuit in the receiver that eliminated signals outside a very narrow frequency band.[92] After American entry into the war, the British government gave the U.S. Navy access to the Royal Naval Signal School, which permitted a knowledgeable officer to take detailed notes on the acceptor-rejector circuit. Conveying an unwarranted hint of skepticism, the officer informed Admiral Sims that British fire control radios could communicate on nineteen different channels within a frequency band of only 180 kilohertz.[93] As the war progressed, other radio experts examined British equipment, using the information gained to improve American apparatus. By the end of the war, the Navy Department had retrofitted most warships operating in European waters with British-style radios for telegraphic communications.[94]

Resolving incompatibilities between British and American wireless apparatus took some time, but one area that received immediate attention was radio security. The Admiralty shared with the U.S. Navy some of the capabilities of the German communications intelligence organization and did what it could to improve the relatively lax security procedures on American warships. Just days after Taussig's destroyers arrived in Queenstown, Royal Navy officers, signalmen, and wireless operators descended upon them to teach the British system of signal flags and radio codes. In June 1917 the Navy Department began distributing mechanical encryption machines for shipboard use, and in October it printed a comprehensive instruction on codes and signals. By the end of the year, the two navies even were jointly issuing some security publications.[95]

Although American warships operating in European waters generally adopted British radio equipment and techniques, there was a notable exception: radiotelephony. According to the scholar who most closely has studied Royal Navy wireless in the early twentieth century, the Admiralty deliberately retarded development of radiotelephony while awaiting improvements in vacuum-tube technology. Another scholar suggests the "apparent British backwardness" in this area likely stemmed from the Royal Navy's knowledge of German intercept capabilities, which led the service to focus on light and flag signals.[96] To be sure, well-trained operators could use such signals quickly and effectively when weather conditions permitted. Yet therein lay two potential problems: the weather was not always good; and once convoy operations commenced, ship-to-ship communications had to take place not only between warships but also between warships and merchant vessels.

The radiotelephone potentially offered American and British naval commanders an effective tool for coordinating ships' movements during anti-submarine warfare patrols and convoy escorts. Such a device certainly would have aided Joseph Taussig during one early patrol, when his destroyer nearly sank a British warship before the latter could signal via searchlight to cease firing.[97] After the war, Taussig recounted the initial difficulties created by escort duty: "Communications were a particularly difficult problem at first. Some of the ships carried only one radio operator, and no signalmen . . . It was usually necessary in such cases to deliver a message by word of mouth through a megaphone. And this was not always easy."[98]

To be effective, any system of wireless telephony had to be reliable and easy to use. Fortunately for the Allies, the Western Electric radiotelephone met both criteria. With some modifications, Western Electric pushed into production the type of set it had provided to the U.S. Navy in early 1917. Officially designated the CW 936, the new equipment could operate at five discrete frequencies, any of which could be selected with the flick of a switch. First installed on American subchasers (110-foot wooden patrol craft built specifically to counter the U-boat), the CW 936 was rugged and simple to operate. It occupied less than two cubic feet of space and weighed slightly over fifty-five pounds, making installation on any class of vessel fairly straightforward. By the end of the war, approximately one thousand CW 936s had been produced and installed on American, British, and French vessels.[99]

The chief of the Bureau of Steam Engineering later declared Western Electric radiotelephones "of inestimable value in the antisubmarine campaign."[100] Subchasers employed them to coordinate attacks, and although U-boats might intercept such communications, their low silhouettes meant that German submariners almost always saw anti-submarine patrol vessels before they themselves were spotted by Allied lookouts.[101] For this reason, an Allied subchaser reporting a U-boat sighting rarely provided the submarine commander with premature tactical warning. For convoys, on the other hand, communications intercepts could lead to disaster, so escort commanders tried to maintain radio silence whenever possible. Cruising instructions provided by one escort commander in late 1917 are both typical and informative: "All vessels are enjoined not to use telephone except for official business . . . There will be no UNNECESSARY COMMUNICATION."[102] Yet if the worst happened and a convoy fell victim to an attack, the radiotelephone became an invaluable asset. Later describing some wartime uses of the radiotelephone, no less an authority than William Sims wrote:

A submarine chaser under way in 1918. Subchasers employed American-manufactured CW 936 radiotelephones to help coordinate attacks against German U-boats. Rugged and simple to operate, the CW 936 also saw service with British and French naval forces. *Naval History and Heritage Command photograph NH 83617.*

> Running conversations were frequently necessary between destroyers and the ships which they had been detailed to escort. "Being pursued by a submarine" . . . cries of distress like this were common. Another message would tell of a vessel that was being shelled; another would tell of a ship that was sinking; while other messages would give the location of lifeboats which were filled with survivors and ask for speedy help . . . At times the surface of the ocean might be calm . . . yet the air itself would be uninterruptedly filled with these reminders of war.[103]

Along with subchasers and destroyers, Allied forces used another important anti-submarine platform during World War I, the airplane. When the United States entered the war in April 1917, the French government expressed a strong interest in American support for airborne anti-submarine patrols; within weeks the Navy Department dispatched a detachment of men to France for this purpose. A flurry of construction took place soon thereafter, and in mid-

November a Tellier seaplane crewed by three U.S. naval personnel took off from Le Croisic, France, to execute the first airborne anti-submarine patrol flown from of an American base. The mission ended prematurely, however, when the craft experienced engine trouble and landed in rough seas. Certainly the crew would have liked to radio back for help, but it had a problem. The Tellier did not have a working radio. Fortunately, though, the plane carried messenger pigeons, and a French destroyer eventually came to the rescue.[104] On another patrol several months later, a pilot in a different type of aircraft spotted a surfaced U-boat and dove in for an attack, releasing a bomb that barely missed its target. Without a radio (or messenger pigeons) to alert other forces, the pilot returned to base, where he jumped into a new plane and flew back to the spot of the encounter. Needless to say, the U-boat escaped.[105]

Long before incidents such as these, Stanford Hooper had recognized the potential value of aircraft radio, which was why he had pressed to establish the Naval Aircraft Radio Laboratory. From the start, the officer in charge of the laboratory took a holistic approach, stressing not only transmitters and receivers but also the human element, because "the success of [aircraft radio] development depends as much upon the competency of the personnel as upon the efficiency of the apparatus."[106] Aiding the cause, in early 1917 Atlantic Fleet commander-in-chief Henry T. Mayo ordered all subordinate commands to solicit volunteer radio operators for assignment to aviation duty.[107] Many of these volunteers later found themselves headed to France and England, where they supported allied anti-submarine warfare operations.

Throughout 1917, material improvements to airplane radio equipment fell mainly under the supervision of civilian radio aide Benjamin Miessner. Even before American entry into the war, the former enlisted electrician had spearheaded the development of an aviation telephone system that enabled clear communication between pilots and crew members. The system's high per-unit cost initially precluded the service from buying it in large quantities, but fiscal constraints dissipated with wartime mobilization.[108] In May, the Bureau of Steam Engineering held a conference to establish research and design priorities for aircraft radio equipment, and Hooper and Miessner both attended. The latter quickly returned to Pensacola, where he worked with the officer-in-charge to expand laboratory facilities and hire new personnel. The bureau designated airplane transmitters a top priority, seeking devices with a maximum weight of one hundred pounds and a minimum range of one hundred miles.[109] Five different companies, including those headed by Lee De Forest

Lieutenant Commander William M. Corry Jr. gives final instructions to a pilot about to depart Brest, France, on an anti-submarine patrol in 1918. Because of the potential unreliability of aircraft radio, the aviator takes with him a basket of messenger pigeons. *Naval History and Heritage Command photograph NH 52846.*

and Guglielmo Marconi, claimed to have apparatus that could meet the navy's specifications, and in September the Naval Aircraft Radio Laboratory tested each firm's equipment. Although De Forest submitted the only continuous wave set, it did not perform nearly as well as a spark gap transmitter manufactured by Emil J. Simon.[110] Simon's apparatus, a two-channel device that weighed eighty-five pounds, cost almost a thousand dollars apiece. Even so, the bureau accepted Simon's price and placed a large production order. The service shipped many of these transmitters to both France and Britain.[111]

In order to locate the Naval Aircraft Radio Laboratory closer to the Atlantic Fleet, the Navy Department moved it to Hampton Roads, Virginia, in early 1918. Miessner did not make the move; curiously enough, pro-German statements allegedly made by Miessner's wife and mother-in-law led to his dismissal from government service.[112] Miessner eventually cleared his name, but without a strong guiding hand, the pace of work slowed.[113] To rejuvenate research activity, the radio division dispatched a reserve lieutenant commander, A. (Albert) Hoyt Taylor, to Hampton Roads, where he took charge of the service's development efforts with respect to aircraft radio. A former physics professor from the University of North Dakota, Taylor had joined the naval reserve just days before Congress declared war on Germany.[114]

When Taylor arrived in Hampton Roads, he found "active work in progress on the testing of various transmitters and receivers, both on the bench and in flight."[115] The most common type of transmitter was a bureau-designed version of the Simon apparatus that had been acquired the previous year; the service by then also had created a standard aircraft receiver. Designed by the Radio Test Shop in early 1918, the new receiver contained vacuum tubes and had two-stage signal amplification.[116] Its performance surpassed that of previous receivers but electrical interference remained problematic, for even as the laboratory modified equipment to enhance reception, the navy adopted more powerful aircraft engines that generated higher levels of interference.[117] Taylor and his colleagues pressed forward with solutions that brought about incremental improvement, including a redesign of the radio helmet originally developed by Miessner in Pensacola.[118]

Voice communications from naval aircraft presented yet another challenge, as rushing wind rendered the type of phone used on the CW 936 completely ineffective (by way of analogy, imagine a person talking on a cell phone with his or her head hanging out the window of a moving car). Possibly influenced by Miessner's work, Magnavox succeeded in constructing a microphone mounted in such a way that its diaphragm was exposed on both sides to external noise vibrations. This arrangement meant that the sound waves emanating from an operator's voice were the sole acoustic impulses acting on the diaphragm.[119] According to one contemporary radio expert, throughout the war Magnavox's apparatus was the only kind "at all satisfactory for aircraft use in the Navy service."[120] General Electric designed a series of successful aircraft radiotelephones around the Magnavox microphone; these sets began to enter service shortly before Germany agreed to the Armistice that ended the First World War. Aviators were excited to receive voice radios, but even as late as the fall of 1918 patrol aircraft sometimes still had to rely upon pigeons to get messages through.[121]

The naval war fought by the United States in 1917–18 was certainly not the type of conflict anticipated a few years earlier. The potential severity of an unrestricted U-boat campaign seems to have escaped most American policy makers, including naval officers, and prewar efforts to improve shipboard command and control focused largely on systems that would aid commanders during a fleet engagement. Nonetheless, from an operational perspective American naval command and control was adequate during the war. Four factors contributed to this outcome. First, the U.S. Navy benefited from close collaboration with experienced allies, particularly the Royal Navy. This circumstance lessened an otherwise steep learning curve. Second, the service profited from a robust, preexisting infrastructure in radio research and design. The U.S. Navy created the world's first laboratory dedicated exclusively to aircraft radio, hired knowledgeable experts to staff it and complementary facilities, and drew heavily upon the capabilities of domestic firms like Western Electric and Magnavox. Third, a relative lack of fiscal constraints after American entry into the war permitted the service quickly to install new command and control technologies like the CW 936. In peacetime, the navy had to replace such technologies at a measured pace; in wartime, the cost of radios was merely a drop in the bucket. Finally, while anti-submarine warfare required massive coordination at the strategic level, at the operational level it presented fewer command and control challenges than the type of fleet-on-fleet action witnessed at Jutland. With the U-boat seemingly defeated, efforts to improve shipboard command and control systems returned to the prewar paradigm.

Indeed, Jutland noticeably influenced the mind-set of senior naval officers in the years after the First World War. The battle had been a strategic victory for Britain, but American naval officers were particularly interested in exploring why the Royal Navy, with its numeric superiority, failed to win at the tactical level. Then as now, much of the focus centered on the loss of three British battle cruisers and on the decision making of participating naval commanders.[122] Yet American officers also identified command and control as a key aspect of the battle. Writing in 1921, Captain Alfred Hinds criticized the British for their limited scouting away from the main fleet, taking care to emphasize the importance of such activity for the U.S. Navy. He posited that the biggest "stumbling block" for the British was "lack of accurate information." According to Hinds, "The necessity for efficient tactical scouting, prompt and accurate reports and the charting of all incoming information seems to be far and away the most important of our lessons from Jutland."[123] Four years later, Captain Thomas Hart examined matters from the German perspective. He argued that the High Seas Fleet made the same basic information-sharing mistakes as the Grand Fleet, a "coincidence" that prompted him to wonder whether blame should rest "upon individuals or upon naval systems." Hart never directly answered his own question but left little doubt where he stood on the matter: the U.S. Navy needed better methods of command and control.[124]

One way to acquire such systems was to improve existing shipboard facilities and equipment. Both during the war and after, U.S. naval officers took care to document the Royal Navy's best shipboard communications practices. Three things generated much interest. First, in 1918 the British began adopting a new antenna arrangement in which every receiver had its own aerial. Such aerials were shorter and therefore less sensitive, but the Royal Navy partially offset lost sensitivity through signal amplifiers. Commenting on the arrangement, an American officer wryly remarked: "This idea has been developed, mainly, to render unnecessary the carrying of spare aerials which (as experience in action tends to show) cannot readily be renewed during an engagement . . . to send men aloft during an action to renew aerials would be to seriously reduce [one's] valuable operating staff."[125] A second practice that interested American naval officers was the Royal Navy's use of dedicated radios for fire control. The U.S. Navy had no equivalent to the British Type 31, a vacuum-tube set with very sharp tuning and a range of up to twenty miles. American observers were somewhat surprised at the Royal Navy's reluctance to use radiotelephony for fire control but marveled at watch standers' ability to work "at phenomenal speeds with no interference and no trouble." Lastly, U.S. personnel took a significant interest in the plotting facilities and interior communications systems of British warships.[126]

Identifying the Royal Navy's most successful practices with respect to shipboard command and control was relatively straightforward; implementing these and other improvements was more difficult. Two critical issues the service had to address after the fighting ended were personnel losses and cuts in naval funding. The Navy Department anticipated both but underestimated their magnitude. Probably nowhere was attrition more acute than in radio. Ironically, the vast expansion of U.S. naval training during the war had provided, for the first time, an adequate supply of operators. The service trained nearly half of all its radiomen at Harvard, where more than twelve thousand men went through a course of instruction lasting from sixteen to twenty weeks. Under the direction and guidance of Arthur Hepburn, who in early 1918 was serving as Naval District Base commander in New London, Connecticut, the service also established a radiotelephone school where hundreds more received training.[127]

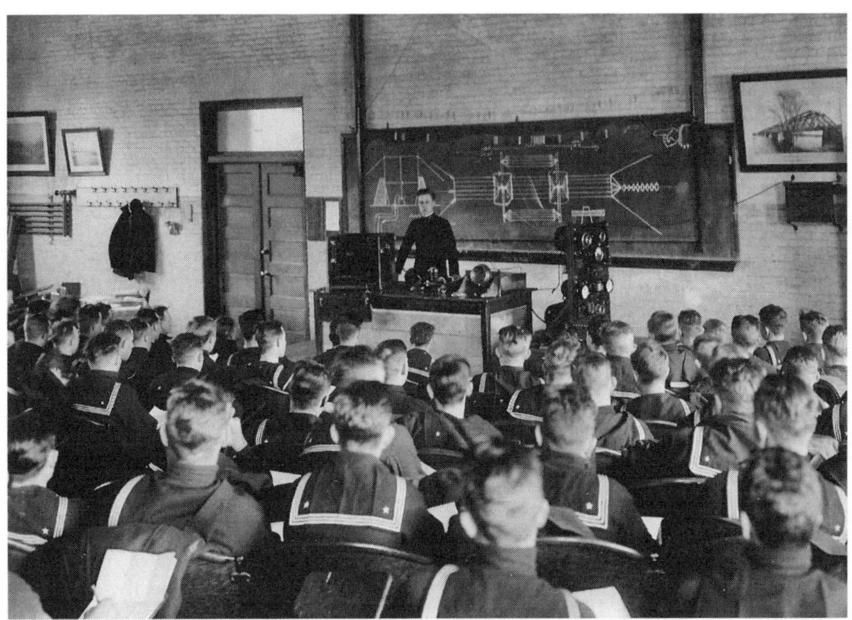

Radiomen receiving instruction in transmitting at Harvard University, most likely in 1918. The vast expansion of U.S. naval training during World War I provided a plentiful supply of operators, but after the Armistice the navy had great difficulty retaining skilled radiomen. *Naval History and Heritage Command photograph NTC-Cam 1907 (NR&L Files).*

Naval leaders had recognized that there would be postwar attrition of radiomen, but the exodus was more severe than expected and created a manning headache for the fleet. The cynical words of one officer succinctly capture what was clearly the view of many: "Navy personnel not expert . . . cushy jobs for commercial concerns attract [the fleet's] good men."[128] Expressing his concern, Atlantic Fleet commander-in-chief Henry Mayo informed Secretary Daniels that there existed "a serious shortage of men available for radio, either with previous experience or for training," a situation brought about "by discharge of practically all the prewar and war personnel [and by] the high wages paid radio operators on merchant ships." Mayo went on to point out that the average battleship was manned at only about three-fifths of its full radio complement and that fewer than half these men could "read and transmit messages at the Fleet speed."[129]

The navy's inability to pay competitive wages to radio operators was just one issue in the service's return to an environment of relative fiscal austerity.

In fiscal year 1922, federal spending on commissioned warships totaled more than $192 million. Five years later that figure had dropped by one-quarter, and a comparable amount would not again be spent until 1933.[130] The five-power Washington Naval Treaty (signed by diplomatic representatives of the United States, Britain, Japan, France, and Italy on 6 February 1922) fueled this decline, and Congress would choose not to fund a fleet equal to treaty limits until 1934. Although most navalists opposed shipbuilding limitations on the grounds that a powerful fleet was essential for guarding U.S. possessions and interests in the Pacific, Congress and three consecutive Republican presidents disagreed.[131] American naval personnel therefore again turned their attention to improving the operational capabilities of the vessels they possessed. Nevertheless, budgetary constraints were real and influenced what could be accomplished. In practice, fiscal limitations frequently hampered changes to shipboard command and control systems.

Tight budgets most affected two aspects of shipboard command and control: equipment and facilities. During World War I, the navy spent more than $3.3 million on arc transmitters, yet by 1918 this type of apparatus already was being superseded.[132] At the time of the Armistice, many spark sets still remained in service, so the postwar fleet consisted of vessels carrying three different and generally incompatible types of transmitters: spark, arc, and vacuum tube. The radio division warned of its inability to fix the problem in late 1921, when it informed fleet officers that the Bureau of Engineering was unable to authorize expenditures for many of the equipment improvements then needed.[133] Complaints from the fleet were endemic, with senior officers penning comments like "The modernization of Fleet radio equipment should be pushed vigorously," "The Commander-in-Chief desires not *more* equipment so much as better equipment," and "Any further increase in efficiency must come from the addition of modern equipment."[134] Nor was the budget problem limited to radio apparatus. The war had demonstrated the lasting importance of visual signaling systems, and exercises revealed that expensive high-powered searchlights were the best available means for long-distance communication when the operational situation warranted radio silence.[135] In 1925, however, the commander-in-chief of the United States Fleet felt compelled to tell the CNO that, after reviewing the subject of visual signal equipment, he found the methods and devices employed "still very crude," adding, "No real change has taken place for many years."[136]

Inadequate shipboard plotting facilities were just as problematic as outdated communications equipment. Major warships of the era already had designated plotting areas for fire control, a subject covered well by historians Jon Sumida and David Mindell.[137] Yet, as the battle space expanded in both distance and dimension, commanders discerned a need to picture visually their surroundings. Royal Navy gunnery expert Frederic Dreyer proposed plotting the approach of an enemy fleet in May 1914; a few months earlier Stanford Hooper had tinkered with what possibly was the first plot to track opposing exercise forces.[138] Later that same year, American inventor Hannibal Ford applied for a patent on an instrument called the Battle-Tracer. Ford's invention mechanically plotted on paper the course and speed of both "own ship" and a target vessel.[139] Originally designed for incorporation into a fire control system, the device's ability to provide a bird's-eye view of the battle space was so valuable that naval officers began to consider using it for other purposes. Beginning with flagships and extending to other vessels, the Navy Department eventually installed plotting facilities on nearly all warships.[140] Shipboard modifications were expensive, though, and work proceeded slowly. Throughout the 1920s, naval officers expressed much displeasure over the lack of adequate shipboard facilities from which to exercise command.

In 1924, for example, one of the service's most senior officers argued that to arrange forces properly a naval commander "must have information of enemy and own forces plotted on a chart where he can clearly visualize their relative positions and determine his own disposition and courses of action" and lamented that he could not obtain "any valuable information regarding the enemy forces from the instruments provided [to me]."[141] Battleship division commander Louis M. Nulton, who had commanded one of the battleships attached to the Grand Fleet during World War I, was even more blunt. Of his flagship, Nulton wrote:

> The NEW YORK is not designed, nor at present arranged, to meet the requirements of a Commander of mixed forces . . . A man on a bicycle is as well fixed to compete with a man on a motorcycle as the material flag arrangements are filled to meet present screen (and fleet) Sector Commander's work . . . The upper flag plot is filled with overflow radio material and apparatus so congested that operators practically fall over each other . . . lower flag plot [is] inaccessible for rapid transit to and from the bridge and has *no* communications . . . I consider the present lack of facilities on board the NEW YORK the best asset the enemy has.[142]

To ensure his superiors did not misconstrue such comments as superficial whining, Nulton then stated:

> If the paragraphs relating to the lack of facilities and personal inconvenience of the Sector Commander are read as merely being an expression of individual lack of ease, the entire fundamental principle involved will have been lost by the reader. The point involved is that if 80% or 90% of one's physical and mental energy is spent in overcoming overhead or an obstacle supplied by lack of facilities there remains but a small fraction of the total available to be applied to the legitimate work of the problem in hand.[143]

Although fiscal constraints impeded upgrades to equipment and facilities, naval commanders possessed a freer hand in other areas. One of these was shipboard communications policies and procedures, changes to which required much thought but few capital expenditures. For example, in mid-1918 the Atlantic Fleet began to investigate new procedures for handling radio messages, and in early 1919 fleet commander-in-chief Henry Mayo formally adopted the "Force Wave System."[144] The purpose of the new system was to align radio channels (waves) with the fleet's operational organization. Before Mayo's reform, the Atlantic Fleet had a designated "calling wave" on which any ship wanting to communicate would transmit another ship's unique call sign. The receiving vessel would acknowledge the call, after which both vessels would shift to a "working wave" in order to exchange message traffic. Mayo recognized the impracticality of this process for a sizable fleet, writing: "The number of ships operating together which can exchange calls on a single wave is limited . . . with the present number, it is impossible." Under the Force Wave System, the commander-in-chief gave each fleet component its own channel—for example, 1,512 meters (198 kilohertz) for the cruiser force, 378 meters (793 kilohertz) for the submarine force—while he took control of the old calling wave to ensure a means of "instant communication with all of his ships."[145] The Force Wave System was an improvement over its predecessor, and although naval personnel would modify Mayo's plan, its basic elements remained in effect until the fleet adopted a sweeping new radio frequency plan in April 1927.[146]

Of course, successful changes to fleet communications procedures had to be not only well conceived but also practiced. As previously discussed, such practice was a part of American naval exercises dating back to the Key West maneuvers of 1874. The deployment of the Squadron of Evolution in

1889 had inaugurated a new operational mind-set, and by the early twentieth century fleet maneuvers took place with increasing regularity. During the 1900s and 1910s, these maneuvers were rarely all-inclusive, however, as diplomatic missions and armed interventions frequently kept elements of the fleet from participating. While a large portion of the fleet remained in Hampton Roads during World War I, after the Armistice the United States divided its naval forces more or less equally between the Atlantic and the Pacific. This arrangement lasted until 1922 when President Warren Harding, at the CNO's urging, moved the nation's battleships to the Pacific.[147] Precipitating this shift was the five-power Washington Naval Treaty, which cut the navy's inventory of battleships in half and prohibited the United States from fortifying territories in the Pacific.[148]

The terms of the Washington Naval Treaty contributed to a major reorganization of the fleet, formally instituted in December 1922. Under the new organizational structure, most American naval forces combined to form a single United States Fleet, itself consisting of a battle fleet, commanded by an admiral, and a scouting fleet, led by a vice admiral.[149] Home-ported on the West Coast, the Battle Fleet consisted primarily of the navy's battleships and (as they became available) aircraft carriers. The Scouting Fleet, initially stationed on the East Coast, consisted mainly of cruisers and maritime patrol aircraft. In time of war, the two fleets, along with an assemblage of submarines, destroyers, auxiliary vessels, and Marine Corps expeditionary forces, would come together to take the fight to the enemy.[150] Needless to say, integrating these forces created major command and control challenges.

Even so, American naval officers recognized that the new organizational arrangement presented not just challenges but opportunities as well. Relative international calm during the 1920s and early 1930s meant the fleet had ample opportunity to train, and train it did. The training cycle coincided with the fiscal year, with the first half of the year devoted to elementary professional skills (e.g., torpedo or gunnery practice) and the latter half dedicated to more advanced training. The culmination of this cycle was a "fleet problem," routinely held in late winter or early spring, which brought together all available ships as part of a simulated wartime scenario. The fleet problems gave naval personnel, especially senior officers, the most realistic possible training short of actual combat. Preplanned but unscripted free maneuvers in which each side had specific a mission to execute, fleet problems usually related in some way to the navy's war plan against Japan.[151] Each problem had rules to determine the results of engagements, along with umpires to enforce the rules and make reasonable estimates of simulated damage. Significantly, these problems gave American naval commanders a chance to gain valuable experience at a time when two technologies, the airplane and the submarine, were making the coordination of seaborne forces ever more difficult.

Black and Blue

In February 1923, the new United States Fleet conducted its first major exercise, a simulated attack on the Panama Canal involving a majority of American naval forces. The *New York Times* heralded the event as "manoeuvres on the most extensive scale ever planned for the American navy," a claim that certainly captured the grandeur of the moment.[152] Without question the maneuvers were large, with nearly 150 ships and more than thirty-nine thousand personnel participating.[153] In this inaugural and appropriately named Fleet Problem Number One, vessels of the Scouting Fleet operated as a friendly "blue" force tasked to protect the canal against attack by an enemy "black" fleet comprised of warships from the Battle Fleet. Broadly speaking, the objectives of Fleet Problem One and its successors were to practice fleet maneuvers; to test war plans, operational instructions, and tactical doctrine; and to allow commanders and their staffs to plan operations and to make time-critical decisions in a simulated combat environment.[154] Unfortunately, command and control difficulties were a recurring theme in all three areas during Fleet Problem One.

The chief umpire for navy's first fleet problem was U.S. Fleet commander-in-chief Hilary P. Jones, whose responsibilities included the monitoring and evaluation of participating forces. In his final report for the exercise, Jones noted numerous deficiencies related to issues of command and control, going so far as to write that fleet communications "as carried through in this problem show a similarity, on a small scale, with the British Fleet before and during the Battle of Jutland."[155] The admiral pointed out that the number of available radio channels had been inadequate, creating congested circuits and ineffective communications.[156] According to Jones, the fleet lacked sufficient procedures for prioritizing messages or, as he tersely stated: "Important information was delayed by the transmission of relatively unimportant information."[157] This point had been emphasized by the blue commander-in-chief, who in his own report appended a copy of every message sent or intercepted by his flagship along with the wry comment that the amount of tactically relevant information could have been "reduced to about six messages."[158] Finally, Jones reported that because commanding officers generated so many duplicate reports commanders had considerable difficulty maintaining a coherent tactical picture during the exercise.[159] In a nutshell, Fleet Problem One demonstrated to the navy's leadership that commanders did not yet possess the necessary means to manage and process available information.

Jones's observations were supported by and based upon feedback from his subordinate commanders and individual commanding officers. For example, the commanding officer of one battleship remarked that his biggest problem "was one of communications," lamenting, "The average length of time to get through a contact report was about two hours."[160] A blue destroyer squadron commander reported the same prolonged delays, noting that these could lead to "great confusion."[161] Other officers were even more blunt. The commander of the blue air patrol stated that radio communications between his aircraft and ships of the fleet "were very unsatisfactory,"[162] while destroyer commanding officers found ship-to-ship communications "almost impossible," "far from satisfactory," and "inadequate."[163] One destroyer skipper even expressed his opinion that the whole communications system had "practically failed completely."[164]

Nor were the command and control troubles experienced in Fleet Problem One limited to radio. The Black Fleet, which maintained radio silence during most of the exercise, had significant problems executing long-range visual signals. Screening vessels in particular had difficulty maintaining contact with the main body, leading one squadron commander to state, "Present means of destroyer communications by visual for great distance are very unsatisfactory," and one division commander to suggest that the navy had underestimated greatly "the importance of visual communication at long range."[165] Message delays often exceeded two or three hours, with the result that warships in the screen had tremendous difficulty maintaining position.[166] One commanding officer aptly summarized the matter by noting that "accurate and efficient communication is imperative in order that the scouts may be kept informed as to the movement of the fleet. The whole basis of the communication system is embodied in one principle—the dissemination of information. If the information received is faulty and incorrect it does more harm than good."[167]

Two battleships under torpedo attack during Fleet Problem One (February 1923). One torpedo has missed the closest battleship astern but the other torpedo has scored a hit just below the furthest battleship's forward turrets. Fleet Problem One highlighted the difficulties of commanding and controlling geographically dispersed forces against a dimensionally diverse opponent. *Naval History and Heritage Command photograph NH 100448.*

The navy's senior leaders realized these issues needed to be addressed as quickly as possible. In July 1923, CNO Edward W. Eberle appointed a special committee to consider the design of future fleet problems and to suggest improvements.[168] Minutes and handwritten notes from this committee reveal significant concern over the demonstrated unreliability of ship-to-ship communications, especially those between a fleet's main body and its scouting vessels. Committee members agreed that new exercises should continue to test rigorously the fleet's command and control systems. Several officers suggested that scenarios for future maneuvers should require commanders to employ radiotelephones or tactical radio direction finding, and one even argued that the next fleet problem should include "simulation of a complicated action, in which, during the smoke and confusion of battle, the O.T.C. is faced with the necessity of communicating by radio with some of his own units . . . whose identity is uncertain and whose flagship and unit organization is destroyed or crippled."[169]

That Eberle would initiate action to address the command and control problems laid bare by Fleet Problem One is hardly surprising. Although Eberle's sole biographer labels him a gunnery officer at heart,[170] a close look at Eberle's career reveals that he had interest and experience in command and control systems. As an ensign Eberle was a vocal proponent of naval messenger pigeons, positing that warships, "by dispatching pigeons, may be able to keep shore stations informed."[171] In late 1897 he published a remarkable professional article describing how messenger pigeons carried on board cruisers could aid a naval commander trying to keep tabs on a Spanish fleet in or near Cuban ports.[172] Wireless rendered this idea moot, of course, but several years later the young officer found himself pursuing the same ends with new means. From 1903 to 1905, Eberle was flag lieutenant of the Atlantic Fleet; in that job, he helped John Hudgins and others incorporate radio into fleet operations. During World War I, Eberle was superintendent of the Naval Academy, and in 1922, while serving as the Pacific Fleet commander-in-chief, he convened a special board to explore the modernization of capital ships' radio installations.[173] Most critically, he had been the Black Fleet commander during Fleet Problem One.

As much as any senior naval officer, Eberle appreciated the difficulties inherent in the command and control of large fleets. Yet even in his role as CNO, Eberle could do only so much. Fiscal constraints pressed hard, limiting the navy's ability to buy modern communications equipment and build better command and control facilities. The CNO understood that immediate improvements had to derive from more efficient policies and procedures, but he needed the right men for the job. Fortunately, Eberle and the service found them in two veteran hands, one an admiral ready to assume fleet command, the other an experienced radio expert. The former was Samuel S. Robison, who became the Battle Fleet commander-in-chief in the summer of 1923. The latter was Stanford Hooper, who took over as the United States Fleet's radio officer just weeks later.

Curiously, Samuel Robison is an officer largely overlooked by historians. In the 1920s, however, he did more than any other individual to improve the fleet's system of command and control. A Pennsylvania native, Robison graduated from the Naval Academy in 1888 and sailed with the Squadron of

Evolution in the early 1890s. During the Spanish-American War, he fought in the Battle of Manila Bay (1 May 1898), earning a performance rating of "excellent" from his commanding officer.[174] Like Eberle, Robison then gained valuable experience working with radio during its early stages of development. In 1904 the Navy Department gave Robison orders to the Bureau of Equipment, where the bureau chief put him in charge of the newly established radio division. Why Robison received this particular assignment remains a bit of a mystery since he appears not to have had any prior background in electricity or electrical systems. Most likely, the chief simply recognized Robison as a highly capable officer with the ability to learn quickly. And learn quickly he did. Robison took the late John Hudgins's *Instructions for the Use of Wireless-Telegraph Apparatus* and expanded the work greatly, producing a highly successful manual on wireless telegraphy for naval electricians. When Robison completed his tour, the bureau chief took care to identify him as "a most excellent Electrician."[175]

Robison served on a pair of armored cruisers before the navy sent him back to the Bureau of Equipment in 1909, where he was placed in charge of the design, installation, and upkeep of shipboard electrical systems. He worked closely with the radio division, once again earning acclaim for his "wide knowledge in electrical matters" and his superior expertise in "Electrical Engineering in all its branches [including] Wireless Telegraphy."[176] Robison went on to command three warships and served as the Atlantic Fleet submarine force commander from July 1917 to September 1918. After the First World War, he burnished his already strong professional reputation by performing well in several high-profile assignments, including eighteen months as the military governor of Santo Domingo and six months on the navy's General Board. On the eve of Robison's appointment to command of the Battle Fleet, navy secretary Edwin Denby called him "an exceptionally able officer" and praised his work as "very well done indeed."[177]

Robison's knowledge of shipboard communications systems surpassed that of any previous fleet commander in the U.S. Navy. And with Stanford Hooper serving as the radio officer of his immediate superior, U.S. Fleet commander-in-chief Robert Coontz, Robison could be confident that efforts to improve communications efficiency would receive support from above. One of Robison's first actions was to direct his staff to study "the best way to coordinate visual, radio and sound," specifying that he wanted the most efficient possible means for coordinating his forces.[178] Robison lamented the fleet's lack of

Battle Fleet commander-in-chief Samuel S. Robison, c. 1925. Largely overlooked by historians, Robison was instrumental in improving the fleet's systems of command and control. *Naval History and Heritage Command photograph NH 80541.*

modern communications equipment, at one point pleading with the CNO for more high-intensity searchlights, at another exasperatedly informing Coontz: "WITH PRESENT RADIO EQUIPMENT THE BATTLE FLEET IS NOT READY FOR BATTLE."[179] Yet Robison knew that complaining to the Bureau of Engineering, which agreed that existing equipment did not meet fleet requirements,[180] was like preaching to the choir. He thus focused heavily on organizational reforms, one of which was to create shipboard communications departments. This reform ensured a single officer on each warship held responsibility for coordinating all available methods of communication: flags, lights, semaphores, radio, and underwater sound signals, the last of which was then still experimental. Another key reform pushed through by Robison was the elimination of radio strikers, a personnel practice that allowed individuals with no shore-based schooling to become radiomen exclusively through on board training. To Robison, the complexity of contemporary radio operations made strikers both unwise and uneconomical.[181]

Seeking to lead by example, the Battle Fleet commander pushed his subordinates to improve the efficiency of their radio organizations. He achieved notable success. In 1923 the time required to deliver a radio dispatch within the fleet averaged more than half an hour. By June 1924, the average had decreased to just over five minutes.[182] The following month, Robison instituted a record-keeping system for radio traffic and began publishing a monthly report showing the average transmission times for all subordinate commands. By April 1925, the mean time to transmit a message was less than four and one-half minutes, with Robison's flagship leading all commands at an average transmission time of two minutes, fifty-eight seconds. These improvements prompted the satisfied admiral to write: "From an organization which was used without confidence . . . the radio organization has become the most important carrier as regards both quality and quantity of communications in the Battle Fleet."[183]

Robison's improvements to shipboard command and control extended beyond organizational reforms to include tactics as well. When Robison commanded the Atlantic submarine force during World War I, he became impressed with the work of a lieutenant commander named Chester Nimitz. Some five years later, just as Nimitz was completing his studies at the Naval War College, Robison sensed an opportunity. He pulled some strings to get Nimitz, by then a full commander, assigned to his flagship and made him the tactical officer. At the War College, Nimitz had written a thesis on Jutland;

he also had participated in several war games. These experiences gave him an idea for how a naval commander could more efficiently control a large, geographically dispersed fleet. In the latter part of 1923, Nimitz pitched his idea to Robison, proposing a new type of cruising formation in which screening vessels positioned themselves in concentric circles around a fleet's capital ships.[184] This so-called circular formation differed from the traditional scouting line, a rectilinear formation in which echelons of vessels sailed ahead of a fleet's capital ships.[185] In the circular formation, vessels occupied sectors around the main body, thereby minimizing the likelihood of a surprise attack. The formation's circularity also obviated the need for complex maneuvers to change direction and allowed a fleet to shift more rapidly from a cruising to an approach formation.[186]

Despite such theoretical advantages, many naval officers believed the circular cruising formation would make station keeping prohibitively difficult, particularly at night.[187] Robison knew the only way to test the new formation was to try it at sea, something he did on a limited basis in early December.[188] This small-scale trial went relatively smoothly, so Robison proceeded with a comprehensive test during Fleet Problem Two, a simulated advance across the Pacific through a barrier of hostile submarines. A few weeks later in Fleet Problem Four,[189] a movement to an advanced base hypothetically located several hundred miles from Japan, Robison further experimented with the new cruising formation. Because many of the communications issues encountered in Fleet Problem One remained unresolved, however, these exercises did not provide an unambiguous verdict on the circular formation. One senior commander reported that on two occasions screening vessels "found themselves very much out of position," and he complained that with existing means "it seems practically impossible to devise any effective methods of communication with vessels of the distant screen."[190] Conversely, another senior officer declared the circular cruising formation "the best thing of the kind thus far devised" and looked forward to the day when "a broadcast message would be hardly less certain of immediate accurate receipt by every vessel of the Formation, than would be the case if the same message were delivered by spoken word to the Commanding Officers of those vessels assembled in the O.T.C.'s Cabin."[191] Robison concluded that while communications difficulties remained, the circular formation "has successfully met such tests as it has been possible to apply."[192] Tellingly, he employed it again during the following year's maneuvers.[193]

Other officers shared Robison's belief that ship-to-ship communications were a critical and increasingly important determinant of success in battle.[194] Before Fleet Problem Five (1925), for example, the Blue Fleet commander claimed that the exercise was "as much a problem in communications as it is a scouting problem and the success or failure of the whole maneuver will depend largely upon the radio personnel of the fleet."[195] Several months after Fleet Problem Seven (1927), the U.S. Fleet commander-in-chief noted that "the primary function of the Communications system is the tactical and strategical control of the Fleet"; a year later his successor wrote: "Efficiency of communications is indispensable to the fullest efficiency in the exercise of command. Particularly does tactical command demand rapid and accurate communication."[196] Reflecting back years later, Robison's protégé Nimitz stated simply that "effective coordination and ultimate operational success depends upon efficient communications."[197]

Of course, American naval commanders long had recognized the importance of command and control, but the fleet problems threw a Klieg light on the issue. As such, senior officers became increasingly vocal about the fleet's lack of modern communications equipment. Robison was less strident than others, perhaps because he had spent considerable time in the bureaus and understood the budgetary constraints they faced. By 1926, however, even Robison had become so frustrated that he took the unorthodox step of telling navy secretary Curtis Wilbur how the Bureau of Engineering and the navy yards should reallocate their spending priorities in order "to put Fleet Radio on a Satisfactory war basis."[198] Others complained as well. Nine months after relieving Robison as commander-in-chief of the Battle Fleet, Charles F. Hughes reported that no receivers and a mere fourteen transmitters were "satisfactory for use on a combatant ship."[199] In mid-1927, the commander of the Battle Fleet's destroyer squadrons complained that his ships carried "antiquated" direction-finding equipment and still used CW 936s for radiotelephone communications, which after years of wear and tear were "unstable in frequency" and generally "unreliable."[200] A few months later Hughes directed his ire toward the Bureau of Construction and Repair, criticizing it for the slow building pace of new searchlight platforms. He pointed out that the Bureau of Engineering had developed "a satisfactory signal searchlight" but that the current pace of construction would necessitate "eight years more before this project is completed."[201]

The Bureau of Engineering was sensitive to such complaints, often responding directly to criticisms from fleet commanders.[202] Yet the bureau faced a dilemma not unlike the ones faced by the Signal Office in the late 1870s and the Bureau of Equipment at the turn of the twentieth century. In the former instance, the service delayed the purchase of large quantities of Very signal flares so that it could obtain more reliable flares. In the latter, bureau chief Royal Bradford adopted a wait-and-see approach toward naval radio after Guglielmo Marconi failed to demonstrate his wireless system could reliably prevent interference. In both cases, the fundamental dilemma was whether or not to provide a new communications technology to the navy's warships. If the new technology were unreliable, then the fleet would be stuck with ineffective equipment that would need to be replaced or refurbished. On the other hand, a technical breakthrough not introduced promptly enough might put American naval forces in peril during a conflict. During the 1920s, the Bureau of Engineering faced just this dilemma with two communications technologies: underwater sound signaling and high frequency (HF) radio.[203]

An important organizational change facilitated work in these two areas. In 1923 the navy consolidated all radio and sound research at a single facility, the Naval Research Laboratory.[204] This consolidation brought together in one location personnel from the Naval Radio Research Laboratory (formerly the Naval Radio Telegraphic Laboratory), the Radio Test Shop, the Naval Aircraft Radio Laboratory, and the Naval Engineering Experiment Station in Annapolis.[205] The new facility initially had just two divisions, one devoted to radio and the other to underwater sound. A. Hoyt Taylor, the former physics professor and naval reserve officer who had helped develop aircraft radio during and after World War I, led the radio division. Harvey Hayes, a leading sonar expert who too had been a physics professor, took charge of the sound division. The navy's goal in establishing the Naval Research Laboratory was both to economize and to provide a more collaborative environment for research and experimentation.[206] As with any scientific lab the researchers experienced successes and failures, but in aggregate the new facility aided the navy in obtaining better shipboard command and control systems.

Personnel at the Bureau of Engineering relied upon the expertise of scientists and engineers at the Naval Research Laboratory to help them decide when a new technology should be provided en masse to the fleet. Unfortunately, in the case of underwater sound signaling the shore establishment furnished American warships with a technology that was not operationally ready. One of the problems the bureau was trying to solve was that of communicating with submarines.[207] The cramped interiors of these vessels precluded installation of large transmitters, and the repeated cycle of diving and surfacing wreaked havoc on antenna arrangements.[208] Visual communications were no less problematic. As Samuel Robison described the problem in 1924, "Submarines can use flags only at expense of quick diving, consequently the use of flags on submarines is impracticable. No proper searchlights and no other method of visual day communication have been developed, although experiments have been tried with smoke bombs. These may be developed but will probably be of value chiefly to attract the attention of surface craft so that they may come close enough for short range communication."[209]

By the time Robison wrote these words, underwater sound signals could be received reliably at a distance of up to one thousand yards.[210] The Bureau of Engineering apparently believed this range could be expanded to a tactically significant level with the devices then available, so in fiscal year 1925 it acquired underwater sound transmitters and receivers for a large portion of the fleet.[211] This equipment served a dual purpose. First, it could be used to detect other vessels, surfaced or submerged, by sending out sound waves and listening for a response (i.e., active sonar). Second, ships could use it to communicate via Morse code. At first, sporadic "phenomenal results," such as when a submarine tender copied sound signals from a battleship fifty miles away, encouraged naval personnel. Yet, because such results were rarely reproducible, Robison eventually concluded that underwater sound equipment served "little useful purpose." Even worse, it diverted valuable funds away from radio modernization. Robison suggested further "experimental development work" at the Naval Research Lab but recommended removal of nearly all sound apparatus on surface ships.[212] While the Bureau of Engineering did not go quite that far, for several years thereafter it provided the fleet with considerably less equipment and focused on more limited trials aboard selected vessels.[213]

In contrast to underwater sound signaling technology, the bureau's adoption of high-frequency radio equipment was quite successful. From the start, the Naval Research Lab ran an active research program in HF radio. A. Hoyt Taylor later recalled that "in the late fall of 1923 we did not realize the tremendous possibilities for the use of high frequencies in the field of Naval Communications [but] we did see that they would certainly be extremely valuable, provided we could sufficiently stabilize transmitters."[214] The lab helped solve this problem by establishing an on-site facility to evaluate and grind quartz crystals. Such crystals, when properly cut and installed, provided more than adequate frequency control. In 1924 the Naval Research Lab installed an experimental HF transmitter and receiver on the new rigid airship *Shenandoah*, which used these devices on its famous transcontinental voyage that autumn. *Shenandoah*'s apparatus performed well, and the following year U.S. Fleet flagship *Seattle* received an HF radio for shipboard trials. Fleet Radio Officer Hooper was initially skeptical, but after working with it under operational conditions, he became a strong proponent of the new technology.[215]

The Naval Research Laboratory went to work fixing the few problems experienced during the *Seattle* trials, such as overly temperamental receivers, and the Bureau of Engineering started supplying the fleet with HF radio equipment in the latter part of 1926.[216] Within months, naval commanders were complaining. This time, however, the complaints were not about inadequate apparatus. Instead, senior officers were imploring the Navy Department to expedite deliveries of HF radios to the fleet.[217] According to one radio officer, operators were so excited about the new technology that some of them actually assembled "home made" HF transmitters "for use both afloat and ashore."[218]

Enthusiasm over HF radio ran high for several reasons. To begin, there was less static in the HF band. Next, because the ionosphere refracted HF radio waves back to earth (so-called sky waves), long-range radio communications were possible with low-power transmitters. Taylor later reminisced that "it was possible to get long range communication with sets of ridiculously small power."[219] Less power meant lower costs and, most critically, smaller antennas and more compact equipment. For submariners these characteristics were especially welcome. In early 1928, the Bureau of Engineering equipped two of the fleet's newest submarines with HF transmitters. Installation was somewhat problematic because the bureau accidentally gave the manufacturer, General Electric, incorrect dimensions for the submarines' hatches; workers were thus surprised to discover that they could not lower the new transmitters into the boats. The senior submariner in the Pacific badly wanted those radios, though, so he authorized the use of blowtorches to make necessary cuts to get the equipment in place. The apparatus performed well once installed, probably a fortunate outcome for that officer's future career.[220]

One can reasonably argue that although World War I was fought chiefly on land, it was won at sea. Neither the Marne nor the Somme, with all the battlefield carnage, inflamed the people and politicians of the United States like

Lusitania and *Sussex*. Indeed, American neutrality would end only after Germany launched an unrestrained submarine offensive designed to force Britain, the nation with the world's most powerful navy, to sue for peace. Largely because the Royal Navy's industrial demands so thoroughly disrupted Britain's merchant shipbuilding industry, Germany's strategy nearly succeeded.[221] Allied adoption of the convoy system helped to stave off disaster, and the U.S. Navy played an important (although probably not decisive) role in that system.

In many ways the U.S. Navy was unprepared for anti-submarine warfare. In the realm of command and control, however, the service transitioned relatively smoothly from predicted fleet battles to actual anti-submarine operations. Before American entry into World War I, naval officers Arthur J. Hepburn and Stanford C. Hooper spearheaded efforts to improve shipboard command and control systems. Tangible results from those efforts included the adoption of new equipment, such as arc transmitters, heterodyned receivers, and wave meters; the creation of more efficient training programs for enlisted personnel; and the establishment of a larger and more effective research and design program for naval radio. To be sure, unanticipated problems arose when war arrived. But the existence of facilities like the Radio Test Shop and the Naval Aircraft Radio Laboratory, as well as collaborative relationships with companies like Western Electric and Emil J. Simon, allowed the navy to address problems quickly and efficiently.

Nineteen months of wartime cooperation with experienced allies benefited the postwar U.S. Navy, particularly in the area of communications security. The service adopted numerous Royal Navy techniques and shipboard arrangements, and American naval officers spent considerable time studying the command and control problems that occurred during the Battle of Jutland. Yet the biggest issue faced by the navy after the war was the federal government's embrace of fiscal austerity. As in other areas, economy measures adversely affected shipboard command and control. And while American naval commanders often complained about obsolete equipment, poor facilities, and insufficient numbers of trained personnel, they also sought improvements in areas requiring limited expenditures, most notably communications policies and procedures.

In 1923 the navy began rigorously testing the ability of its warships to operate as a coordinated body during large-scale annual maneuvers. Unfortunately, during the service's inaugural fleet problem numerous command and control issues arose, ranging from absurdly long message delays to poor information management. Many of the navy's senior leaders recognized that these problems needed immediate attention, and they acted accordingly. Fleet commander Samuel S. Robison, a knowledgeable and highly capable officer, led the way by significantly improving the fleet's communications efficiency and introducing important changes in the way the fleet operated. The Navy Department also consolidated most of the research related to command and control systems at a single facility, the Naval Research Laboratory. The Naval Research Lab was instrumental in bringing HF radio to the fleet, an important technical advance that would prove enormously valuable in the 1930s and beyond.

Ironically, the Great Depression of the 1930s eased some of the fiscal pressures faced by the navy. High unemployment made the recruitment and retention of skilled personnel easier, and significant deflation during the administration of Herbert Hoover (1929–33) actually boosted the navy's purchasing power. For example, in fiscal year 1929 the service spent $2.8 million on radio apparatus; three years later it spent the same amount, but in deflation-adjusted terms that sum equaled approximately $3.6 million.[222] The election of Franklin D. Roosevelt in November 1932 and growing congressional concerns over Japan's intentions in the Pacific augured well for future naval spending. Even so, throughout the 1930s budgets remained tight. Yet the biggest issue for naval commanders during that decade was an emerging sublime technology, the aircraft carrier, which added a new level of complexity to the already difficult problem of how best to command geographically dispersed and dimensionally diverse seaborne forces.

CHAPTER FOUR

A Most Complex Problem

Demanding Information

> Due to aircraft, tactics have become more complicated. The complications are increased further by the submarine.
>
> Thomas P. Magruder, 1930

> The difference between a good officer and a poor one is about ten seconds.
>
> Arleigh A. Burke, 1943

Information

Olive Wilbur hit the bottle hard across the bow. Glass shattered, and champagne flew. The date was 16 November 1927, and the wife of navy secretary Curtis Wilbur had struck the blow that officially commissioned USS *Saratoga*. A similar ceremony took place the following month for another aircraft carrier, USS *Lexington*. The United States would commission five more carriers before December 1941, but until the mid-1930s the only ones present in the fleet were *Saratoga*, *Lexington*, and the experimental *Langley*. Products of the five-power Washington Naval Treaty, *Lexington* and *Saratoga* had been laid down as battle cruisers; as such, they were big and fast. The service would employ one or both in every fleet problem conducted after 1928.

Anticipating the arrival of *Lexington* and *Saratoga* into the fleet, the Navy Department used *Langley* to develop and improve carrier flight operations. The officer in charge of this task was Captain Joseph M. Reeves. An Illinois native, Reeves had entered the Naval Academy in 1890. In his junior and senior years Reeves played right tackle on the football team, and he may well have invented the football helmet, a piece of equipment worn by him during the army-navy game in 1893.[1] As an ensign Reeves participated in the Battle of Santiago de Cuba (3 July 1898), but he missed the cruise of the Great White Fleet and remained stateside during the First World War. Probably not on track for flag rank, Reeves's career received a boost in 1924 when he became head of the tactics department at the Naval War College. His performance in that assignment helped him secure follow-on orders to the naval aviation observer's course in Pensacola, Florida, and in the fall of 1925 Reeves became commander of the Battle Fleet's aircraft squadrons.[2] His main focus over the next several years was on improving the operational capabilities of the navy's sole aircraft carrier.

When Reeves assumed command of the Battle Fleet's air squadrons, *Langley* had been in commission for more than three years. Yet the carrier was hardly an efficient fighting machine. To a large degree, this situation derived from *Langley*'s status as an experimental vessel. Reeves did not believe *Langley*'s role as a test platform precluded maximum operational efficiency, however, and he wasted little time imparting this vision to his subordinates. Within months, he had increased the number of planes the carrier could successively launch from six to sixteen. To expedite aircraft recovery, Reeves embraced the idea of organizing the flight deck crew into color-coded cohorts, each with a specific task to perform (e.g., brown for crew chiefs, blue for airplane pushers, yellow for directors). One of Reeves's biggest innovations was the deck park, a technique in which crewmen moved landing planes to the back of the flight deck and then raised a wire barrier so aircraft still aloft could come aboard as quickly as possible.[3]

As *Langley*'s personnel implemented reforms, the carrier played an increasingly aggressive role in the service's annual fleet problems. In the maneuvers of 1924 and 1925, *Langley* participated but accomplished little. In Fleet Problem VI (1926), Reeves sent out *Langley*'s aircraft to "attack" an enemy cruiser, and in Fleet Problem VII (1927) he used them to strafe the opposing fleet's destroyers.[4] By April 1928, Reeves had managed to squeeze forty-two planes on board *Langley*, up to thirty-six of which could be sortied in a single launch sequence. During Fleet Problem VIII (1928), the carrier provided aerial scouting for the Battle Fleet, and in May *Langley* became the centerpiece of a joint army-navy exercise designed to test Hawaii's defenses by having a large part of the fleet conduct a simulated shore bombardment and aerial attack on Oahu. Early on the morning of 17 May, the carrier launched thirty-five planes in just seven minutes; they arrived over the island at daybreak, attacking Wheeler Field and several other army installations.[5] This simulated attack was in many respects a harbinger.

Even as Reeves worked to improve *Langley*'s performance, the Navy Department placed two more carriers under his command. Reeves, by now a rear admiral, was enthusiastic yet apprehensive. On the one hand, *Lexington* and *Saratoga* were more than twice as fast and nearly three times the size of their predecessor. On the other, Reeves would be sailing in uncharted waters, developing new tactics and doctrine for coordinated carrier operations. Expressing his concerns to William Moffett, chief of the Bureau of Aeronautics, Reeves wrote that "we have made progress in the work of operating one carrier in Fleet operations, but with the advent of several carriers this work becomes many times more difficult and more complex. In this tactical work there is no precedent to guide us and I am at times almost appalled by its importance."[6]

Perhaps Reeves's anxiety reached a peak during the early morning hours of 26 January 1929. Fleet Problem IX was well underway, with Reeves aboard the black carrier *Saratoga*. His mission was to destroy the Panama Canal. Reeves's superior, Black Fleet commander-in-chief William V. Pratt, was many miles away in his flagship. At Reeves's urging, four days earlier Pratt had detached *Saratoga* and an escorting cruiser to make a circuitous high-speed run toward the designated launch point. Around 3:30 a.m., Pratt received an intercepted message indicating that *Saratoga* was preparing to launch aircraft. This news was both good and bad. On the positive side, it meant *Saratoga* had arrived within raiding distance of the canal's locks. Unfortunately, it also meant that blue forces knew *Saratoga*'s geographic position. Pratt undoubtedly checked his plot, perhaps breathing a sigh of relief when he saw that his battleships would arrive at a prearranged rendezvous point in time to protect *Saratoga* from blue forces. What Pratt did not know was that the plot on his flagship was wrong; his forces were in reality thirty-five miles too far away. *Saratoga* would have no additional protection once daylight arrived.[7]

A clever ruse had helped Reeves even get to that point. On the afternoon of 25 January, a blue scout destroyer discovered a large aircraft carrier but could not positively identify it. Reeves later recounted the story, revealing that "the bright young men of my staff did a bit of quick thinking and radioed an order to take 'station astern.'"[8] The destroyer commanding officer, now convinced that what he saw was blue carrier *Lexington*, complied. When the destroyer got close enough, Reeves opened fire with *Saratoga*'s eight-inch guns, making quick work of the opposing vessel.[9] Despite his staff's quick thinking Reeves knew *Saratoga* was in grave peril, for the blue destroyer had radioed back information regarding the carrier's position prior to being sunk.[10]

Early the next morning, Reeves gave the order to prepare to launch planes. Crewmen pushed aircraft into place; sparks flew as engines warmed up. A bright moon made the task easier, but Reeves credited the crew's perfect execution to months of rigorous training: "Even under the difficult light conditions, and the imperative necessity of the greatest haste in launching the planes, squadron after squadron went off at the tick of the watch, just as in practice."[11] Shortly before 5:00 a.m., the first pilot in line pushed his throttles to full power. A plane roared down *Saratoga*'s flight deck and into the air. Sixty-nine more followed, primarily fighters, light bombers, and torpedo planes. Also in the mix were four observation aircraft, the only ones to carry radios. Rather remarkably, given the advance warning available to blue, *Saratoga*'s aircraft arrived unopposed over the canal's southernmost locks. Multiple loads of simulated ordnance quickly turned these locks into imaginary rubble.[12]

Back on *Saratoga*, developing events tempered enthusiasm for the bold raid. Reeves had anticipated a rendezvous with Pratt's forces so *Saratoga* would have some protection during aircraft recovery. He scanned the seas, "expecting to sight [Pratt's battleships] at any moment."[13] A destroyer appeared on the horizon, then several larger vessels. Much to Reeves's dismay, these were not the ships he was expecting. At the very moment the first squadron of black aircraft was dropping simulated ordnance on the Panama Canal, three blue battleships opened fire on *Saratoga*. According to the official exercise observer these vessels needed just five minutes to destroy the unprotected carrier.[14] If *Saratoga* really had been sunk, in one fatal blow the navy would have lost not only an aircraft carrier but also more than 60 percent of all officer pilots trained for carrier operations.[15]

Scholars have pointed to *Saratoga*'s strike on the Panama Canal in Fleet Problem IX as a seminal moment in American naval history. Craig Felker calls *Saratoga*'s air raid "truly significant," and Charles Melhorn uses the event to mark the point where "the carrier came into its own . . . [with] its first great tactical triumph."[16] Other historians downplay the significance of Fleet Problem IX but admit that it boosted aviators' morale and gave credibility to the use of carriers as operational units.[17] Some scholars choose to portray Fleet Problem IX as a turning point forsaken; in this interpretation, conservative battleship admirals failed to recognize that the fast carrier task force was the future of naval power.[18] Often cited by these historians are the words of U.S.

Fleet commander-in-chief Henry A. Wiley, who in his post-exercise critique stated that "there is no analysis of Fleet Problem IX fairly made which fails to point to the battleship as the final arbiter of Naval destiny."[19]

In hindsight, Wiley's comments seem to lend credence to the idea that conservative officers failed to appreciate the full potential of carrier aviation. This interpretation has some steadfast proponents. Historian Robert O'Connell argues that a majority of American naval officers were irrationally prejudiced toward the battleship, even going so far as to depict *Lexington* and *Saratoga* as "unwanted prototypes" that were "something of an embarrassment" to conservative interwar naval officers. Clark Reynolds highlights "the relentless opposition of the Gun Club" against aviators like John Towers, and William McBride characterizes pre–World War II U.S. naval officers as members of a "thought collective" that embraced a "battleship technological paradigm."[20] A common theme in the work of these scholars is the conviction that naval officers identified with a particular type of warship. According to O'Connell in particular, senior naval officers held an undue reverence for the battleship, "the single most important artifact of their professional existence."[21]

Was the battleship important? Yes. But to claim that this particular artifact dominated the thinking of American naval officers is to neglect considerable evidence to the contrary. For example, a generation ago Thomas Hone revealed that the percentage of U.S. naval expenditures devoted to battleships declined after 1927, dropping sharply even after the final treaty limiting capital ship construction expired at the end of 1936.[22] More recently, Craig Felker emphasizes the service's concerted efforts to test and develop submarine tactics and doctrine during the fleet problems,[23] while John Kuehn's work on the General Board highlights the role of the Washington Naval Treaty's fortification clause in shaping American naval innovation during the interwar period.[24] To date, however, no one has explored the critical operational problem arising from the aircraft carrier's introduction into the fleet: how best to exercise command over forces more numerous and geographically dispersed than ever before.

Fleet Problem IX offers an excellent point of departure for exploring shipboard command and control during the 1930s. To be sure, efforts toward this end built upon the previous work of American naval personnel, from Foxhall Parker's new signal book to the improved efficiencies achieved by Samuel Robison in radio communications. Yet Fleet Problem IX marked a break from the past in several ways. It was the first time the service conducted annual maneuvers with multiple aircraft carriers, and it was the first use of a detached fast carrier task force. The total number of aircraft involved, more than 260, was many times greater than in any previous fleet problem.[25] There was another new development in Fleet Problem IX, though, that has gone largely unnoticed by historians. For the first time, the navy's official report for a fleet problem contained a substantial entry under the heading "INFORMATION."[26]

Why did U.S. Fleet commander Henry Wiley and his staff single out this topic in their critique of Fleet Problem IX? The most proximate reason was that the information management practices of both sides had been poor. Wiley was especially upset over the mishandling of important tactical information. Regarding black's inability to interdict the blue battleships that sank *Saratoga*, he believed such a disaster never should have occurred, declaring: "It should have been possible to establish the positions of the forces concerned with sufficient exactness to make such a failure impossible. There is no more vital thing in naval operations."[27] Others echoed similar sentiments, including former radio division chief Arthur J. Hepburn, who called black's failure "the most serious and mortifying blot on a record of very fair achievement."[28] Wiley did not hesitate to invoke the Battle of Jutland, claiming that it furnished "a classic example of the importance of correct navigational positions in cooperating forces."[29]

For all his displeasure with black command and control, Wiley directed his harshest criticism at the Blue Fleet for its inability to communicate securely via radio. During Fleet Problem IX, a special group of intelligence officers stationed on board the Black Fleet flagship intercepted and decrypted a large percentage of blue's messages,[30] prompting the assigned observer to comment that there had been a "complete breakdown of Blue radio communications so far as secrecy is concerned." The observer went on to describe the success of black's radio intelligence organization:

> For all practical purposes it can be said that all Blue despatches by radio were translated into original English text and made available to the Commander in Chief, Black Naval Forces in time, in most cases, for appropriate action to be taken. With all due regard to the cleverness and efficiency of the Decrypting Unit in the Black Force Flagship, it is desired to stress as emphatically as is possible that this complete lack of secrecy was not due so much to the Decrypting Unit as to the simplicity of the task presented

> to the Unit, due to our own general and serious lack of knowledge of communications.[31]

Wiley was so upset about this breakdown that he recommended making the misuse of codes and ciphers a punishable offense equal in severity to that of providing codes and ciphers directly to the enemy.[32]

Wiley's explicit focus on information drew attention to what was arguably the most abiding concern of operational commanders in the decade before World War II. To command naval forces effectively, they required information that was accurate, timely, and secure. Of course, commanders in previous eras had demanded information with the same characteristics. But by 1929 a confluence of developments had coalesced to magnify considerably the disadvantages of inaccurate, late, or compromised information. The navy's early fleet problems already had illuminated a range of command and control issues, from poor shipboard plotting facilities to the pervasive presence of old communications equipment. Yet nothing exacerbated the command and control difficulties faced by naval commanders as much as the aircraft carrier. Such vessels were limited in number and highly vulnerable, raising significantly the value of any information about their location at sea. Furthermore, as growing numbers of aircraft joined the fleet (including both seaplanes and dirigibles) commanders had to control more discrete units. Finally, because aircraft were many times faster than ships, naval commanders often had much less time to make critical decisions. Thus, as the service worked to integrate aviation into the fleet, naval personnel frequently found themselves in situations of unsettling complexity. Simplifying this complexity required better systems for the management of information. Not until the service developed such systems could the full potential of carrier aviation be realized.

Carrier Vulnerability

At the core of the U.S. Navy's efforts to integrate the aircraft carrier into fleet operations was an intractable issue: carrier vulnerability. Most officers had anticipated the danger one aircraft carrier posed to another. Joseph Reeves, for example, remarked after Fleet Problem IX that he honestly had feared only the other fleet's carrier.[33] As just recounted, however, *Saratoga*'s demise actually resulted from an unexpected encounter with three opposing

USS *Saratoga* conducting flight operations in the early 1930s. Carriers were highly valuable but also truly vulnerable assets that exacerbated the command and control difficulties faced by naval commanders. *Naval History and Heritage Command photograph NH 95043.*

battleships. This occurrence was no anomaly. Throughout the 1930s, surface ships and submarines hypothetically sank the navy's carriers again and again, in fleet problem after fleet problem. Over time, a pattern emerged with respect to these sinkings: they generally derived from incorrect or, more often, incomplete, information. Such experiences help to explain why effective shipboard command and control systems were an indispensable element of carrier warfare.

From 1930 to 1940, the navy conducted twelve fleet problems. In nearly all of these, surface warships hypothetically damaged or sank some or all of the opposing carriers.[34] In Fleet Problem X (1930), for example, enemy cruisers found both flattops assigned to the blue forces, "badly damaging" one of them. The other carrier sped away to report the attack but provided the blue commander-in-chief with contact reports that were "both incomplete and inaccurate."[35] In Fleet Problem XI (1930), surface warships engaged and impaired both *Lexington* and *Saratoga*, the latter of which was accidentally sunk by friendly fire. This incident occurred in part because an important dispatch had been "confused with another that had been received at the same time."[36] During Fleet Problem XII (1931), another simulated attack on the Panama Canal, *Lexington* twice encountered enemy surface ships, both times escaping only after escorts laid down smoke screens to cover the carrier's retreat.[37]

Fleet Problem XIII (1932) played out much like a carrier duel between *Saratoga* and *Lexington*, but ultimately torpedoes from a nighttime destroyer attack were the blows that sent *Saratoga* to the bottom.[38] The following year in Fleet Problem XIV (1933), *Lexington* was the more unfortunate carrier. Bad weather the night before a planned air raid caused *Lexington* to lose contact with several escorting black cruisers. The strike group commander described what happened next: "Efforts were continued to contact the cruisers by radio and at 0554 course was set for the launching point . . . The planes were in the process of re-warming when at 0605 a BLUE battleship was sighted close aboard at an estimated distance of 4,500 yards. The battleship opened fire and the carrier tried to escape but in a few moments came under fire from a second battleship."[39] The umpire ruled *Lexington* out of action before even a single aircraft had taken off.[40]

Drawing on lessons learned from these fleet problems, American naval officers sought to develop ways to reduce carrier vulnerability. Their efforts bore but limited fruit as the decade progressed. During Fleet Problem XV, an extensive exercise carried out during the spring of 1934, every participating carrier was lost. Most dramatic were the events of 6 May. That morning more than one hundred carrier-launched gray aircraft conducted a successful attack on two blue carriers, sinking one and inflicting heavy damage on the other.[41] Just as in Fleet Problem IX, however, the exhilaration of success turned quickly to dismay when the launching carriers themselves came under fire. The U.S. Fleet commander described the situation in dramatic prose, writing: "Just as the GRAY squadrons were returning to [their carriers], flushed with this decisive victory over BLUE, an embarrassing situation arose in turn for GRAY. While these planes were about to land, there appeared over the horizon [three BLUE cruiser divisions], consisting of eight heavy cruisers."[42] These eight cruisers, in combination with six dozen blue planes launched just prior to the destruction of their own carriers, made quick work of the gray flattops. The following day gray cruisers returned the favor, sinking the lone remaining blue carrier at close range with a barrage of simulated torpedoes.[43] Despite gray's success in finding and attacking the last blue carrier, the commander of the gray cruiser force informed his professional colleagues: "Unless one has actual experience in seeking information concerning a well screened force, it is hard to realize the difficulties of doing so."[44]

In contrast to Fleet Problem XV, during Fleet Problem XVI (1935) opposing forces sank no aircraft carriers. In part, this resulted from a decision to suspend all exercise-related air operations at a critical point so available aircraft could search for a downed patrol plane (unfortunately, the plane and its crew of six were never recovered).[45] Other factors played a role as well, including a large exercise area and "perfect radio silence" on the part of black forces.[46] Participants were unsure what to make of this development. One senior officer found it "interesting to note that although more carriers were present during this problem than ever before, it was the first time during any problem that no carrier was accurately located, bombed, or torpedoed."[47] Joseph Reeves, who by then had risen to become the U.S. Fleet commander-in-chief, praised each side's ability to keep its carriers undetected from the other. Yet he also cautioned against criticizing the forces tasked to locate the enemy's carriers, stating that the "desire for information" should not "obscure the fact that the ultimate objective is battle and the infliction of damage . . . or the prevention of damage to one's self."[48]

Opposing forces once again engaged carriers in Fleet Problem XVIII (1937), heavily damaging all three flattops operating with a black fleet tasked to fight its way through white territory to relieve a besieged island garrison. As in Fleet Problem XV, hypothetical damage was extensive. The carnage began when a white submarine torpedoed both *Langley* and *Lexington*. Hours later the two ships, which were operating together as part of a black strike force, came under attack by white cruisers.[49] The next day, aircraft flying from the only white carrier, *Ranger*, found black carrier *Saratoga*, dropping nineteen hundred-pound bombs. Black patrol planes soon responded in kind, attacking *Ranger* and putting the flight deck out of commission.[50] Meanwhile, *Lexington* suffered further. Another submarine found the unfortunate carrier, launching a simulated barrage of four torpedoes from just a thousand yards away.[51] Adding insult to injury, in the waning hours of the exercise three white battleships combined to concentrate fire on the beleaguered vessel.[52] According to the Black Fleet umpire, *Lexington* suffered theoretical damage of 92 percent, two-thirds coming from torpedo hits and one-third from gunfire.[53]

The pummeling absorbed by the navy's aircraft carriers during Fleet Problem XVIII generated much criticism against Admiral Claude Bloch, the Battle Force commander-in-chief.[54] Particularly vocal were those most experienced in carrier operations. In something of an understatement, future chief of naval operations (CNO) Ernest King claimed that the problem "was not a conspicuous success," and the commander of the Black Fleet's air forces remarked that "a carrier tied down to a slow formation is quite certain to be put out of action."[55] Modern scholars have been no less critical. Albert Nofi characterizes Bloch as "conservative" and among "the least imaginative of the Navy's senior officers in the interwar period."[56] The latter depiction is not without merit, but Bloch was not so much conservative as he was inflexible and controlling. In general, he believed the proper location for carriers was near the fleet's main body, a disposition that permitted tight control over such forces. Bloch justified his placement of black's carriers by pointing out that they "were able to contribute their full and proportionate share," and that since "all [warship] types shared in the damage sustained, the carriers may be said to have borne no more than their part of the brunt of the battle."[57] His remarks were somewhat disingenuous, as the loss of a gunnery turret on a battleship or a cruiser was certainly less damaging than a gaping hole in the middle of a carrier flight deck. World War II would show that a key to successful carrier employment was tactical flexibility and that reliable command and control systems enhanced such flexibility. Perhaps not surprisingly, Bloch appears to have had less interest in these systems than his predecessors had.[58]

Even so, Bloch never claimed his decision to locate aircraft carriers near the main body was absolute. He wrote that there was "no infallible rule for the stationing of carriers" and that each problem "must be solved upon its merits" because "the variables are many and a change in the value of any, or several, results in a different tactical treatment to obtain a proper solution."[59] Other senior naval officers expressed similar sentiments. For example, after Fleet Problem XXI (1940) a different Battle Force commander-in-chief opined that "it would be a mistake to formulate a hard and fast rule covering the question as to whether the carrier should be separated or kept with the disposition. The situations that may confront a commander are too varied. The best location must be determined by the attendant circumstances."[60]

Of course, such circumstances were ever changing. By 1940 the threat posed to aircraft carriers by surface ships, although still genuine, had declined.[61] During Fleet Problem XX (1939), the first in which new fleet carriers *Yorktown* and *Enterprise* participated, opposing commanders positioned their flattops as they saw fit, depending on the tactical situation.[62] Except for *Enterprise*, which opposing planes sank during the exercise, the heavy damage suffered in Fleet Problems XV and XVIII was not repeated. *Yorktown* emerged from the problem essentially unscathed, and the harm inflicted on other carriers came mainly from air attacks.[63] Aircraft also generated most of the damage endured by carriers during Fleet Problem XXI.[64] Nevertheless, surface warships engaged flattops during both problems. In Fleet Problem XX, a cruiser fired on *Lexington* from a range of fourteen thousand yards, and in Fleet Problem XXI surface ships attacked *Yorktown* twice, first with eight-inch gunfire and later with destroyer-launched torpedoes.[65] The threat was waning, but it remained real.

A big reason carriers increasingly had less to fear from surface vessels was that aircraft performance improved significantly during the latter half of the 1930s.[66] Speed, endurance, and payload all improved substantially. For example, the service's main fighter in the mid-1930s was a biplane (the Boeing F4B-4) that possessed a maximum speed of 188 miles per hour and a combat radius of about 120 miles. This aircraft was followed by a series of Grumman biplanes up to 40 percent faster. These Grummans (F2Fs and F3Fs) had much greater endurance, with a combat radius of 300 miles or more. Monoplanes entered the scene in the late 1930s, and the navy began to replace its biplanes with the Brewster Buffalo. Soon thereafter, the more agile Grumman F4F Wildcat superseded the Buffalo. Both fighters possessed a combat radius similar to the best biplanes but had maximum speeds well over 300 miles per hour.[67]

Improvements in aircraft payload led to an important new type of plane, the specialized dive bomber. The origins of dive-bombing—a technique where aircraft dive on a target from high altitude at a very steep angle—remain obscure but date to the 1920s. According to one account, a Marine Corps pilot learned the tactic from the army and suggested it to his navy brethren. In the fall of 1926, a squadron of fighters simulated this type of attack on ships of the Battle Fleet, and by the early 1930s dive-bombing was a regular part of the fleet's annual maneuvers.[68] The tactic was hard to defend against and more accurate than level bombing; however, the amount of ordnance a carrier aircraft could bear was limited. The Grumman F3F could carry two hundred-pound bombs of the kind that hypothetically damaged *Saratoga* during Fleet Problem XVIII, but plans for a plane that could conduct a diving attack with ordnance of up to one thousand pounds had been under development since 1929.[69] After a few fits and starts, in 1936 the fleet finally began to receive significant numbers of dive bombers capable of dropping thousand-pound bombs.[70] Five years later, the Douglas SBD Dauntless would become the fleet's primary dive bomber. It had superior carrying capacity and greater range than any previous aircraft of its type and would serve with distinction during World War II.

Better aircraft reduced the likelihood of an encounter between surface ships and carriers for two reasons. First, scouting aircraft with longer ranges could locate enemy ships farther away. In an ideal world, this gave commanders time either to maneuver away from the threat or to move their own forces into an interdicting position. Of course, the ocean environment was rarely ideal; weather could be bad, and scouting planes sometimes broke down. Moreover, navigational errors or human mistakes could lead to faulty reports. After Fleet Problem XVI, for example, one commander voiced his frustrations at just such an occurrence: "Various reports were received concerning enemy main body which proved to be our own forces anchored off Midway . . . It is better to be deployed than destroyed, but misinformation could bring about a mal-deployment with its attendant handicaps."[71] Also, in an era before radar or night flight operations, a carrier always risked a morning encounter with enemy surface vessels that had been hundreds of miles away the evening before.[72] All in all, however, better scouts aided the survivability of aircraft carriers.

The other reason surface warships became less of a threat to aircraft carriers was because carrier aircraft became a significant threat to surface warships. A few bombs weighing one hundred pounds might send a destroyer to the bottom, but they were unlikely to sink an armored vessel. Bombs weighing one thousand pounds were an entirely different matter. Furthermore, a much-improved carrier-based torpedo bomber, the Douglas TBD Devastator, entered service in the fall of 1937.[73] Although the torpedo bomber was easier to defend against than the dive bomber, a coordinated attack by both types of aircraft could be especially problematic. Battleships and cruisers became increasingly dependent on friendly fighters for protection, which only magnified further the value of carriers. Of course, carriers too were susceptible to coordinated air attack. Thus, even as the threat posed to carriers by enemy surface ships declined, the potential lethality of opposing air attacks escalated.

While attack from above may have been the biggest concern of American naval commanders, another problem lurked below. The service's inaugural fleet problem had revealed that U.S. submarines were mechanically unreliable and too slow to keep up with the battle line.[74] Troubled by these and other submarine-related deficiencies, in 1926 CNO Edward Eberle created a committee of experienced submariners to advise him and the secretary of the navy on such matters. Members of this group advocated building submarine boats with long endurance, improved habitability, and superior submerged performance. These ideas quickly gained acceptance, and by the late 1920s few commanders thought subs should be employed near the battle line. In 1932, after much design work, the fleet received the first U.S. submarine intended specifically for offensive patrol operations in the Pacific. By the end of the decade, twenty-nine such boats were in service.[75]

Submarines presented a real yet modest threat to surface warships during the early fleet problems, but that threat increased markedly over time as submariners gained valuable operating experience and eventually better boats. By the late 1930s, the navy's annual maneuvers had revealed that submarines posed a danger to aircraft carriers second only to the one posed by other carriers. As recounted previously, for example, submarines achieved notable success against participating carriers during Fleet Problem XVIII (1937). Two years later in Fleet Problem XX, one of the navy's newest submarines put several torpedoes into *Ranger*, prompting a senior officer to comment that the subs then entering service clearly increased "the vulnerability of carriers to submarine attack."[76] The officer was prophetic. The following year in Fleet Problem XXI two of the new boats torpedoed *Lexington*, and a third fired four torpedoes at *Saratoga* from virtually point blank range (twelve hundred yards).[77]

To make use of an analogy, the carrier of the 1930s was like a prizefighter with a devastating hook and a glass jaw. By the end of the decade, carrier aircraft could sink the most powerful warships afloat, but carriers themselves remained highly vulnerable to attack. And while American naval personnel worked diligently to integrate carriers into fleet operations, experience gained in the service's annual fleet problems highlighted a recurring dilemma: naval commanders too often lacked the information they needed to keep their carriers out of harm's way, and even when such information was available, they had limited ability to defend against attacks. In fleet problem after fleet problem, carriers were hypothetically torpedoed, bombed, and shelled. From the perspective of U.S. naval commanders, these platforms thus introduced an unprecedented level of uncertainty into warfare at sea. As one senior officer opined after the 1932 annual maneuvers: "Fleet Problem XIII should call to the attention of all officers associated with it, the idea of the swiftness of events in a modern sea battle. Enemy battleships almost over the horizon . . . bombers, torpedo planes, cruisers . . . destroyers assembling behind a smoke screen for attack, submarines unseen—these indicate the necessity for quick perception, quick response to a situation which may be partially hidden, and a thorough knowledge of the tactics of all arms."[78] Joseph Reeves described the situation in more colorful terms, arguing that opposing carrier forces were like "blindfolded men armed with daggers in a ring. There is apt to be sudden destruction of one or both. If the bandage over the eyes of one is removed the other is doomed."[79] Improved systems for the command and control of naval forces came directly out of the navy's efforts to remove these bandages.

Timely and Secure

For centuries, navies have been highly capital-intensive organizations. The early twentieth-century U.S. Navy was no exception, and personnel devoted much of their time and effort to developing new and better technologies for warfare at sea. Many of these technologies, among them airplanes, dirigibles, battleships, cruisers, destroyers, submarines, and aircraft carriers, were impressive to see. Yet, for all their grandeur, such platforms achieved maximum potential only through coordinated operations. This task was more difficult than some have appreciated, but it was one fully understood by American naval officers. As interwar fleet commander William V. Pratt aptly described the matter, "The coordination of activities of each type with every other type is a serious problem requiring concentrated study and long experience before high efficiency can be attained. The coordination of the air force effort as a whole, with that of the surface [and submarine] forces, then remains as the most complex problem of all."[80]

Key to the coordination of naval forces was information about enemy and own forces. Beginning in the early twentieth century, such information increasingly came from electronic devices, which meant not only that there was more information available but also that it often had to be filtered through intermediaries. To manage this situation, naval personnel both requested and developed systems that could provide them with relevant information in the quickest and most reliable manner. As one commander informed his subordinates in 1939, "An O.T.C. must have early, accurate, complete and continuous information of the *enemy* . . . [and also] a current appreciation of *our own* situation in terms of disposition, movements, and remaining offensive and defensive strength." He then gave an order that "no information in the possession of a subordinate should be withheld unless it is known without any doubt whatever that the O.T.C. is already in possession of the information. If there is any doubt, the information should be supplied."[81]

A prominent theme in the exercise reports of American naval officers during the interwar period is the demand for timely information. In the Battle Fleet commander's report on Fleet Problem VI, for example, future CNO Charles Hughes complained that there had been unacceptable delays in the receipt of important information. He curtly stated: "Delayed communications are of little or no value."[82] Once *Lexington* and *Saratoga* entered the fleet, officers expressed comparable sentiments with regularity. After Fleet Problem X, one commander remarked, "Timely information prevents a tactical surprise," while another stated that "intelligent action demands good and timely information."[83] Before Fleet Problem XI, the Black Fleet commander-in-chief informed his subordinates that success would "depend primarily upon quick dissemination of information and orders to own Force with a timely knowledge of the enemy resulting from accurate information and contact reports."[84] A few years later, *Langley*'s commanding officer bemoaned the information delays that had prevented his aircraft from crippling *Saratoga*, noting that even "a minute's delay in launching an attack has a serious effect on the results. Timely information is vitally important."[85] Comments similar to these are found in the records of nearly every fleet problem held during the 1930s.

One of the biggest obstacles to the rapid transmission and receipt of tactical information was communications security. During the First World War, British naval officers had impressed upon their American counterparts the risks associated with the transmission of unsecure messages, but this danger had to be balanced against the need for timely information. The service's annual fleet problems stressed the need for rapid and secure radio communications, and U.S. naval personnel took steps both to protect the navy's own and to exploit the opposing side's radio traffic. Fleet Problem VIII, held in the spring of 1928, appears to have been the first in which dedicated personnel attempted to intercept and decrypt the opposing fleet's wireless messages. In that problem, the navy assigned a special decrypting unit of three officers and an enlisted assistant to the orange fleet. Although this group intercepted only one tactical message during the exercise, naval commanders clearly grasped the potential benefits of such a unit.[86] Accordingly, the navy assigned a decrypting team to the black fleet for the following year's maneuvers, this time stationing it on the flagship of the commander-in-chief.[87]

Unfortunately for blue, the work of the black decrypting unit in Fleet Problem IX was nothing short of outstanding. Over the course of the problem, this group logged more than five hundred intercepted messages, many of which provided vital tactical information to black commander-in-chief William Pratt. According to one member of the decrypting team, intercepted messages provided significant details about the opposing side's plans and dispositions, and in most instances important information was available to Pratt and his staff less than three hours after transmission by blue.[88] U.S. Fleet commander-in-chief Henry Wiley called the "complete breakdown of secrecy" a "serious matter," while Pratt noted simply that "the importance of denying information to the enemy was emphasized and the value resulting to our own forces when enemy information was obtainable was prominent."[89] Still, Wiley recognized the inherent trade-off between speed and security. He informed the CNO that "secrecy, even if delay cannot be eliminated, is essential . . . [but] there comes a time when no delay can be tolerated . . . [sometimes] plain language must be used."[90]

Various factors contributed to the success of the black decrypting group during Fleet Problem IX. First was the excellent work of the decrypting team, which included several individuals who later would be instrumental in the navy's code-breaking exploits against Japan during World War II.[91] Second was the relatively high volume of radio traffic between blue platforms, which not only made black's task of breaking the blue codes and ciphers easier but also meant that more information was available for the taking. Third was simple carelessness by blue personnel, who in numerous cases violated established procedures by relaying previously encrypted messages in plain language. Finally, members of the black decrypting unit benefited from their familiarity with the navy's enciphering equipment.[92] While most officers retained faith in the service's supposedly "excellent" codes and ciphers, at least one commander openly questioned their value during a real war.[93]

The fleet problems also drew attention to the slowness of the service's encryption and decryption processes. Delays of several hours were routine, an unacceptable hindrance for most air operations. As one knowledgeable officer wrote after Fleet Problem IX, "Communications with aircraft . . . brings up the subject of, when to use code or plain English. In this operation if code had been used the delay in transmission might have caused the aircraft squadrons to entirely miss their objectives . . . [I am] of the opinion that the importance of the time element in any aircraft operation is greater than that of secrecy."[94] After the ensuing fleet problem, another naval officer was even more emphatic: "The use of plain language by the fleet air observer when action is imminent now appears to be a matter of necessity. Coding entails undue delay."[95]

Nor was the dilemma limited to aircraft communications. Naval commanders frequently complained about "coding lag," the time delay created by the encryption and decryption of messages, and one commanding officer was only partly jesting when he remarked after Fleet Problem XIV that radio "may be put out of action or the ship sunk during the time required to code messages."[96] A vocal proponent of the notion security should be sacrificed to timeliness when warranted by the tactical situation was future CNO William H. Standley, who in April 1933 wrote that, while there was "only a single advantage" to such a idea, it was "a tremendous one in that extremely valuable time is gained at the moment when it is most valuable."[97] Several years later as part of an ongoing dialogue between the Office of the Chief of Naval Operations (OPNAV) and the fleet over communications security, CNO Standley stressed to new U.S. Fleet commander-in-chief Arthur Hepburn that "plain language radio messages will necessarily be resorted to in the stress of battle for the transmission of information of immediate value."[98]

Recognizing that the navy's existing system of tactical radio communications could not provide information that was both timely and secure, naval commanders called upon the shore establishment to provide "a more rapid system of cryptography."[99] At the same time, they continued to use communications intelligence as a means for acquiring information about opposing forces. In Fleet Problem XI, for example, the Black Fleet commander organized three separate decrypting units, instructing them to make every possible effort to break the enemy cipher.[100] Because blue had a more advanced cipher than black, however, black's radio messages were the ones compromised during the exercise.[101] As such, throughout much of the problem, the Blue Fleet commander-in-chief had valuable information about his opponent. As one staff officer later described the situation,

> The BLUE Commander, by breaking the BLACK codes, was in the earlier stages of the problem advised of the movements and plans of the enemy's single vessels and small forces in time to act against them. In the final stages of the problem positive information relegated BLACK's plans from the realm of surprise to that of precisely known facts. The information thus obtained was of far more value to BLUE than all the combined information he did obtain or could have obtained from his entire scouting forces and it was obtained without the expenditure of a gallon of fuel or risk of a single vessel.[102]

The success of early shipboard decrypting teams prompted operational commanders to expand the presence of such units in the fleet. In Fleet Problem XIV, for example, the Black Fleet commander ordered the establishment of at least one decrypting unit in every task group under his command.[103] The same fleet problem also appears to have been the first in which the service placed special "Combat Intelligence Units" on board every major warship.[104] Existing records shed little light on how shipboard cryptanalysts and intelligence specialists worked with one another, if at all, but clearly important information had to be sent back to the force flagship, via either visual signals or radio transmissions. This inevitably increased the volume of communications, which led both to congested circuits and to enhanced opportunities for the cryptanalysts and intelligence specialists of the enemy.[105] Ironically, the burgeoning skills of American naval personnel added a degree of urgency to the push for more rapid and secure communications systems.

Operational commanders took two tacks in their efforts to develop systems with these capabilities. First, they sought to improve existing methods and procedures for radio security. One of the simplest ways to enhance security was to decrease the amount of radio traffic between naval forces. CNO William Standley emphasized this fact to the U.S. Fleet commander-in-chief when he wrote, "The degree of security desired must govern the volume of communications . . . the question as to when, and how to communicate . . . is really a function of command."[106] Yet establishing the parameters for such a vision was easier said than done. The fleet's communications plan contained dozens of radio frequencies, and commanders typically gave their subordinates considerable discretion in choosing how and when to use these circuits. Even radio silence was not absolute. After Fleet Problem X, for example, the Blue Fleet commander-in-chief acknowledged that, despite his strong preference for visual communications, "radio was at times imperative."[107]

Largely because American naval commanders so ardently promoted tactical flexibility and individual initiative, subordinates sometimes were too willing to break radio silence.[108] This happened in 1934 during Fleet Problem XV. A gray radio intelligence unit reported all transgressions to the officer in tactical command (OTC), but he appears not to have disciplined violators. The official exercise report commented simply: "The O.T.C. himself maintained radio silence until 1000. By that time so many other commanders were making frequent use of radio that further radio silence on his part was useless."[109] Enforcing communications security without squelching initiative was a challenge. The service addressed the issue primarily by stressing the dangers of ill-advised radio transmissions.[110] As the black commander-in-chief informed his subordinates before Fleet Problem XX, "Each [commander and commanding officer] must weight the advantage to be gained by transmission of information by radio against the loss of security . . . USE OF RADIO MUST BE REDUCED TO A MINIMUM, ACCORDING TO THE CARRYING OUT OF THE MISSSION AND ASSIGNED TASKS."[111] Prior analyses had shown that by the late 1930s subordinates took such guidance seriously. For example, an OPNAV study on ship-to-ship communications after Fleet Problem XIX reported that both sides had maintained "strict supervision of radio circuits" and an appropriate "volume of traffic" throughout the exercise.[112]

A key procedural change that minimized the amount of traffic between operating naval forces was a new method of radio communications in which a commander would broadcast messages out to his forces. First implemented in April 1930, this "no receipt" method relieved subordinate commands of the requirement to acknowledge receipt of messages sent out by the commander-in-chief. In addition to reducing the amount of traffic between vessels, the broadcast method also prevented an enemy from ascertaining the size of opposing forces by recording the number of responses to transmitted messages. An obvious disadvantage to such a system was that a commander had no positive assurance that any specific unit had received a message. Conversely, individual commanding officers never knew if they were in receipt of all the information their superiors had sent. The navy soon arrived at a solution to this dilemma by giving each message a serial number so subordinate commands could ascertain whenever a message had been missed.[113]

The service also reduced the volume of tactical radio traffic by developing a common operational doctrine. After Fleet Problem One, CNO Edward Eberle issued a significantly revised version of the navy's *War Instructions*. According to these instructions appropriate doctrine entailed a mutual understanding of the intentions and plans of the commander-in-chief or OTC. Ideally, this understanding would enable the coordination of forces even in the absence of detailed tactical orders. As the *War Instructions* expressed the idea, "Subordinate commanders shall exercise the initiative within their respective spheres of action, but always in loyal support of the intentions and general plan of their senior commanders. Otherwise confusion and failure will result, instead of coordination and success."[114] While such prose sounded good, the *War Instructions* of 1923 offered few details on how to achieve this vision. A big step toward this goal took place in 1930 when fleet commander William Pratt introduced a new and extremely effective set of battle instructions.[115] These instructions gave OTCs enhanced flexibility by providing an outline for tactical employment of the fleet. Unless an OTC provided orders to the contrary, subordinates were expected to take initiative in accordance with the overall plan. OTCs only needed to transmit broad changes to the original plan because subordinates could be expected to do their utmost to fulfill the previously declared intentions of the commander-in-chief.[116]

Praise for the new instructions was widespread. The Battle Fleet commander found them "simple to understand and to use in operations," while the Scouting Fleet commander remarked that they provided "the greatest single advance in fleet tactics I have known in my years of service . . . [they afford] to the O. T. C. an extraordinary increase in the flexibility of control from the beginning of tactical scouting through the general engagement, and until the final dispersion of the enemy."[117] Pratt's tentative instructions underwent several revisions before the service finally codified them as a tactical publication in early 1934.[118] By then most officers considered the fleet's tactical doctrine fundamentally sound. After Fleet Problem XV, held in the spring of 1934, U.S. Fleet commander-in-chief David Sellers noted that he had given "no important orders to any of the detached forces," pointing out that these forces "conducted their operations entirely by means of the current Battle Instructions, thus proving the great progress made in the indoctrination of the fleet."[119] Nevertheless, the CNO reminded Sellers that "even the best of doctrines will not cover all situations and the provisions for a senior to communicate with his subordinate must remain. Communications are the servant of command, and, therefore, must provide the tools with which command is exercised."[120]

In 1933 the service took another step toward improving communications security when Sellers organized a group of specially trained officers and tasked them to look for security violations in the fleet's message traffic. These specialists scrutinized all encrypted messages originated in or addressed to units of the fleet and advised the commander-in-chief on how to enhance communications security. They also worked to educate communications personnel throughout the navy via "personal contact, correspondence, and lectures on the dangerous possibilities attendant upon the careless use of codes and ciphers."[121] The CNO attached high value to the work of this group, noting that it was "of inestimable value in training and indoctrinating our officer personnel with regard to the military weaknesses and dangers which result from radio messages which have been poorly or carelessly prepared and encrypted."[122] Senior operational commanders repeatedly praised this "Fleet Security Unit," pointing to measurable progress in the coding of messages and the employment of cryptographic aids.[123] Indeed, the unit was so successful that in 1939 the U.S. Fleet commander actually decreased by two-thirds the number of officers on his staff responsible for monitoring communications security.[124]

The focal point of American naval commanders in their second tack toward achieving rapid and secure communications was encryption technology. During World War I, the service had adopted its first mechanical encryption machine, the Naval Cipher Box. Designed by a junior officer, this device reportedly remained uncompromised throughout the war.[125] In the mid-1920s the navy adopted a cylindrical cipher used by the army, but naval officers complained that it was cumbersome to use.[126] In order to encrypt a message, an operator had to assemble twenty-five aluminum disks in a specific order as determined by the keyword(s). The method for so doing involved a time-consuming, multistep process.[127] Evidence also suggests that the service's communications experts knew cylindrical devices were, by the nature of their design, inherently unsecure.[128] As such, the Navy Department devoted substantial effort toward developing a quicker, more secure encryption machine.

One promising possibility was a new type of machine developed by civilian inventor Edward Hebern in the early 1920s. Hebern's electromechanical device had rotors that enabled polyalphabetic encipherment, which theoretically made it more secure than existing machines. Confident of future riches, the inventor built a sizable factory in Oakland and solicited orders from both commercial firms and the armed services. Unfortunately for Hebern, legendary U.S. Army cryptanalyst William Friedman was able to decrypt messages enciphered by his machine, casting doubts on its efficacy. Yet this was not the inventor's biggest problem. In March 1926 he was convicted of securities fraud, and his company went bankrupt a short while later.[129] Hebern organized a new firm, but the navy was unsure who actually owned the rights to his encryption devices—the new company or the old one, which remained in receivership for some time.[130] Facing serious financial hardship, Hebern attempted to sell his intellectual and physical property to the navy, including all "patents issued or pending, models, drawings, tools, dies, jigs, and manufacturing parts and supplies on hand." The service declined, but OPNAV suggested to Hannibal Ford (the inventor of the Battle-Tracer) that he might wish to pursue a business relationship with Hebern.[131] In other words, while the Navy Department saw potential in Hebern's machines, it wanted to deal with a financially sound firm.

Ford and Hebern never joined forces, but the latter was able to raise new capital and continued to develop his encryption devices. Despite Friedman's success at breaking Hebern's enciphered messages, officers in OPNAV seemed most concerned about the mechanical reliability of the models provided for testing. One naval officer wrote "how sorry" he was that the "crude" mechanical features of the device prevented the fleet from acquiring "a machine on which you could write out your plain language message the same as you would on a typewriter and have it come out in cipher."[132] Hebern was persistent, however, and in the spring of 1928 he exhibited a new model to two of the fleet's most knowledgeable cryptographic experts. One or possibly both of these individuals wrote OPNAV to recommend further investigation,[133] and several weeks later the CNO ordered a thorough analysis of "Mr. Hebern's device."[134]

In January 1929 the Navy Department acquired several machines for testing, providing them to the Battle Fleet for operational trials. During a ten-week period, operators sent and received more than 320 messages using the new device. Battle Fleet commander-in-chief William Pratt reported that only 13 of these messages were indecipherable upon receipt, an error rate he found acceptable "in view of the great need for a high speed cipher of high security." Pratt recommended that "the Hebern Cipher Machine be adopted and issued to at least all capital ships and administrative commands afloat," and although U.S. Fleet commander Henry Wiley concurred he wanted Hebern to make some minor electrical and mechanical modifications.[135]

This Hebern did, delivering two new prototypes to the service in the spring of 1930. Pratt, by then the U.S. Fleet commander-in-chief, examined the new model with an eye toward its potential suitability for naval service and reported to the CNO that "the Hebern Cipher Machine, as modified, will attain a very satisfactory performance." Director of Naval Communications Stanford C. Hooper proposed purchasing thirty machines at a cost of fifteen hundred dollars apiece, which he believed was the smallest number necessary to provide "a good and thorough test." He also suggested making a large production contract as soon as these devices had proven their worth, mainly on the grounds that such action would reduce the cost per machine by 60 percent.[136] In June 1930 the Navy Department awarded Hebern's firm a production contract for thirty-one devices, of which OPNAV hoped to take final delivery by August of the following year. Hebern actually finished production ahead of schedule, and in the spring of 1931 personnel at the Naval Research Laboratory conducted acceptance tests on two machines.[137] The lab found some minor material problems but filed a report leading Hooper to conclude that the Hebern machine was likely to "perform with a relatively high degree of reliability." In September OPNAV began distributing the new devices to all senior operational commanders, along with orders to employ them both before and during Fleet Problem XIII.[138]

In another example of the big responsibilities assumed by junior personnel as the service adopted new command and control technologies, the Bureau of Engineering gave more than fifty enlisted petty officers and several junior officers special training on the Hebern Cipher Machine. The bureau expected these individuals to keep the machines in working order while underway and to make minor repairs as necessary.[139] They did an excellent job. At the end of Fleet Problem XIII, the Battle Force commander singled out two enlisted personnel for their "ingenuity" in making important alterations to his Hebern cipher machines, and U.S. Fleet commander-in-chief Frank Schofield informed the CNO that the devices assigned to him "stood up exceedingly well despite the heavy traffic experienced. There were no derangements which were not remedied within an hour."[140] Elsewhere Schofield enthusiastically reported: "The improved Hebern machine has been found generally satisfactory and represents the greatest advance in the field in many years."[141]

Schofield's enthusiasm notwithstanding, Hebern's cipher machines were far from perfect. The typewriter was sluggish and the wiring arrangements less than ideal; as one senior commander reported after Fleet Problem XIII, "The greatest number of failures . . . have been of an electrical nature . . . the machine becomes grounded whenever in operation its base comes in contact with other metal."[142] One of the device's biggest drawbacks was that it operated on battery power and was therefore incompatible with standard shipboard electrical distribution systems. For these reasons the navy placed no further production orders, choosing instead to work with Hebern in fixing such faults before buying more electromechanical cipher machines.[143]

Even as the navy worked with Hebern to develop a rotor-based encryption machine, several officers tried to create a device more efficient than the cumbersome cylindrical cipher. Early in 1931 Lieutenant (j.g.) Morris Smellow began working with several others to build a new encryption device, which OPNAV soon designated as the "strip cipher." The idea behind this device was to move letters from the cylindrical cipher's alphabetic disks onto strips of paper.[144] An operator would sequentially arrange the strips in a special holder (in an order determined by the cryptologic key), sliding them left or right to align vertically the first twenty-five letters of a message. To encipher these, the operator would then substitute letters from a different column randomly selected. To decrypt a message, recipients would merely reverse the process. Such a system was theoretically faster than existing cylindrical ciphers and unquestionably more secure, because the Navy Department could supply new paper strips much more frequently than it could replace alphabetic disks.[145]

Morris Smellow was yet another of the junior personnel—akin to Edward Very, Albert Niblack, and Benjamin Miessner—who made important contributions in American naval command and control before following a different career path. This observation is notable because it stresses how improvements in command and control systems derived from the activities of a variety of personnel, not just from the work of dedicated proponents like Stanford Hooper. Smellow, who grew up in Atlantic City and graduated from the Naval Academy in 1923, eventually became a supply officer, but in the late 1920s and early 1930s his primary focus was cryptography. His interest in that area appears to have begun after the Navy Department assigned him to the Office of Naval Communications in late 1928.[146] In the summer of 1929 Smellow proposed a new type of cylindrical cipher and, later that year, submitted detailed drawings for an "electric typewriter coding machine." According to Smellow, the biggest advantage of his machine was that it would be "much faster than any code in use at present."[147] Although the service ultimately adopted neither device, several high-ranking officers took note of Smellow's interest in cryptographic apparatus.[148]

As such, in early 1931 the director of OPNAV's code and signal section asked Smellow to investigate the feasibility of a strip cipher device. The young officer embraced the assignment, taking his work home with him night after night and completing a preliminary design in less than a month. The Bureau of Engineering allocated funds to construct a prototype, and after a series of modifications by Smellow and several others, it finally concluded that the device was ready for operational testing. In April 1932 the CNO directed several ships and shore activities to employ the new device and report upon its practicability for use in naval service.[149]

Feedback was positive. The commanding general of the Marine Corps' second brigade reported that the strip ciphers provided to him were "in all respects practicable for general issue and could very advantageously be used," while the commander of the naval squadron tasked to test the new cipher wrote that "the Device has proved very satisfactory . . . It is urgently recommended that its use be continued." These officers also reported the new apparatus to be surprisingly rugged and less prone to garbles; most critically, it was also about twice as fast as its predecessor.[150] A week after receiving this feedback, OPNAV directed the manufacture of some two hundred strip ciphers for use during Fleet Problem XIV.[151] Experience gained during the problem revealed that while a few minor defects remained, the strip cipher was indeed faster than its predecessors.[152] Officers in the fleet proposed various remedies to these defects, and OPNAV incorporated several of their suggestions into an improved design.[153] For his "valuable contribution" in helping to develop the strip cipher, Morris Smellow received a written commendation from the CNO.[154]

The army also adopted the strip cipher, but as a manual encryption system, its maximum speed was still insufficient for many tactical situations. Naval commanders recognized the speed they desired would have to come from an automatic device like the Hebern Cipher Machine. Unfortunately, Hebern ran into problems when he redesigned the model that had been acquired by the service in 1931. Although naval officers had welcomed those machines, they also called for more reliable apparatus.[155] Heeding such calls, in 1932 the navy purchased one of Hebern's latest machines and subjected it to a series of comprehensive tests. For reasons that remain unknown, the new device failed these tests.[156] By 1934, many naval officers had lost faith in Hebern's ability to construct a reliable machine, and operational commanders again began to stress the need for more rapid encryption systems.[157] Writing in mid-1934, Admiral David Sellers captured the sentiments of many when he informed the CNO: "The Commander-in-Chief feels that adequate security with adequate speed will not be obtained until the Fleet is provided with apparatus by which a message can be simultaneously encoded and transmitted, and received and decoded . . . He realizes that cryptographically secure apparatus cannot be developed overnight, but it must be developed before Fleet communications can be said to be adequately prepared for war."[158]

Even as Sellers wrote these words, the Navy Department already was pursuing a new electromechanical encryption machine, one eventually designated the ECM (electric cipher machine). Scholars know surprisingly little about the ECM, although it appears to have been developed jointly by the Bureau of Engineering and the Teletype Corporation, a wholly owned subsidiary of Western Electric.[159] First tested on a wide scale during Fleet Problem XVII, operational commanders hailed the new machine as an improvement over previous cryptographic devices.[160] Yet they also noted some unwelcome flaws; as the Battle Force commander summarized the matter: "The ECM has proven the greatest single advance in [electromechanical cryptography], but it suffers from a confusing complexity of settings and occasional mechanical defects in operation."[161] These deficiencies eventually were resolved, but the navy never procured large numbers of ECMs because by the late 1930s it had an even more secure electromechanical device under development, the ECM Mark II. Although the name implies an improved version of the ECM, the Mark II actually employed an entirely different design in which a circuit arrangement called the "stepping maze" made the code wheels move in a seemingly "irregular and erratic manner."[162] In early 1940 the navy disclosed details of the machine to the army, and the ECM Mark II entered joint operational service just four months before the Japanese attack on Pearl Harbor. More than ten thousand of these machines saw service during World War II. The ECM Mark II apparently remained uncompromised for the entirety of its service life, which ran through the late 1950s.[163]

Of course, not even the best cryptographic equipment could prevent an enemy from detecting and obtaining a line of bearing to a transmitting ship or airplane. If an opposing force could obtain two or more lines of bearing, it could plot a cross-fix to determine the geographic position of a transmitting platform. While this concept was theoretically simple, it was hard to execute. Early radio direction finding (RDF) equipment was difficult to operate, and the accuracy of any given cross-fix depended upon both range and relative position. Most critically, even fairly small bearing errors could result in highly inaccurate plots. Despite these obstacles, the potential advantages of a good RDF system were such that operational commanders repeatedly called for better equipment from the shore establishment. When this equipment finally became available, fleet personnel developed innovative ways to employ the new technology.

Find without Being Found

American naval officers recognized the potential value of radio direction finding only a few years after the service acquired its first wireless sets. Around 1905 inventor Reginald Fessenden informed the Bureau of Equipment he had developed a device that could determine the direction of an incoming radio signal within thirty degrees,[164] and the following year the bureau actually tested a shipboard radio direction finder.[165] Neither of these RDF devices was operationally practical, but the bureau's radio division kept tabs on the work of civilian inventors, and in 1913 division chief Arthur Hepburn decided the time was right to reevaluate matters. He acquired direction finders from two European companies and had them tested in the Atlantic Fleet. The results were disappointing. Even Fleet Radio Officer Stanford Hooper, who was generally optimistic about the new technology, became discouraged and recommended against keeping the experimental equipment then installed.[166]

Hooper faced the issue from a new perspective when he took over the Bureau of Engineering's radio division in the spring of 1915. With war raging in Europe, he made the study of RDF equipment a top priority, and in early 1916 the bureau purchased some new direction finders from the American Marconi Company. These also failed to perform adequately. Hooper again became discouraged, but shortly thereafter he received an unexpected phone call from Frederick A. Kolster, a young engineer working at the National Bureau of Standards. Kolster told Hooper he had developed an instrument that could determine the bearing to a radio transmitter and asked the division chief if he would like a demonstration. Although Hooper was skeptical, he trekked over to Kolster's lab, where he found a promising new device.[167] Within weeks of Hooper's visit, the Bureau of Engineering agreed to pay Kolster twenty thousand dollars in exchange for exclusive rights to his "Kolstermeter."[168]

Shipboard installations began in late 1916, accelerating rapidly after the United States entered World War I the following spring. The Bureau of

Engineering also constructed a series of RDF shore stations along the Atlantic coast—for planned use against any German U-boats that might venture near the United States—but none of these sites became operational before the Armistice.[169] Many warships deploying to European waters had RDF equipment on board, but with a typical bearing error of more than two degrees, Kolstermeters were not accurate enough to aid substantially in the fight against the U-boat.[170] Nevertheless, direction finders were useful for convoy duty because, whenever the need arose, duly equipped escorts could more easily locate their convoy, even at night or in bad weather. According to one official report, such occurrences were common.[171]

For nearly a decade after the war, operational forces struggled to employ RDF equipment that was too inaccurate for most tactical purposes. Naval commanders made progress in training operators and achieved a few successes, such as when blue forces obtained RDF bearings to opposing black vessels during Fleet Problem V; however, into the mid-1920s radio direction finders generally remained "erratic," "liable to excessive errors," and "far from satisfactory."[172] The Bureau of Engineering worked diligently to improve the accuracy and reliability of shipboard RDF systems.[173] Gradually, its efforts began to pay dividends. During the fleet's annual maneuvers in 1927, a division of battleships used RDF bearings to intercept opposing forces, and the following year U.S. Fleet commander-in-chief Henry Wiley noted that Fleet Problem VIII had "well demonstrated the value of [RDF] tracking."[174]

No sooner had the fleet successfully begun to employ shipboard radio direction finders than available apparatus became obsolete. Like the U.S. Navy, most of the world's navies adopted high frequency (HF) radio in the late 1920s and early 1930s. Yet existing RDF equipment could not detect HF radio waves. American naval personnel were quick to notice this fact and simply stopped transmitting on lower frequencies during fleet problems. In August 1930 U.S. Fleet commander-in-chief William Pratt requested that HF direction-finding equipment be given top priority, and the Bureau of Engineering shortly thereafter took steps "to expedite the development of efficient direction finders for both the intermediate and high frequency bands."[175] Over the next two years the bureau tested several different HF RDF systems. None of them performed well, leading the navy's senior at-sea commander to comment: "The lack of high frequency direction finding equipment remains the most serious deficiency in Fleet radio material. The Commander-in-Chief realizes the problem in under consideration and urges the highest priority for it."[176]

The Bureau of Engineering continued to pursue an HF RDF system for fleet use, finally equipping a limited number of platforms with improved equipment before Fleet Problem XVI in 1935.[177] *Saratoga* was one of those to receive the new HF direction finder, and operators employed it with success during the exercise. They obtained and plotted bearings to many of the opposing side's vessels and provided critical locating information that led directly to attacks on three enemy submarines. The latter development prompted one senior commander to observe that the "use of radio is made exceedingly dangerous by this new development, and submarines operating against our Fleet during an overseas passage will be unable to use radio or [will] expose themselves to great danger as a consequence."[178]

World War II would demonstrate that submariners did in fact have much to fear from HF direction finders, but in the mid- to late 1930s the navy still had several difficulties to overcome with respect to the employment of RDF technology. HF direction finders were more dependent on operator skill than those that detected lower frequencies, and obtaining bearings to HF sky waves could be particularly difficult.[179] Operational commanders called on the shore establishment to provide better equipment, which ultimately arrived around 1940 in the form of the loop direction finder.[180] Yet one difficulty inherent to the operation of RDF systems could be solved only by fleet personnel—namely, the establishment of methods and procedures to correlate the bearings obtained by operators on different platforms.

At first, seaborne commanders took a decentralized approach. Operators on each RDF-equipped vessel plotted bearings to intercepted radio transmissions, informing their individual superiors whenever they obtained pertinent information.[181] This approach provided commanders with bearings to transmitting platforms; however, it did nothing to establish range. To determine range, a cross-fix was necessary, which required plotting one ship's RDF intercept with that of another. Other factors being equal, the ideal cross-fix resulted when two intercepting units had perpendicular lines of bearing to a target. This geometry usually occurred only when sizable distances were involved, which meant that a fleet maintaining radio silence generally could not correlate intercepted bearings. This is exactly what happened to the senior officer present on *Saratoga* during Fleet Problem XVI, who had been unable to provide important information to the OTC.[182]

Fleet Problem XVI thus highlighted for naval commanders the desirability of a more centralized approach to radio direction finding, or, as one senior officer argued, "It has been found impracticable satisfactorily to decentralize intercept watches to a number of vessels. Centralization . . . is necessary."[183] Yet without a system of ship-to-ship communications unlikely to be detected by the RDF systems of an opposing force, such an approach was risky. One possible solution was to create shore-based control and tracking centers. The purpose of these centers was to correlate RDF intercepts and to provide commanders with a picture of the operating environment through a "no receipt" broadcast. Although this arrangement turned out to be relatively unsuccessful in practice, it highlights the ways in which American naval officers pursued innovative methods for managing the growing amount of information generated by new electronics technologies.

Throughout the latter 1930s one of the biggest advocates for centralized RDF facilities was Arthur J. Hepburn, who from 1935 to 1938 served first as the Scouting Force commander and then as the commander-in-chief of the United States Fleet. Hepburn had been chief of the Bureau of Engineering's radio division for three years in the early 1910s and had commanded the American submarine chasers operating out of Queenstown during World War I. After the war he held a variety of billets that prepared him well for fleet command, including one tour as director of naval intelligence (1925–26) and another as commander of the U.S. Submarine Force (1931–32).[184] In Fleet Problem XVI, Hepburn was the Black Fleet commander, and in that role he saw opposing white forces bomb three of his submarines after gaining valuable information from RDF intercepts. No surprise, then, that for the remainder of his tour as Scouting Force commander Hepburn pushed to advance the training and readiness of his RDF operators.[185]

After he became U.S. Fleet commander-in-chief, Hepburn made the further integration of HF direction finding into fleet operations one of his top priorities. He believed the best way to accomplish this end was to create "an efficient shore network with a control and tracking center." According to Hepburn, lessons learned from this arrangement could then be applied toward establishing similar facilities on board ships.[186] Especially after the outbreak of the Second Sino-Japanese War in July 1937, the navy pursued Hepburn's vision aggressively. Working in tandem with several different naval districts, the Bureau of Engineering tested a variety of radio direction finders to identify the most effective equipment and to create a small cadre of skilled operators. By early 1938, CNO William D. Leahy believed the service was ready to test its shore-based RDF system. His office drafted detailed plans for that purpose.[187]

Fleet Problem XIX, conducted in the spring of 1938, offered an ideal environment in which to execute these plans. During the problem, the navy placed a shore-based control and tracking station at the disposal of each force commander. These "central control stations," located in San Francisco (for black) and Pearl Harbor (for white), were to be linked by "landwire or cable" to a series of shore-based direction finders. A main purpose of the stations was to coordinate intercepted radio transmissions and to provide seaborne commanders with information on opposing forces' "areas, positions, movements [and/or] lines of bearing." According to Leahy, "This forwarding of *exact* information . . . is considered of the greatest importance."[188] In addition, the Navy Department provided the black commander with a separate direction-finding network exclusively for patrol aircraft. Designated "Air Net," this system consisted of three HF direction finders linked to a central control station in San Diego. Operators at this station plotted lines of bearing to all transmissions from black patrol planes, deriving cross-fixes whenever they received multiple simultaneous intercepts.[189] Air Net served a dual purpose by providing corroborating information on enemy contact reports and also by creating a datum for search operations in the event of a lost aircraft. Because flying for extended periods over the ocean created considerable navigational difficulties for aviators, naval commanders saw tremendous potential in Air Net.[190]

Unfortunately, neither Air Net nor the other direction-finding networks established for Fleet Problem XIX performed as well as expected. The commander responsible for the white direction-finding network reported "negligible" results, and the Air Net control officer noted that only 50 percent of cross-fixes were "reasonably accurate."[191] Hepburn, whom the navy placed in charge of the black direction-finding network, remarked that the performance of his primary RDF station had been "wholly inadequate."[192] This situation derived partly from a lack of adequately trained operators at the station, but Hepburn nevertheless discerned several positive trends in the service's inaugural test of centralized RDF facilities. He noted that operators had improved their skills over the course of the exercise and pointed out that they obtained excellent cross-fixes whenever accurate estimates of transmitting frequencies were available ahead of time. Especially encouraging, according to Hepburn, was the fact that RDF operators sometimes could obtain multiple simultaneous bearings even on relatively short radio transmissions.[193]

Encouraged by these tangible if somewhat limited achievements, CNO William Leahy once again directed the use of shore-based RDF networks in

Fleet Problem XX. As before, the Navy Department provided black forces with two separate direction-finding networks: one to plot, analyze, and broadcast white radio transmissions; another to track black patrol aircraft (i.e., Air Net).[194] Significantly, the Black Fleet commander also established an RDF control and tracking center on his flagship. Like Hepburn before him, Vice Admiral Adolphus Andrews believed that centralized RDF facilities could give naval commanders an advantage over their adversaries. He argued that such facilities, if operated properly, would provide critical information in a timely manner. Expressing these sentiments to his superiors shortly after the 1939 maneuvers, Andrews remarked:

> The OTC or the Fleet Commander with all his immediate sources of information, such as aircraft, surface and sub-surface scouting . . . may find that quick but sound estimates made from a radio analysis *right at hand*, will be of definite assistance. The ability of the ship to hear transmissions not heard on shore at some distant point, the question of the value of minutes and seconds when determining courses of action, the responsibility of the Fleet Commander in evaluating all sources of information at first hand, these considerations must be weighed . . . Remember that we want every bit of information concerning the enemy which we are able to obtain. Some little detail which might appear insignificant at first, might very well prove to be an important clue in determining just what the enemy is doing.[195]

Despite such high-level support, the navy's direction-finding networks performed only slightly better in Fleet Problem XX than they had during the previous year's maneuvers. The chief radioman at one direction-finding station reported that his operators had trouble intercepting transmissions during daylight hours, while the commander in charge of Air Net lamented his inability to obtain multiple simultaneous bearings on a regular basis.[196] The control and tracking facility on Andrews's flagship did little better, and like their shore-based counterparts, the operators within it had considerable difficulty obtaining accurate cross-fixes. Because of this, plots of intercepted white transmissions generally provided black personnel with little more than "the approximate area in which the enemy was located."[197]

The original hope of Hepburn, Leahy, and others—that a series of dispersed RDF receivers could feed lines of bearing to a central station where personnel would plot accurate cross-fixes—turned out to be a chimera given the accuracy of RDF technology and the vast geography of the Pacific. Andrews recognized this situation when he noted that even one degree of bearing error could result in a highly inaccurate cross-fix, while another senior officer complained that faulty information from the Black Fleet flagship's RDF center had led him to place forces "some 1200 miles" out of position.[198] Yet operational experience also revealed a silver lining. As articulated by Andrews, "The best information was obtained [when] all bearing lines were plotted on the chart, no crosses even attempted. In this way, area limits were determined . . . this system of plotting single bearing lines and taking the area of greatest density will give the OTC his best information."[199]

After Fleet Problem XX, the navy appears not to have pursued further the idea of shipboard RDF control centers, returning instead to a decentralized approach for seaborne radio intercepts. From a tactical standpoint, this arrangement worked well, as a ship or aircraft receiving an intercept could proceed either down or, if the tactical situation warranted, away from a line of bearing. Indeed, during the Second World War HF direction finders would prove of great value in the fight against the U-boat, especially after effective shipboard equipment entered operational service in the spring of 1943.[200] The navy also constructed an extensive network of HF direction-finding stations in the Pacific. This "Mid-Pacific Net" provided key information to operating forces both before and during the war.[201] Yet there was a flip side to the maturation of RDF technology, and it lay in the danger that American forces would themselves be victimized by the HF direction-finding capabilities of an enemy. Radio silence was a practical if inefficient solution to this problem, but what naval commanders really wanted was an electronic communications system akin to the porridge that Goldilocks deemed "just right"—in other words, a system that would enable communications at tactically adequate distances but that would not propagate too far, lest an adversary might be listening.

In actuality the fleet already possessed a limited-range communications system: visual signals. The problem with most visual signaling systems, however, was that their maximum range was too short. After Fleet Problem One, for example, several commanding officers reported that they could not read flags, semaphores, or blinker-tube signal lights at distances greater than about five nautical miles.[202] The visual signaling method with the furthest range was the electric searchlight, which the navy had begun experimenting with in the late 1870s. By the 1920s, the maximum effective range at which signals could be conveyed with this device was around fourteen miles, but only larger vessels employing newer equipment under ideal weather conditions could achieve that range. Rough seas made the searchlight's narrow beams hard to read; as one senior officer described the matter, "Due to rolling . . . it becomes a matter of great difficulty to 'keep on' the transmitting vessel, causing errors in reception, requests for repeats and consequent delays."[203]

Although the service could do relatively little about this environmental limitation, it took other steps to improve the fleet's visual signaling methods. In the aftermath of Fleet Problem One, the Navy Department implemented competitive drills to stimulate interest in the subject and promulgated a revised general signal book that Samuel S. Robison believed was "a great improvement over any previous issue."[204] A few years later, Battle Fleet commander Charles Hughes introduced a record-keeping system for visual signaling similar to the one Robison previously had adopted for radio, declaring that fleet personnel could not allow improvements in radio to lead to the "neglect and desuetude of visual communications."[205] Yet procedural changes such as these could not solve the biggest problem related to searchlight signaling—namely, unreliable and insufficiently luminous searchlights.

During World War I, the standard U.S. Navy searchlight had been twelve inches in diameter. For the most part these devices were reliable, but personnel found their maximum range inadequate during anti-submarine warfare patrols and convoy escort duties, let alone for the type of fleet operations being practiced by American naval forces in the 1920s. The Bureau of Steam Engineering solicited bids for more powerful apparatus, eventually purchasing twenty-four-inch signal searchlights from the Arma Engineering Company and installing them on larger warships. Operational commanders liked the new light's superior luminosity but chafed at its unreliability. They pleaded with the shore establishment for a more dependable twenty-four-inch searchlight. In the early 1920s General Electric answered the call, and the Navy Department slowly replaced Arma's searchlights with ones manufactured by General Electric.[206] The searchlight quickly became the fleet's preferred method of daytime signaling, garnering praise from commanders who variously described it as the "most useful of all visual communications," "a splendid means of daylight signalling at long range," and "invaluable."[207]

Despite this widespread approbation, the searchlight was hardly a cure-all. Submarines continued to carry the less luminous twelve-inch version, and although airplanes could receive searchlight signals, they could not send them.[208] In addition, like all visual signaling techniques, messages sent by searchlight could not be read through heavy fog or smoke, the latter of which likely would be present during battle. At night, however, an opposite problem existed. The beam from a searchlight could easily carry too far, extending well beyond the horizon.[209] While friendly forces could not read over-the-horizon messages, light beams potentially could alert an enemy. For this reason, American naval personnel usually avoided signaling with searchlights at night.[210]

Another limited-range communications system pursued by the navy during much of the interwar period was underwater sound signaling. In the early 1920s a number of officers believed this technology eventually could supplant, or at least supplement, visual methods of tactical communication. Writing in 1924, for example, future fleet commander Henry Wiley told Battle Fleet commander-in-chief Samuel Robison that sonar was "a fertile field for desirable improvements. Equipment designed for directive signalling would result in practically secret communications between ships of a battle fleet."[211] The Naval Research Laboratory and the Bureau of Engineering put considerable resources into developing "more effective intercommunication between vessels by the use of underwater sound waves,"[212] but in the mid-1920s the bureau unintentionally issued ineffective equipment to the fleet. Even though Robison wanted to remove acoustic signaling apparatus from most warships, Wiley remained optimistic. While serving as U.S. Fleet commander-in-chief in the late 1920s, he continued to express his desire for "some method of underwater sound signalling of sufficient range to permit tactical handling of the battle line."[213]

Despite Wiley's support, the navy's shore establishment simply could not develop a practicable system of underwater sound signaling. A big problem was that the faster a ship moved through the water, the harder it was for operators to pick up acoustic signals. Through the mid-1930s, ship speeds greater than about ten knots simply created excessive signal-to-noise ratios.[214] Furthermore, acoustic propagation in the ocean environment was both little understood and difficult to predict. These issues, when combined with promising trends in the development of an alternative technology, eventually led the service to abandon underwater sound signaling as a prospective means of tactical communication.[215]

That alternative technology was limited-range radio. American naval officers long had recognized the tactical advantages of radio equipment that would surpass visual communications ranges but that could not be heard beyond the horizon. As early as 1905 the Bureau of Equipment expressed an interest in limited-range wireless, and before American entry into World War I the navy's General Board advocated the development of a radiotelephone that was "for short distance communication embracing a limited tactical area (assured communication to 20 miles)."[216] Throughout the 1920s, U.S. naval commanders pleaded repeatedly for limited-range radios. Following the fleet's annual maneuvers in 1924, the service's senior seaborne commander emphasized the pressing need "to reduce transmission range to an approximately known distance," while after Fleet Problem Five one of the exercise observers remarked: "The best signal force in the world will not do away with the necessity of radio communication, or underwater communication, and it is necessary . . . [to ensure] that such radio waves we must send out shall not extend more than 50 or 60 thousand yards from the point of origin."[217]

One possible way to achieve limited-range radio was to employ low-power transmitters. To some extent the Western Electric radiotelephone adopted during World War I, the CW 936, already provided limited-range equipment. Its maximum dependable range, however, was only about fifteen miles, a distance insufficient for large-scale fleet operations. Furthermore, because the relatively low frequencies at which the CW 936 operated easily carried over the horizon, vessels employing it might inadvertently give away their location.[218] The CW 936 was no anomaly; at existing operating frequencies, naval personnel could never be sure how far any low-power signal would travel. And once the navy adopted HF radio in the late 1920s, relatively low-power transmissions routinely traveled thousands of miles through sky-wave propagation. Indeed, this very capability was the driving force in the navy's decision to adopt of HF radio. In short, low-power limited-range radio was not the answer.

By the mid-1920s, though, many officers felt a solution was in sight, with one admiral informing his colleagues that "it is not as hopeless as it looks" with respect to limited-range radio.[219] This optimism undoubtedly stemmed from the fact that the Naval Research Laboratory already was experimenting with "super-frequencies," or what today would roughly equate to the lower end of the "very high frequency" (VHF) band.[220] Unlike frequencies below about fifteen hundred kilohertz, which can follow the curvature of the Earth, or HF waves, which can refract through the ionosphere, super-frequency transmissions propagated down a line-of-sight trajectory. Barring atypical atmospheric conditions, only the Earth's curvature and the heights of the sender and receiver determined the maximum distance at which such waves could be detected. Super-frequency radio thus offered great promise as a means of secure tactical communications.

The fleet's first exposure to super-frequency radio came in the mid-1920s, although nothing tangible developed out of early trials with some experimental equipment.[221] The Naval Research Lab attempted to persuade two different commercial companies to take up super-frequency radio development, but neither firm was willing to do so. As such, in 1928 engineers at the lab built two crystal-controlled super-frequency transmitters capable of operating at up to seventy-five megahertz.[222] Stanford Hooper, the recently appointed director of Naval Communications, saw great promise in these prototypes and encouraged operational tests as soon as practicable. Hooper went on to note that super-frequency radio, if successfully developed, would "better serve the purpose of communication [at critical times] and during periods of radio silence than underwater sound signalling."[223]

In 1929 the Naval Research Laboratory installed these experimental sets on the flagships of the navy's two most senior operational commanders. Shipboard personnel tested the sets thoroughly, obtaining mostly positive results, and the following year the lab provided an experimental super-frequency radio for aircraft. From the fleet's perspective, the biggest drawback to the new apparatus was that it operated only in telegraphic mode.[224] Engineers at the Naval Research Lab resolved the matter by devising a way to incorporate voice modulation into existing equipment, and in early 1931 the Bureau of Engineering indicated that it was ready to place all super-frequency radio testing afloat into the hands of the Battle Force commander-in-chief.[225]

Shortly thereafter, the Battle Force commenced trials on super-frequency radio. Progress was slow. Although super-frequency transmitters performed adequately, receivers were not yet ready for operational use.[226] Other problems included short-lived vacuum tubes, radio interference, and excessive energy loss between antennas and transmitters/receivers. The service previously had addressed all three issues, but the move to super-frequencies caused them to reemerge. The Bureau of Engineering and the Naval Research Lab worked in concert to address the problems, leveraging the expertise of commercial firms whenever possible.[227] Senior commanders voiced their enthusiasm for limited-range radio and awaited better equipment, if not always patiently. In the fall of 1935, for example, U.S. Fleet commander-in-chief Joseph Reeves reminded the CNO that lack of super-frequency equipment was "one of the most keenly felt needs of the Fleet."[228]

Reeves did not have long to wait. In early 1936 the fleet finally began to take delivery of production-model super-frequency equipment in the form of the CXL radio.[229] Initial feedback was guarded yet hopeful; in his annual report for the year, Battle Force commander William Leahy remarked, "Although it is premature to arrive at definite conclusions regarding the usefulness of this equipment, the general impression obtained has been very favorable."[230] Reeves was a bit more optimistic, writing:

> By using limited range radio, capable of being restricted to the limits of a Fleet disposition, the Commander-in-Chief will have available the most satisfactory method, (even superior to flag hoists and other systems), for the instantaneous dissemination of signals, orders and information to all ships of the disposition without danger of intercept by an alert enemy intelligence organization . . . encrypting of signals will not be necessary, relays will be eliminated, and 24-hour operation, unaffected by darkness or low visibility, will be available.[231]

Over the next few years, Reeves's vision slowly took shape. The service installed CXL radios on ships of the battle line, where they performed "in a generally satisfactory manner."[232] Operational experience revealed that the CXL worked best in telegraphic mode, however, so officers in the fleet pushed the shore establishment to provide a new line-of-sight voice radio, preferably one capable of operating at even higher frequencies.[233] The Naval Research Laboratory needed little encouragement, as it already was working on improved super-frequency equipment. One project involved a radio that could operate at frequencies up to five hundred megahertz, well into what later would become the ultra high frequency (UHF) band. The navy began testing this apparatus on ships and aircraft in 1937, and even before UHF equipment became operationally practicable, naval officers began debating how best to integrate it into the fleet.[234]

These debates would carry into the Second World War, but the U.S. Navy's principal limited-range radio throughout that conflict was something different, the TBS radio transceiver. Designed by the Naval Research Laboratory and manufactured by the Radio Corporation of America (RCA), the TBS was simply a superior voice radio. Its origins trace to early 1938, when the Navy Department contracted with RCA for a line-of-sight radio that would operate at frequencies between sixty and eighty megahertz. RCA delivered acceptance models in late September, which the Naval Research Laboratory ran through a series of comprehensive tests. The lab's engineers were impressed, reporting the performance of the TBS radio to be "generally excellent" and commenting that the new equipment represented "a definite step forward in the design of [super-frequency] communication systems."[235] Although battleships retained their CXL radios for several more years, destroyers and submarines received TBS installations starting in fiscal year 1940.[236] Fleet personnel embraced the new equipment, at some unknown point in time bestowing upon it the moniker "Talk Between Ships." The TBS radio would go through eight different upgrades and served as the fleet's primary system of tactical voice communications throughout the Second World War.[237]

Fleet Problem IX marked an inflection point in American naval history. It was the first fleet problem in which more than one aircraft carrier participated, and it saw the first use of a detached fast carrier task force. Not coincidentally, Fleet Problem IX was also the first time the fleet commander-in-chief's official report contained a thorough entry under the heading of information. By singling out that particular topic Admiral Henry A. Wiley was drawing attention to an issue his predecessors—naval commanders such as A. Ludlow Case, Albert S. Barker, and Edward W. Eberle—had grappled with for decades.

Wiley and his contemporaries recognized that a change not just of degree but also of kind was taking place. The aircraft carrier was powerful yet vulnerable; in theory and in practice, it reduced considerably a commander's margin for error. During the navy's fleet problems, surface ships and submarines hypothetically sank opposing carriers again and again. Some naval officers argued then, and some historians agree now, that the improper employment of carrier forces led to these sinkings. Perhaps so, but aircraft carriers were vulnerable no matter how commanders employed them. The early part of World War II offers disheartening evidence in support of this conclusion. Of the six American carriers assigned to the Pacific in 1942, Japanese naval forces sank four and damaged the other two.[238]

The flip side of carrier vulnerability was the power a carrier's aircraft could bring to bear. In the early 1930s such aircraft were relative weaklings, but by the end of the decade their brawn had become evident. All-metal monoplanes like the Devastator and the Dauntless possessed longer ranges than their

Destroyers on maneuvers with signal flags flying, October 1936. Not until the arrival of TBS in 1940 did the fleet finally gain a reliable system of limited range voice radio. *Naval History and Heritage Command photograph NH 60270.*

predecessors and carried heavier ordnance, while new fighters like the Grumman F4F Wildcat allowed carriers better to defend themselves against enemy air attack. During World War II, the service gradually replaced most of these first-generation carrier monoplanes with more powerful aircraft, such as the Grumman Avenger and the Vought Corsair.[239]

Increased brawn, in combination with the pervasive risk of catastrophic loss—perhaps from battleships emerging out of the fog or from submarines lurking below—truly created a "most complex problem" for American naval personnel. One way senior officers addressed the difficulties of command in an environment of geographically dispersed and dimensionally diverse naval forces was to create a common operational doctrine that emphasized initiative and tactical flexibility. Yet the days when a battle could be won with a single signal at the start of the action, a la Horatio Nelson at Trafalgar, had long since passed. Practical experience revealed that coordinated naval operations were extremely difficult and that superior command and control gave combatant commanders a decisive advantage.

As such, the 1930s witnessed aggressive efforts by U.S. naval personnel to create systems that could provide senior officers with accurate, timely, and secure information. The tangible results of these efforts included quicker and more secure cryptographic machines, valuable experience with RDF networks, better searchlights, and a superior VHF radio for limited-range communications. One should not, however, view the story as a narrative of unalloyed progress. Efforts to create shipboard RDF control and tracking centers were unsuccessful, and the service pursued a technological dead end in the case of underwater sound signaling. Yet these failures are as relevant as the successes in laying bare the mentalité of interwar American naval officers, who as a group understood the critical importance of shipboard command and control systems and embraced technology as a way to improve such systems. Far from being technologically conservative, this group was genuinely progressive in its attitude toward command and control technology.

Significantly, as fleet personnel worked with the navy's shore establishment to improve shipboard command and control, they gradually created environs different from those used by earlier generations. In these new surroundings, successful commanders needed not only courage and good seamanship but also an ability to manage information in time-critical situations. Yet even with the numerous new command and control systems developed by the service before and during the 1930s, an operational commander or a ship's commanding officer usually could count on having a few minutes to think over important tactical decisions. Naval radar would alter this state of affairs, moving more and more decisions from the realm of minutes into one of seconds.

CHAPTER FIVE

Creating the Brain of a Warship

Radar and the CIC

> How complex a captain's life used to be, and how relatively simple it can be now . . . the vast number of items that once went only to the captain, who had to weigh each bit of data himself and decide whether to use it, discard it, or file it in the back of his mind for future use . . . the Combat Information Center is now the agency whose primary function is to filter and evaluate nearly all of this material for him. The captain receives the information he needs when he needs it; and he is free to concentrate on his decisions and to carry the burden of command.
>
> "CIC Handbook," 1943

> Tallyho from Sam One
> Sam Two's also here
> It won't take long to splash'm
> So hand me some cold beer.
>
> Verse from an anonymous author, 1944

Something Completely Out of the Ordinary

Lieutenant Commander William W. Outerbridge saluted smartly and said, "I relieve you, sir." The day was a Friday. The place was Pearl Harbor, Hawaii. And the date was 5 December 1941. Outerbridge had just assumed command of USS *Ward*, a destroyer originally commissioned during the First World War. Numerous thoughts undoubtedly crossed Outerbridge's mind, but he would have little time to reflect upon them. The following morning, *Ward* got under way for what everyone on board must have thought would be routine patrol duty in the waters just south of Pearl Harbor. By the end of the weekend the entire crew would be serving in a wartime navy.[1]

On the morning of 7 December, Outerbridge arose at four o'clock when *Ward* received a message, sent by a nearby minesweeper via visual signals, reporting a possible submarine contact. Although skeptical, Outerbridge initiated a search of the area. Finding nothing, the new commanding officer contacted the minesweeper by voice radio to obtain clarification. Outerbridge discovered that his ship had been searching in the wrong direction, so he reversed course and ordered his crew to continue the search. Thinking the minesweeper must have spotted a buoy or some flotsam, Outerbridge returned to his cabin to catch a little more sleep.

Soon thereafter dawn arrived, illuminating the verdant volcanic mountains north of Pearl Harbor. The commanding officer of a navy cargo ship returning to port later recalled that the morning was "especially beautiful." Any thoughts of liberty in Hawaii were abruptly interrupted, however, when that same officer noticed something unusual off his starboard quarter. Was it a submarine? He reported the "suspicious object" to *Ward* by radio, which steamed over for a look.[2] A few minutes later the pilot of an U.S. Navy patrol plane flying overhead thought he saw an American sub floundering near the surface. He informed *Ward* and dropped smoke pots to provide the destroyer with a datum on the distressed vessel.[3]

Meanwhile *Ward*'s helmsman spotted an oddly shaped buoy in the distance, which he pointed out to the officer of the deck. The two men kept an eye on the alleged buoy; after a few minutes, they realized it was moving. The officer of the deck quickly called his commanding officer, and *Ward* went to general quarters. Outerbridge had orders to attack *any* submarine in his patrol area. What if the pilot was right, though, and the mysterious vessel really was an American submarine? Someone, possibly the ship's executive officer, asked Outerbridge what he was going to do.[4] If Outerbridge hesitated in answering this query, then he did so for only a moment. *Ward* closed range and opened fire at 6:45 a.m. The first salvo missed, but the second found its mark. A four-inch shell pierced the conning tower of a Japanese midget submarine, which sank to the bottom. *Ward*'s kill would be confirmed six decades later when underwater archaeologists finally discovered the submarine's barnacled hull.[5]

Not knowing that *Ward*'s gunfire had mortally wounded the Japanese vessel, Outerbridge continued the attack with depth charges. At 6:51 a.m. he reported to the local district watch officer via radio message that *Ward* had attacked a submarine. Two minutes later, he elaborated on his initial report: "We have attacked fired upon and dropped depth charges upon sub operating in defensive sea area."[6] Because common knowledge held that anti-submarine patrols sometimes reported false contacts, Outerbridge sent the second message to stress that he had actually seen the submarine. Why was he not more explicit? The answer lies in the fact that *Ward*'s messages were encrypted, probably in code but possibly by strip cipher. Either way, Outerbridge could send out a clarifying report more quickly by simply adding a few words to the original message, which is exactly what he did.

The station receiving *Ward*'s report delivered it to the watch officer in charge of Pearl Harbor's anti-submarine nets and booms, Lieutenant Commander Harold Kaminski. Kaminski did not receive the message until 7:12 a.m., a delay of nearly twenty minutes likely resulting from slow decryption by shore personnel. Yet once he had the message in hand, Kaminski acted with alacrity. He ordered the ready duty destroyer to get under way immediately to assist *Ward* and began making a series of phone calls. The first person Kaminski reached was the Pacific Fleet duty officer;[7] the second was the local naval district's chief of staff, who questioned *Ward*'s report and told the duty officer to get further confirmation from the destroyer. Kaminski attempted to do so but placed a higher priority on alerting other naval personnel. He was on the phone when he first heard, then saw, a Japanese aircraft pass overhead. The noise of the ensuing attack was deafening, with Kaminski later recalling: "It was impossible to use the phone during this time because of the noise."[8]

Unbeknownst to Kaminski, ten minutes before he received *Ward*'s report two U.S. Army soldiers operating a mobile radar set on the northern tip of Oahu had picked up what one of them later described as "something completely out of the ordinary."[9] What the two men saw, in the form of spikes on an oscilloscope, was a wave of Japanese aircraft headed for Pearl Harbor. After some hesitation, the operators reported their contact, via telephone, to an "information center" in Fort Shafter, located about four miles east of Pearl Harbor. Theoretically, this center stood at the heart of Oahu's air defense system. Within it, personnel could plot information received from any of six mobile radar sites positioned around the island's perimeter. In the event of an enemy air raid, a control officer and his staff were supposed to direct pursuit planes to intercept the attacking aircraft. While such a raid could not be stopped completely, naval officer William E. G. Taylor, a veteran of the Battle of Britain who in 1941 was one of the country's few experts in air defense, posited that effectively directed defending aircraft would have been able "to break up, to a large extent, a raid of the sort that came in on December 7."[10]

Of course, no intercepting defenders were there to meet the attacking Japanese aircraft that fateful Sunday morning. Instead, a comedy of errors led to tragic failure. Most critically, the information center's manning was inadequate. Only two individuals were on duty when the radar operators phoned in their report, and the senior officer present, U.S. Army lieutenant Kermit Tyler, had stood only one previous watch in the information center. Tyler thought a group of B-17 bombers was due to arrive from the mainland sometime that morning, so he told the operators "not to worry about it."[11] Figuring any inbound planes had to be friendly forces, Tyler felt no need to telephone the operations officer at the pursuit wing tasked to defend Oahu. After the attack began, a hodgepodge of personnel flooded into the information center, where chaos reigned. Taylor, who arrived at about 8:30 a.m., later described the scene:

> The information center was in pretty great confusion. In order to man all the necessary positions, the air warning officers had drawn on mess cooks, linemen, and every man that they could lay their hands on—all of whom were inexperienced . . . The main plot had a paper overlay, ripped off the table, making the scale of the plotting table too large for accurate plotting. [Reports] coming in from the various radar stations were in such confusion it was impossible to determine what was going on.[12]

In defense of Tyler, he was part of a system almost predestined for failure. Personnel within the information center at Fort Shafter had no way of determining whether radar reports corresponded to friendly or hostile forces, largely because the army had established neither aircraft approach lanes nor a movement reporting system. Inadequate lines of communication between the two services meant that no one in the army knew of *Ward*'s early-morning encounter with a Japanese submarine, critical information that might have led Tyler to think twice before so casually dismissing a report about unidentified inbound aircraft. Finally, even if Tyler had phoned the pursuit wing operations officer, as he should have done, the army's existing alert condition fundamentally precluded a rapid sortie of defending aircraft.[13]

One of the few things that actually performed well on the morning of 7 December 1941 was the army's mobile radar equipment. The set that had detected the inbound Japanese aircraft was a long-wave, air-search radar manufactured by the Radio Corporation of America (RCA). Prior tests by the Signal Corps indicated that it could detect approaching aircraft at distances of up to 150 miles, and operators picked up the inbound Japanese planes very near that maximum range. The two operators then tracked the raid for almost half an hour, until it was within twenty miles of Oahu.[14] Simply stated, the failure of Oahu's air defense system resulted not from inadequate equipment but from other factors, including a lack of information sharing, poor preparedness, insufficient training, and bad judgment. Institutional friction also contributed to the disaster, as the Signal Corps' narrow focus on technical performance had led it to forestall Interceptor Command's prior requests to assume operational control of Hawaii's radar equipment.[15]

Unlike the army, the navy's adoption of radar generated little bureaucratic infighting. In fact, the opposite was true. As with Very's signal flares, wireless telegraphy, and electromechanical encryption machines, the service's shore-based technical infrastructure worked closely with operating forces to adopt the new technology. Nonetheless, two factors unique to warfare at sea challenged the service's ability to employ radar effectively. First, the physical facilities used to process radar information (i.e., the naval equivalent of the information center at Fort Shafter) had to be created within the confines of already crowded warships. Second, air defense was just one aspect of ship defense. For naval commanders, radar helped to protect against not only air but also surface and submarine attacks.[16]

Further complicating matters, radar information came in the form of images, not words. Operators had to translate and integrate these images into existing shipboard systems in order to make radar truly useful to commanding officers and senior commanders. This process required personnel to observe electronic representations (frequently referred to as blips), convert them into numeric form, and transfer them to a plot, where decision makers could view radar information in a more easily intelligible form. Needless to say, this conversion process appreciably increased the possibility that critical information might be distorted or lost in translation.

These issues greatly complicated the U.S. Navy's efforts to design and construct command and control facilities for the employment of radar. Nevertheless, by late 1941 the service had made considerable progress in developing radar technology and integrating it into the practice of warfare at sea: training programs had been initiated, new tactics had been developed, and better equipment was on the way. Yet solutions to the problems created by shipboard space constraints and three-dimensional threat axes were not easy to achieve, and much work remained before American naval commanders could exploit fully the potential of seaborne radar. The navy may have been ahead of the army at the time of Pearl Harbor, but naval personnel still had a steep learning curve ahead of them.

Curiously, the history of the shipboard facilities built to manage radar information has not attracted the interest of historians. This marks a sharp contrast with the history of naval radar, for which some excellent scholarship exists.[17] Yet the technical aspects of American naval radar development are just one part of a larger story, a story that includes important contributions by now largely forgotten individuals. At the heart of their efforts were shipboard spaces in which officers and sailors worked with specialized equipment to collect, organize, process, evaluate, and disseminate radar information. The U.S. Navy initially called this type of facility the Radar Plot, but combat experience revealed that decision makers had great difficulty integrating information from the Radar Plot with other sources of information, which could include radio messages, visual signals, direction-finding intercepts, sonar contacts, reports by lookouts, and fire control solutions. To address this information management problem, American naval personnel transformed the Radar Plot into the Combat Information Center (CIC), a facility in which information from all sources could be collected and analyzed as expeditiously as possible. By the end of World War II, the CIC had become an indispensable artifact, the brain of the American warship.[18]

Radar Goes to Sea

The origins of radar in the United States date to 1922.[19] In September of that year, A. Hoyt Taylor—the former North Dakota physics professor and reserve officer who had helped develop radio for naval aircraft during and after World War I—and another radio engineer, Leo C. Young, were conducting tests with experimental high-frequency apparatus when they observed that large objects, like buildings and ships, created unusual interference patterns when relative motion existed between object and receiver. Realizing that this phenomenon might have useful applications, Taylor wrote the Bureau of Engineering to request support. The bureau concluded a practical system could not be built with existing electronic components, so Taylor and Young moved on to other projects. For nearly eight years afterward, no one in the United States would seriously investigate systems for radio detection.[20]

Renewed interest in such a system began in the summer of 1930 when Young and another radio engineer discovered that airplanes reflected enough high-frequency radiation to create a noticeable interference pattern in a distant receiver. They informed Taylor of their discovery, who joined them in performing further experiments. Convinced that the Naval Research Lab might be able to build equipment that could detect aircraft and ships at considerable distances, Taylor again wrote the Bureau of Engineering to request support. This time the bureau agreed to sponsor further research, albeit with a minimum of funding. Young and Taylor made little progress, however, until Young recommended shifting from continuous wave to pulsed equipment sometime in late 1933. The former type of system relied on the Doppler effect, whereas the latter used the time delay between transmission of a radiated signal and receipt of its echo to determine the distance to an object.[21] In pursuing this idea Young drew inspiration both from his own prior experiments and from work ongoing in the Naval Research Laboratory's sound division, which was then investigating better ways to display return echoes from sonar pings.[22]

Pressed for time by other commitments, Young and Taylor delegated work on a possible pulsed wave system to Robert M. Page. The son of a Minnesota farmer, Page initially wanted to go into the ministry but discovered a different light in college and chose instead a career in physics. Graduating near the top of his class, he accepted a job at the Naval Research Laboratory and began work there in 1927. For more than six years, Page's research focused on finding ways to reduce radio interference, but he tackled his new assignment with enthusiasm and in late 1934 carried out tests that demonstrated the feasibility of pulsed radar. Soon thereafter a laboratory committee recommended devoting more resources to the radio detection problem, prioritizing it ninth out of fifteen different projects. The Bureau of Engineering balked, probably because bureau chief Harold G. Bowen believed the potential operational value of the new technology warranted a higher priority. The laboratory thus elevated the radio detection problem from ninth to third in importance, which was certainly a leap of faith because Page's experimental apparatus displayed numerous shortcomings. The main problem was with the receiver, which became saturated by high-power pulses and could not recover in time to detect returning echoes. Page spent most of 1935 redesigning his receiving unit, during which time he developed a circuit that reduced feedback and allowed for rapid recovery from high-voltage signals. By the spring of 1936, Page was ready for a practical test of his new system.[23]

That test came the last week of April and surpassed even Page's expectations. The redesigned apparatus revealed clear and distinct echoes from planes up to nine miles away, a distance engineers were able to double after some "fruitful tinkering."[24] In June, representatives from the Bureau of Engineering went over to the Naval Research Laboratory to witness a demonstration of the new equipment. It functioned perfectly. The observers informed Bowen, who requested that any remaining work "be given the highest possible priority." The bureau chief also asked the lab to construct an experimental system for shipboard use as soon as possible.[25]

Bowen's enthusiasm was tempered somewhat by the fact that the equipment Page had developed needed two major modifications before the navy could test it as sea. The first was a shift to a higher frequency. The apparatus demonstrated in the spring of 1936 operated at 28.6 megahertz, a frequency that required an antenna nearly 250 feet long to achieve sufficiently narrow beam widths. Although Page had been aware of this issue from the start, he selected the unusual frequency because the Naval Research Lab already had an existing antenna optimized to operate at that frequency. Once initial tests proved successful, the lab shifted its efforts toward the design and construction of a 200-megahertz set. Page selected that particular frequency because he believed it was the maximum one practicable with existing vacuum-tube technology.[26]

The second reason the existing prototype was unsuitable for shipboard installation was because it required two antennas, one for transmission and one for reception. Page believed this would be an acceptable arrangement once 200-megahertz equipment had been developed, but Taylor wanted a single antenna for both transmitting and receiving. Although Page greeted the idea with skepticism, Taylor persisted. So the young Minnesotan went to work, and remarkably, in a matter of months, he and his team developed something that permitted employment of a common antenna. The new device, called a duplexer, was an electronic switch that protected the receiver from high-voltage transmissions without limiting its capacity to detect reflected, low-energy radio waves. Personnel at the Naval Research Lab tested both it and the 200-megahertz system concurrently in the summer of 1936. The duplexer performed well but the 200-megahertz receiver did not readily pick up reflected signals from aircraft.[27] While work remained to be done, Page and his colleagues nevertheless had assembled the basic elements of a working shipboard radio detection system.

As Page, Young, Taylor, and others at the Naval Research Lab achieved one milestone after another in their development of a 200-megahertz radio detection system, senior American naval officers showed an ever-increasing interest in the new technology. After Bowen himself witnessed a demonstration in the summer of 1936, several other bureau chiefs and Chief of Naval Operations William Standley stopped by to see what the equipment could do. In December, U.S. Fleet commander-in-chief Arthur J. Hepburn came all the way from California to witness a demonstration of the new 200-megahertz apparatus. Personnel at the Naval Research Laboratory were particularly gratified by Hepburn's visit; according to the lab's director, the commander-in-chief's "obvious interest" was "a great help to morale."[28] For his part, Hepburn pushed the lab to conduct shipboard tests at the earliest possible date. This idea bothered Page, who wanted more time to perfect the equipment. According to historian David Allison, Page "understood well the power of successful demonstrations, [but] he also knew the disastrous effect a poor one might have."[29] Perhaps Page also knew the story of Lee De Forest's radiotelephones and the Great White Fleet.

A. Hoyt Taylor dismissed Page's misgivings and moved forward with plans for shipboard testing. Taylor's decision to overrule his subordinate derived from various factors, including two decades of experience and a high measure of respect for Hepburn. Nevertheless, the push to get radar out to sea so hastily was a departure from the peacetime practices most often pursued by the service with respect to command and control technology. Why the difference? Three reasons are germane. First, international affairs were volatile throughout 1936. In February, Japanese military extremists attempted to overthrow the emperor and seize control of the government; the following month, Adolf Hitler violated the Versailles Treaty by occupying the Rhineland. Of particular interest to naval officers, at year's end Japan formally withdrew from the Washington Naval Treaty. While this withdrawal had been anticipated, it was still a noteworthy event. Bowen, Taylor, and many others recognized that any delay, even one to improve performance, would constitute a genuine gamble. Second, although the service was far from flush, it possessed the fiscal resources to move quickly. Appropriations had been on the rise for several years, and the chairman of the Naval Appropriations Subcommittee in the House of Representatives was favorably inclined toward the new technology.[30] Finally—and in contrast to circumstances surrounding the service's adoption of radio—a nearly impenetrable veil of secrecy enveloped the work of every other country working on radio detection.[31] Whereas Royal Bradford could thus wait while his envoys traveled around Europe gathering data on the latest advances in wireless telegraphy, the navy's leaders did not enjoy the same luxury with respect to radar. Hepburn in particular held a strong interest in the ties between secrecy and radar development.[32]

With some apprehension Page followed Taylor's orders, and in the spring of 1937 the Navy Department made available to the Naval Research Laboratory a destroyer for shipboard tests. Page later wrote that he "learned a lot about radar on the ocean during these tests," but his biggest takeaway was the need for a better transmitter if aircraft were going to be detected at anywhere near the distances observed under the more ideal conditions available back on land.[33] Led by Page, engineers at the Naval Research Lab spent the next nine months designing and building a more powerful transmitter, which they did by adopting a new type of vacuum tube and by reconfiguring the transmitter circuitry into a ring design. On 1 February 1938 laboratory personnel successfully tracked planes up to distances of forty miles; two weeks later they extended that range out to fifty miles.[34] Satisfied with the results, Page recorded in his notebook: "This completes the entire 200 mc [megahertz] development of radio echo equipment."[35]

On 24 February 1938 the Bureau of Engineering convened a conference to discuss what steps the service should take next with respect to the new technology. In attendance were representatives from the Naval Research Laboratory, the Office of the Chief of Naval Operations (OPNAV), and the material bureaus. Laboratory personnel informed attendees that work on a 500-megahertz system was in progress, but with the Sino-Japanese War raging in Asia, a consensus emerged that the fleet should receive radio echo equipment at "the earliest practicable date."[36] As such, the Bureau of Engineering directed the Naval Research Lab to freeze the design of its 200-megahertz apparatus and to construct as rapidly as possible a prototype for shipboard trials. In early March the lab's director informed the bureau that his engineers would have a service model ready within six months.[37]

In the meantime, a debate arose over which ship should first receive the new equipment. Bowen proposed the carrier *Yorktown*, writing that he favored installation on a vessel that would garner "comment and observation by the greatest number of experienced officers." Others, including the chief of the Bureau of Aeronautics, expressed concern that the large size of the antenna frame and supporting structure might interfere with air operations. They instead suggested converting the old carrier *Langley* into a special test platform.[38] For his part, Taylor actually wanted the trials to be conducted closer to home. He got his wish after several senior officers in the newly established Atlantic Squadron lobbied to get the new apparatus installed on the squadron's flagship, USS *New York*. In early December personnel at the lab packaged up everything and shipped it to Norfolk, where workers installed what the service by then had designated the model XAF radio ranging equipment. A few days into the New Year, *New York*'s XAF was ready to go to sea.[39]

The Bureau of Engineering was optimistic about the Naval Research Laboratory's 200-megahertz radar, but it hesitated to put all of the navy's eggs into a single basket. As such, shortly after the service froze the lab's design it contracted with RCA for a pulsed radio detection device that operated at 385 megahertz.[40] RCA's discovery of radio echoes had come out of the company's research into microwaves, but only in mid-1937 did the Bureau of Engineering finally share with RCA significant details about the work of Taylor, Young, and Page. Many individuals at the Naval Research Lab felt the sixty thousand dollars paid to RCA would have been better spent hiring additional staff to expedite production of the XAF, but the bureau's decision was not unreasonable given the uncertainty still surrounding the new technology. So that trials would be comparable, the Navy Department installed RCA's model CXZ radio ranging equipment on a battleship as well. Thus, on 6 January 1939—the date a worker at the Norfolk Navy Yard tightened the last bolt affixing RCA's model CXZ to the superstructure of USS *Texas*—the navy had two different radars ready to go sea. For many of those involved, a sense of competition filled the air.[41]

Shipboard trials revealed that the winner in any such friendly competition was the XAF. Partly in response to pressure from the navy, RCA had rushed its radar into production. This decision led to disappointing results. Returns generally were weak and inconsistent, parts deteriorated when exposed to moisture, and the apparatus initially could not withstand the vibrations created by the discharge of *Texas*'s fourteen-inch guns.[42] For surface contacts the maximum range of the CXZ was half that of the XAF, and CXZ operators could not detect aircraft at distances over five nautical miles.[43] Disappointed with the CXZ's performance, *Texas*'s commanding officer remarked that RCA's equipment "would be of very little value in war," while his superior informed the Bureau of Engineering that "the XAF equipment was in all respects (except size) markedly superior to the CXZ."[44]

Indeed, the Naval Research Lab's XAF performed quite well, with a dramatic illustration of its potential coming almost immediately during an exercise that pitted four darkened destroyers against *New York*. As these four vessels approached the big battleship, an XAF operator picked them up at about ten nautical miles. Personnel on board *New York* tracked the destroyers for more than an hour, and once they had closed to within five nautical miles *New York*'s commanding officer pointed his searchlights down the relevant bearing and illuminated them. Page reported that the first light "fell dead on [the] leading destroyer" and later recalled that the destroyer skipper had seemed "a little bit shaken" by the experience.[45]

Other tests provided similarly impressive results. Page reported that most ranges were accurate to within three hundred yards and that no appreciable interference existed between the XAF and other shipboard electrical equipment. To the surprise of many, XAF operators picked up echoes from a partially submerged submarine and easily tracked fourteen-inch shells in flight.[46] With respect to approaching planes, *New York*'s commanding officer observed that they "could be reliably reported, day or night, up to twenty miles distant," and Page himself rated as "outstanding" the XAF's ability to detect aircraft.[47] Page noted with pride that the equipment had "stood up remarkably well," commenting that its performance "exceeded our expectations." The senior officers present during the trials were equally enthusiastic. *New York*'s commanding officer proposed immediate installation on all carriers, and the commander of the Atlantic Squadron informed the Bureau of Engineering that he considered the XAF "the most important military development in radio since the advent of radio itself."[48]

These encouraging reports notwithstanding, the trials also highlighted the inherent difficulties of managing radar information. For example, during a different attack exercise conducted the following month, XAF operators again succeeded in detecting and tracking approaching destroyers. This time, however, *New York* failed to illuminate the attacking vessels until after they had launched a hypothetical barrage of torpedoes. Although no one recorded the precise reason for this failure, apparently the searchlight operators either failed to receive or misunderstood the information passed to them from operators in the XAF control station.[49] The navy thus learned in 1939 what the army would discover at Pearl Harbor: that operational success derived not only from dependable radar equipment but also from effective systems for managing radar information.

News of the XAF's performance spread quickly throughout the service, and on 1 May 1939 the navy held a conference to come up with a course of action for procuring XAF equipment. Present were representatives from OPNAV, the

Naval Research Laboratory, and each of the material bureaus. After discussing briefly the advantages and limitations of the XAF, conference attendees unanimously agreed that ten devices should be procured immediately. They reasoned that the admittedly imperfect XAF was nonetheless ready for service, that operational experience would aid in future development, and that the current "international situation" warranted taking "immediate advantage . . . of every device leading to greater military effectiveness." In order of priority, the group proposed that all five of the navy's commissioned carriers, three battleships, and two cruisers receive the new equipment.[50]

Less than two weeks later, Chief of Naval Operations (CNO) William Leahy directed the Bureau of Engineering to procure the recommended devices for installation on vessels of the United States Fleet.[51] Soon thereafter, representatives from both Western Electric and RCA visited the Naval Research Lab to learn more about the XAF and to discuss potential contract specifications.[52] Similar meetings took place throughout the summer, ending once the bureau awarded RCA a contract to build the navy's first production-model radars. Because RCA stated that it would need between seven and ten months to build these sets, the Bureau of Engineering lowered the initial production run to six units so expected improvements could be incorporated into the next production contract.[53]

In mid-October, the Naval Research Laboratory supplied RCA with a complete set of blueprints for the XAF and turned its attention to other issues.[54] One of the most important of these was the development of an improved visual display system for radar information. The existing display on the XAF's receiving unit consisted of an A-scope, which showed only an amplitude spike calibrated to indicate range, and a mechanical bearing indicator that provided the antenna's azimuth. Fleet Problem XX (1939) had revealed that operators experienced difficulty correlating information from the A-scope and the bearing indicator, so the Naval Research Lab began to investigate the possibility of an indicating system that could provide both bearing and range on the same display.[55] This work eventually led to a device that would provide operators with a 360-degree, bird's-eye view of their surroundings.

Personnel at the Naval Research Laboratory also worked to solve other difficulties related to the employment of radar at sea. For example, Page and his cohorts made minor alterations that increased accuracy and improved detection ranges. In addition, they intensified efforts to develop a recognition system that could discriminate between friendly and enemy units, an issue the bureau saw as one of the biggest weaknesses of the new technology.[56] Finally, laboratory personnel investigated methods for distributing radar information throughout a ship. Existing arrangements for interior communications relied mainly on verbal reports, passed by telephone, voice tube, or messenger. Yet because radar information came in the form of images, not words, American naval officers recognized that new means of information sharing were essential. One senior operational commander believed the information obtained from radar was "of such immediate vital interest to a number of ship and fire control stations . . . that the development of a repeater system is extremely desirable," while another senior officer stated simply that "effective coordination between the [radar] operating station and the conning or communication station is essential."[57] Knowledgeable personnel at the Naval Research Lab and the Bureau of Engineering believed these problems could be solved, but they knew the solutions would take time. They also knew that by June 1940—the month in which RCA finally finished the six production models of what the navy by then had designated the "CXAM (Radio Echo Equipment)"—time was of the essence.[58]

To hasten the development of radio echo equipment, the newly established Bureau of Ships (created when the bureaus of Engineering and Construction and Repair merged in June 1940) adopted a two-pronged approach. The first dealt with industrial relations. In 1940 the Naval Research Laboratory was still the nation's leader in radar development, but it was not a manufacturer and could never bring to bear the resources available in private industry. As such, the bureau sought to establish a highly collaborative relationship with RCA and Western Electric, the two firms that had bid on the initial XAF contract. It also persuaded General Electric to enter the field. The navy's leadership allowed the Naval Research Lab to disclose completely all details of its prior work with radar, believing the dire international situation necessitated the unusual arrangement. Although the bureau recognized the three companies' efforts would overlap, it believed "the disadvantage of any unavoidable duplication will be more than offset by the advantage of commercial competition and rivalry in this new field."[59]

The second part of the bureau's strategy for expediting radar development was to rely on fleet personnel for ideas and suggestions that would shape the new technology into an effective tool of war. Anticipating delivery of the first six CXAMs, in the summer of 1940 the Bureau of Ships arranged for an equal number of enlisted radiomen to receive six weeks of specialized training in the operation and maintenance of the new equipment. This move matched previous practices for other command and control technologies, including post–Civil War signaling methods, wireless telegraphy, and the Hebern Cipher Machine, and was intended to provide a cadre of skilled personnel who could disseminate technical knowledge throughout the fleet. The bureau even specifically suggested to the U.S. Fleet commander-in-chief that each expert be tasked to train five other individuals as operators so that shipboard CXAM watches would always be properly manned.[60]

The first ship to receive the CXAM was USS *California*. Workers at the Puget Sound Navy Yard in Washington needed a little over a month to install the new apparatus, but by the end of August *California*'s radar was ready for sea trials.[61] Around the same time, five other warships—four cruisers and the carrier *Yorktown*—also received CXAMs.[62] Although this distribution differed from prior suggestions to equip the fleet's carriers first, OPNAV apparently decided that cruisers afforded a better type of test platform than carriers, which remained limited in number. Nevertheless, both OPNAV and the Bureau of Ships believed that the service needed to test radar on every type of major warship, which is why *California* and *Yorktown* received installations.[63] As events would unfold, the crews of these two vessels would have a significant impact on the development of techniques, devices, and facilities for radar employment.

California's commanding officer, Captain Harold M. Bemis, and his communications officer, Lieutenant Commander Henry E. Bernstein, began to evaluate how the CXAM fit into existing shipboard command and control systems almost as soon as their ship put to sea with the new equipment on board. The Bureau of Ships had tasked the Puget Sound Navy Yard to convert one of *California*'s staterooms into a CXAM operating room, linking it to eleven key stations throughout the ship (e.g., flag plot, central radio, conning tower, bridge) by means of a newly installed sound-powered phone circuit. The CXAM operator could talk to any station with the flip of a switch; he also had the ability to ring a call bell that immediately alerted personnel on the bridge. The bureau's goal was to integrate fully the CXAM into *California*'s existing system of shipboard command and control. In the end, however, only the crew could ascertain if such arrangements were adequate.[64]

Bemis delegated primary responsibility for making such a determination to Bernstein. A native of Jacksonville, Florida, *California*'s communications officer was a gifted athlete who had graduated from the Naval Academy in

1926. After graduation, Bernstein served on *Lexington* and several other vessels before pursuing graduate studies in radio engineering. He first attended the navy's postgraduate school in Annapolis and then went to Harvard where he finished his master's degree in June 1935. After back-to-back sea tours, Bernstein returned to the Naval Academy, where he served as an instructor in the department of electrical engineering. In many ways, Henry Bernstein's professional background was remarkably similar to that of officers like Charles Maddox, Stanford Hooper, and Arthur Hepburn.[65]

Extant records do not specify why the service assigned Bernstein to *California* in June 1940, but one would be hard pressed to conclude that the timing was a coincidence. The young officer possessed substantial operational experience, was an expert in both radio and electrical engineering, and reported aboard just weeks before shipyard workers began to install RCA's model CXAM radar. Bernstein's technical knowledge and work ethic undoubtedly impressed Bemis, who gave his subordinate wide latitude in supervising tests of *California*'s CXAM. In turn, Bernstein relied heavily upon the expertise of the enlisted radioman specially trained by the Naval Research Lab earlier in the year. Initial trials, which lasted about a month, went well. *California*'s CXAM displayed only minor material problems, could detect aircraft at distances of up to nearly fifty nautical miles, and was accurate to within one hundred yards. While Bernstein provided Bemis with much feedback on the CXAM itself—recommending such items as better electrical safety devices and a more accurate range indicator—he also identified issues related to the exchange of radar information. For example, Bernstein noticed that the CXAM operator was too easily distracted by the sound-powered phone system, which often required him to divert his attention from the A-scope. The communications officer suggested instead a microphone "so mounted as to be in front of the CXAM operator at all times while he is observing the screen." Bernstein also noticed the CXAM operator sometimes received conflicting orders from different stations, so he formulated clear procedures by which one station had to hand off control before another station could give the CXAM operator an order.[66]

Bemis forwarded Bernstein's recommendations on the CXAM to both the U.S. Fleet commander-in-chief and the Bureau of Ships, but *California*'s commanding officer did not feel he needed approval from higher authority to address the issues his communications officer had identified with respect to the management of radar information.[67] As such, Bemis encouraged Bernstein to tackle the problem immediately. The young officer informed his boss that the best way to view radar information was to plot it, so Bemis authorized Bernstein to design and oversee the creation of a makeshift plotting room, one that incorporated a horizontal plotting table, voice radios, and several other appurtenances next to the CXAM console. Probably thinking he would be spending a lot of time in the new facility, Bernstein even installed a bunk for the communications officer.[68] Because sailors had given the CXAM a nickname, the Geep, he called the new facility the Geep Plot.[69]

While Bernstein's Geep Plot was the most prominent change made by shipboard personnel in response to the new technology, it was hardly the only one. Personnel on the other five vessels that received the CXAM also suggested and pursued innovative ways to integrate radar into the practice of warfare at sea. For example, officers on at least two cruisers arranged for a phone connection between the CXAM room and all gunnery stations, while the commanding officer of *Northampton* stressed to one of his superiors the overriding importance of "direct battle phone communication" between the CXAM operator and the searchlight director.[70] *Chester*'s commanding officer sought authorization to install makeshift repeaters at various locations throughout his ship, and *Yorktown* tested different ways of directing aircraft via radiotelephone on the basis of CXAM information.[71] In short, the six warships that possessed CXAMs simultaneously pursued changes while flooding the shore establishment with suggestions for improvement.

Rear Admiral Harold Bowen, former chief of the Bureau of Engineering who in 1939 became the Naval Research Lab's director, recognized that the many alterations and recommendations made by fleet personnel would be of little value unless the shore establishment evaluated them and coordinated their implementation. He also knew the problem would only get worse when the fleet started to take delivery of the CXAM–1, an improved version of the CXAM that would enter service in the spring of 1941.[72] As such, in October 1940 Bowen proposed to Chief of Naval Operations Harold Stark and Secretary of the Navy Frank Knox that a single agency or individual be tasked to coordinate the service's radio echo program "in connection with tactical requirements of the Fleet."[73] The admiral conveniently recommended himself for the job. Stark and Knox agreed, designating Bowen as the coordinator for all "general applications of this new system."[74]

Meanwhile, fleet personnel continued to experiment with the CXAM. Other ships emulated *California*'s Geep Plot, which eventually became known

A Grumman F4F Wildcat fighter takes off from USS *Ranger*, November 1942. Note the CXAM-1 radar antenna atop the carrier's island structure. *National Archives photograph 80-G-30244.*

as the Radar Plot. The most active vessel in this regard was *Yorktown*, which received a working CXAM in October 1940.[75] As had been the case on *California*, *Yorktown* initially had the CXAM operator pass information to other stations via sound-powered telephone. This arrangement was wholly unsatisfactory, so *Yorktown*'s commanding officer, Captain Ernest Gunther, placed an officer in the CXAM room to plot contacts on a maneuvering board. The setup allowed the CXAM operator to devote his full attention to the A-scope but did not resolve a fundamental problem: personnel located elsewhere on board could not envision the information being passed to them.[76]

In February 1941 Gunther turned the problem over to his successor, Captain Elliott Buckmaster. Buckmaster, who came to *Yorktown* after successful tours as *Lexington*'s executive officer and as the Ford Island Naval Air Station commander, considered efficient employment of the CXAM a top priority.

After just six weeks in command, he made abundantly clear to his superiors the pressing need for a separate radar plotting room:

> It has been increasingly apparent that separate and complete plotting facilities must be provided in order that full and intelligent use may be made of the information which is obtainable from radar. As at present installed and operated, a mass of unrelated and heterogeneous ranges and bearings is sent from radar by telephone . . . It is manifestly impossible for any person receiving this information to form from it a mental picture which will show him incipient air attacks or approaching targets for gun fire. Furthermore, such a mass of unrelated information is likely to confuse the picture of the tactical situation obtained from other means such as contact reports and reports from lookouts.[77]

Buckmaster went on to recommend the establishment of radar plotting teams and the construction of appropriate physical facilities where vital information could be both processed and evaluated.[78]

Buckmaster's suggestions were similar to those of many senior officers. For example, the commander of the Scouting Force's cruiser divisions, Rear Admiral John Newton, recommended co-locating A-scopes and plotting facilities so that the "dissemination of radar information" could take place with "minimum delay," while the commander of the Battle Force's aircraft squadrons proposed changes to carriers' island structures in order to facilitate better shipboard information exchange.[79] In mid-July Buckmaster weighed in on the matter again, repeating his previous recommendations and highlighting the advice of three Royal Navy observers (temporarily assigned to *Yorktown*) who had emphasized to him the significant advantages of radar-directed fighter control. The Bureau of Ships was somewhat overwhelmed by the number of recommendations received from the fleet, but OPNAV and several other bureaus put pressure on the chief of the Bureau of Ships to act as expeditiously as possible. CNO Harold Stark ultimately took a leading role, authorizing the installation of radar plots aboard all carriers on 21 August 1941.[80] In so doing, Stark concurred with the navy's Interior Control Board, which envisioned the Radar Plot as "the brain of the organization which protects the fleet or ships from air attack."[81]

Meanwhile, back on *California* Henry Bernstein turned his attention to the issue of how to distinguish between friendly and enemy aircraft on the CXAM display console. The Naval Research Laboratory had been working on a plane-to-ship recognition device since 1937, but the lab's initial production model had a rather limited range.[82] One problem was that the aircraft itself carried the transmitting unit, which necessitated the use of a relatively small transmitter. Fortuitously, the navy's successful development of pulsed radar eliminated the need for a powerful airborne transmitter for aircraft identification. Although Page recognized this fact almost immediately, he initially tried to build a system that placed a rotating radar reflector on friendly aircraft. In Page's own words this concept turned out to be "much too awkward," so the inventor turned his attention to developing a transponder that would amplify and retransmit a distinctive signal back to the radar receiver.[83] Page and his cohorts eventually perfected such a device, but it did not become readily available for fleet use until after the attack on Pearl Harbor.[84]

Bernstein seems to have been aware of the Naval Research Lab's efforts to develop an aircraft recognition device that would work with the CXAM. Nevertheless, by early 1941 his patience was running short, a sentiment shared by other officers. In February, for example, cruiser divisions commander John Newton wrote that aircraft identification was critical both for the dissemination of information and for determining "what objects on the screen should be reported and [tracked]." The following month *Yorktown*'s skipper Elliott Buckmaster stressed to a superior the difficulty of recognizing friendly units during tactical exercises, recommending that "every effort be made to develop and issue to the Fleet a recognition device that will register on radar screens."[85] Bernstein was one of the few individuals in the fleet who possessed the expertise to pursue an interim solution, something he did throughout the spring and summer of 1941. Working with two chief radiomen, *California*'s communications officer designed and built a recognition device that automatically transmitted a signal whenever triggered by the beam of a CXAM. This signal caused an image of a specific width to appear at a predetermined spot on the CXAM's A-scope. By early August, Bernstein's device was ready for production. The Pacific Fleet commander ordered the manufacture and test of eighteen devices, justifying his actions to the Bureau of Ships by noting that "the problem of recognizing friendly vessels and aircraft is preventing full use being made of [available] radar equipment."[86] Although the Naval Research Laboratory's more sophisticated recognition system would supersede Bernstein's interim device within a few months, the work of Bernstein and his men reveals how aggressively operational personnel pursued improvements in command and control technology. Bernstein's himself also benefited, as the Navy Department later awarded him a commendation medal for his efforts.[87]

While individuals like Bernstein and Buckmaster focused on the operational aspects of radar employment, other naval officers concentrated on a different issue: the training of personnel. In the fall of 1940 the chief of the Bureau of Ships, Samuel Robinson, wrote CNO Harold Stark to recommend that radiomen be trained "in the design features, operation, and maintenance of [radar] equipment" beginning in January 1941. Stark agreed with Robinson and assigned the Bureau of Navigation responsibility for such training.[88] The bureau chose to add radar instruction to the Radio Material School then being offered at the Naval Research Laboratory, a prudent choice for two reasons. First, the lab held the service's leading collection of radar experts. Second, laboratory director Harold Bowen had been closely involved with the so-called Tizard Mission, a delegation of British scientists and senior military officers who arrived in Washington in September 1940. Along with the famous cavity magnetron (a high-powered vacuum tube capable of generating microwaves), the Tizard Mission also turned over to the United States substantial information about British radar operations. Bowen and Stark both sought to incorporate this valuable information into the navy's curriculum for radar training.[89]

Even with the new curriculum, however, a single school was not going to provide the service with sufficient numbers of enlisted radar experts. Robinson realized that only some individuals actually needed to know how to repair a radar set, so in the spring of 1941 he proposed separating the navy's training pipeline into two tracks: one for operators and another for maintenance men. All trainees would take a short course in how to operate radar; only about 15 percent of these would go on to attend a longer course in radar maintenance.[90] While helpful, this arrangement hardly alleviated the navy's manning problem. An internal report, written in September 1941, estimated that the fleet would need two thousand maintenance men and more than ten thousand operators by the end of the fiscal year. Yet because the Radio Material School at the Naval Research Lab could produce only about sixty radar technicians per month, the bureau sought to close some of the projected shortfall by sending men to a Royal Canadian Air Force radar school in Ontario. Still, less than half the maintenance men needed in the fleet would be available by mid-1942.[91] With respect to operators, the Bureau of Navigation held that sailors could be partially trained aboard ship while awaiting more formal schooling. Naval officers who had firsthand experience with the CXAM challenged this assumption and got their superiors to push for the immediate establishment of a radar operators school at Pearl Harbor.[92] Based in part on these inputs, in November the Bureau of Navigation acquiesced and agreed to support the establishment of a radar school in Hawaii. The attack on Pearl Harbor accelerated matters, and the new school opened in early 1942.[93]

The navy's approach to radar instruction for officers differed somewhat from its approach to enlisted training. Whereas the latter concentrated on the technical and practical aspects of radar operation, the former focused heavily on how to use radar information effectively. As early as the fall of 1940, a period when shipyard workers were installing the first six CXAMs on board American warships, operational commander John Newton suggested offering a course for officers "to aid them in evaluating [radar] information," and the following spring Samuel Robinson recommended to Chester Nimitz, his counterpart at the Bureau of Navigation, that 150 junior officers be trained in the operation and "tactical application" of radar.[94] Nimitz concurred and looked to the Naval Research Laboratory for help, but lab director Harold Bowen reported that the recent expansion of the Radio Material School left him short of facilities and potential instructors. Taking a page from its World War I playbook, the Navy Department instead opened a basic radar school at a civilian institution, Bowdoin College (located in Brunswick, Maine). The first class, comprised of junior officers who possessed either an engineering background or a Bachelor of Science degree, convened in June 1941.[95]

Even before Bowdoin's first radar class commenced, senior personnel were discussing how best to train shipboard officers in the area many believed would be one of the most mentally challenging aspects of radar employment: fighter direction.[96] In late spring 1941, Nimitz wrote CNO Harold Stark to recommend the establishment of two schools (one on each coast) for fighter director officers, a proposal strongly supported by the Bureau of Aeronautics.[97] Stark quickly approved the idea and directed the founding of schools in Norfolk and San Diego. Such high-level interest ensured that there were few delays, and the service's first fighter director classes convened in each location in the fall of 1941.[98]

In San Diego, the navy placed Lieutenant Commander John Hook "Jack" Griffin in charge of the new school. A native of Charleston, South Carolina, Griffin was the son of a naval officer and a 1925 graduate of the Naval Academy. After graduation he performed back-to-back sea tours before earning his wings and spending most of the 1930s in aviation-related billets. From 1938 to 1940 Griffin served in OPNAV's fleet training division, and in early 1941 he was one of a select few individuals sent overseas to observe British methods of fighter direction.[99] Nearly five years earlier the Royal Air Force had initiated exercises intended to determine the best means for "intercepting hostile aircraft with fighters whose navigation was controlled from the ground,"[100] and by 1940 British personnel were the best in the world at this task.[101] While in Britain, Griffin observed firsthand the interception techniques of both the Royal Air Force and the Royal Navy, making him one of the U.S. Navy's few specialists in fighter direction. Griffin's experiences in Britain, along with his professional background in aviation and naval training, made him an excellent choice to head the new school.[102]

Griffin's first class convened on 1 October 1941. It consisted of twenty-five newly commissioned ensigns handpicked by the Bureau of Navigation from the V-7 Naval Reserve Midshipmen's School at Northwestern University. As a group, the ensigns were surprised and somewhat confused by their orders, which directed them to report to San Diego "for active duty under instruction in Fighter Director Control."[103] The training curriculum in Northwestern's V-7 program did not cover the topic of fighter direction, and the ensigns still had only a vague notion about their immediate futures when Jack Griffin walked into an otherwise empty hanger at the North Island Naval Air Station on the first morning of class. Griffin's words from that day are lost to posterity, but he likely told the assembled ensigns that if and when war came they would have one of the most critical jobs in the fleet.[104]

Griffin had much work to do, as the new school possessed very little equipment and no prescribed curriculum. Lack of space at North Island meant that classes had to be held in an airplane hangar outfitted with makeshift classrooms. Undaunted, the school's inaugural commanding officer tasked his students to track down various items from other bases in the San Diego area, including plotting instruments, official publications, and sound-powered telephones. Griffin also acquired one dozen specially manufactured, adult-sized tricycles to simulate friendly and enemy aircraft. Students practiced on a big field with some students piloting friendly tricycles while others directed them to intercept instructor-ridden enemy tricycles. To enhance realism, each tricycle had a blinder so friendly pilots could not look across the field to facilitate interceptions.[105]

Griffin borrowed the concept of training tricycles from the British. He also incorporated other successful British practices into the school's curriculum, such as a designated symbol and numbering system for incoming raids and a standard vocabulary for fighter direction. In addition, Griffin made sure his students received thorough training in all aspects of radar employment, including theory, navigation, filtering, and plotting.[106] Like many others, he believed well-trained personnel were a key to success: "I feel the operator is an extremely important cog in the system . . . so much depends on proper 'interpretation' of what is seen."[107] Yet Griffin knew as well that the job of fighter director was difficult and that some individuals simply did not have the personal demeanor or mental aptitude to succeed. Out of his first class, which the Bureau of Navigation had vetted heavily, Griffin refused to allow four officers (16 percent) to graduate. Still, a majority of students successfully completed training, and Griffin worked closely with the bureau to send them to ships and shore stations where their new expertise would most benefit the service.[108]

By the end of 1941, then, the Navy Department had opened new schools to train officers and enlisted men in the operation and employment of radar. Although these schools could not provide enough trained personnel to meet the service's needs after the attack on Pearl Harbor, they laid a solid foundation for wartime expansion. Indeed, the existence of high-quality training programs before the war was just one of three factors that allowed the navy to integrate radar into combat operations relatively quickly. Another was the close working relationship between the naval shore establishment and the fleet, an institutional ethos that dated back decades. For radar specifically, this environment led to the creation of devices and facilities that assisted operational commanders in managing the tactical information made available by the new technology. No part of the shore establishment was more important than the Naval Research Laboratory, where civilians like A. Hoyt Taylor and Robert M. Page displayed a remarkable ability to think in both technical and operational terms. After Fleet Problem XX, for instance, *New York*'s commanding officer went out of his way to praise Page and his cohorts, writing: "We found them very agreeable shipmates and cooperative at all times, so much so in fact that I came to look upon them as regular members of the ship's company and called on them for service just as I would any other officer of the ship."[109] Lastly, the navy's operational officers, over time and as a group, had spent years thinking about and experimenting with shipboard systems that addressed the escalating complexity of warfare at sea. To these men, the idea that radar should be employed as part of a command and control system, rather than as an isolated electronic device, was readily apparent. Without such an appreciation, the Radar Plot probably would not have been conceived and adopted as early as it was. For the United States this head start was important, as the Radar Plot experienced many growing pains before yielding to its direct descendant, the CIC.

The Combat Information Center

A few days after the devastating attack on Pearl Harbor, Nazi Germany declared war on the United States, and the nation confronted the two-ocean war many had dreaded. In the Atlantic, the main problem for the U.S. Navy was again that of preventing commerce-raiding German submarines from sinking Allied merchantmen. Indeed, the service had been fighting an undeclared war against the U-boat menace for some time, a situation made painfully evident by the torpedoing of two American destroyers in October 1941. In the Pacific, the sudden loss of seven battleships moored at Pearl Harbor meant an end not only to Pacific Fleet commander-in-chief Husband Kimmel's career but also to plans for a decisive fleet engagement against the Imperial Japanese Navy.[110]

The calamity that befell the nation's battleships only magnified the importance of shipboard command and control. Like a chess player who loses a queen, operational commanders had to employ their remaining assets more effectively. Adept coordination of naval forces was key to this effectiveness. As in World War I, convoy escort duty required reliable ship-to-ship voice communications, and the navy's new TBS radios offered a notable improvement over the CW 936s of the late 1910s. During World War II, warships assigned to convoy duty also had access to information derived from decrypted German messages (Ultra intelligence), as shore-based code breakers passed relevant information via radio to seaborne forces.[111] Complementing Ultra intelligence was shipboard high-frequency direction-finding equipment, the first tactical units of which became available in early 1943, and radar. Airborne microwave sets in particular proved invaluable in fighting the U-boat menace.[112]

In the war against Japan, American submarines quickly deployed forward to take the fight into Japanese waters, and in early 1942 carrier task forces carried out several raids in the southwest Pacific. These raids provided valuable combat experience in fighter direction and radar employment, but they also highlighted a number of problems. Few aircraft carried recognition equipment, and the primary means of voice communications remained high-frequency radio, which Japanese forces could easily intercept.[113] The first determined attack against an American carrier took place on 20 February 1942. USS *Lexington* was operating northeast of the Solomon Islands when the CXAM-1 picked up an unidentified airborne contact. *Lexington*'s fighter director officer, Lieutenant Frank "Red" Gill, worked with his radar operator to plot the potential snooper (an enemy plane searching for or shadowing friendly forces) while six Grumman Wildcat fighters scrambled. Four Wildcats circled the carrier while Gill used unencrypted voice radio to move two of them into intercept position. Despite heavy cloud cover, *Lexington*'s fighters soon spotted a Japanese patrol plane. Minutes later, the snooper was a burning wreck on the ocean's surface.[114]

For Gill, directing *Lexington*'s fighters into position for an attack on a single enemy patrol plane was relatively straightforward. Unfortunately, his job was about to get much more difficult. Alerted to *Lexington*'s presence, the Japanese commander on Rabaul launched a raid against the carrier. Of course, American personnel on board *Lexington* knew nothing of this decision. The first sign of trouble came at 3:42 p.m. when the ship's radar operator detected an unidentified contact seventy-six miles due west. It quickly disappeared from the screen. Possibly it was nothing at all. Still, Gill felt uneasy. *Lexington* had protective fighters circling overhead, but these would soon run low on fuel. The fighter director officer thus informed *Lexington*'s commanding officer, Captain Frederick "Ted" Sherman, that he wanted to launch immediately a new group of freshly fueled fighters. Sherman concurred.

As these fighters were taking off, *Lexington*'s CXAM-1 displayed a large blip that again disappeared. Then, at 4:25 p.m. the A-scope showed another spike, forty-seven miles to the west. Five minutes later, the blip on the range scale was at just twenty-four miles. Was the information accurate? Sherman called off aircraft recovery and rang up a flank bell in preparation for evasive maneuvers. Gill went to work, vectoring in six Wildcats. The attack turned out to be genuine, and Gill once again had positioned his fighters well. At 4:39 p.m. the Wildcat pilots started to disperse the raid, downing four enemy aircraft in four minutes. Sherman turned his ship into the wind and launched more fighters, which rose up to join the fight. Gill directed most of them into the fray but held two fighters above *Lexington* in order to give himself a small reserve. His orders undoubtedly displeased the two pilots, who would have wanted in on the action. In the meantime, their brethren pressed the attack. Japanese gunners shot down two American planes but no enemy bombers got through to *Lexington*, the last one being "splashed" miles west of the carrier.

While all this was happening, *Lexington*'s CXAM-1 picked up another unidentified contact thirty miles distant in the opposite direction. Gill may have doubted the contact was another Japanese squadron, but he appears to have hedged against that possibility when he directed two Wildcats to circle overhead. Shortly before 5:00 p.m., he received a report from a destroyer ten miles astern of *Lexington* that lookouts had spotted circling enemy aircraft. Was this an accurate report? The A-scope showed nothing but clutter. Then it showed a spike nine miles to the east. *Lexington*'s lookouts still saw nothing. Finally, some specks. At 5:02 p.m. the carrier's lookouts reported as many as eight Japanese bombers inbound. Gill ordered the two circling Wildcats to intercept; fortunately for *Lexington*, a young naval officer named Edward "Butch" O'Hare was piloting one of the two fighters. When the guns of O'Hare's wingman jammed, O'Hare was left alone in the air to defend *Lexington*. In an amazing display of aerial marksmanship, he singlehandedly shot down three bombers and severely damaged two more. Aided by anti-aircraft fire and Sherman's radical maneuvering, *Lexington* escaped unscathed.[115] Two months later, President Franklin Roosevelt presented O'Hare with the Congressional Medal of Honor.[116]

That *Lexington* suffered no damage from a nearly simultaneous attack by two squadrons of Japanese bombers approaching from opposite directions is rather remarkable. Credit certainly extends to *Lexington*'s pilots, especially Butch O'Hare, and to Sherman, whose skillful ship handling prevented the bombers from getting an easy look at his carrier. Months of prior training played a key role as well. Yet *Lexington*'s narrow escape also underscored problems relating to radar employment. Specifically, distant contacts on the CXAM-1 faded in and out, the radar operator had a difficult time executing a thorough 360-degree search on the A-scope display, and the fighter director officer could not readily distinguish between friendly and enemy aircraft. Given such limitations, Red Gill's performance that day is all the more astonishing.

Indeed, Gill's actions highlight just how important the mental acumen and decision-making capabilities of relatively junior personnel had become. Gill's first critical decision was to launch freshly fueled fighters at the first hint of an incoming air raid. He had time to run this idea past his commanding

Lieutenant Commander Allan F. Fleming performs duties as fighter director officer on board USS *Lexington* (CV-16), November 1943. During World War II, fighter director officers routinely made split-second decisions that could determine the difference between life and death. *National Archives photograph 80-G-431073.*

officer, who considered it prudent and gave the order to do so. Additional Wildcat fighters were thus aloft in time to play a significant role in the battle. Gill's second key decision was to hold back two fighters in reserve. He gave this order without consulting Sherman, who was on the bridge conning the ship. Gill acted on the basis of decidedly uncertain information, yet he sensed intuitively that the tactical situation warranted such a disposition. Sherman himself believed his fighter director officer's actions saved *Lexington* from damage or destruction, and he went out of his way to praise Gill's work.[117]

Unfortunately, some ten weeks later *Lexington*'s good fortune ran out at the Battle of the Coral Sea (7–8 May 1942). Once again Gill was in charge of fighter direction, this time over a two-carrier task force consisting of both *Lexington* and *Yorktown*. On the second day of the battle, American and Japanese scout planes located the other side's carriers at about the same time. The opposing commanders immediately launched attacks against their counterparts; by midmorning each side had dozens of aircraft headed toward the other's carriers. Gill and his assistants, several of whom recently had graduated from Griffin's Fighter Director School, tried to place *Lexington*'s fighters into position to intercept the incoming raid. *Yorktown*'s fighter directors, over whom Gill retained overall authority, attempted to do the same. Regrettably, the successes of 20 February were not repeated. Despite evasive maneuvers, two torpedoes rocked *Lexington* and dive bombers scored hits on both carriers.[118] After valiant yet ultimately unsuccessful damage control efforts, Sherman gave the order to abandon *Lexington*. He was the last individual to leave, later recalling that the signal flags for "I am abandoning ship" still flew from a yardarm as the doomed carrier slipped beneath the waves.[119]

What accounts for the disparate outcomes of February and May 1942? First and foremost, at Coral Sea the Japanese air attack was considerably more formidable than in the earlier battle. The February attack had consisted of seventeen land-based Mitsubishi G4M bombers, infamous for their tendency to burst into flames when hit. At Coral Sea, enemy pilots executed coordinated dive-bomb and torpedo attacks while protected by fighter escorts.[120] The total number of attacking planes, *not* counting the fighters, was three times as many as *Lexington* had encountered off Rabaul. Making matters worse, highly experienced Japanese pilots were flying the attacking planes. To stop this onslaught, Gill had just seventeen Wildcats under his direction.[121]

With such unfavorable numbers against an enemy whose pilots were also highly skilled, the U.S. task force's air defenses would have had to function perfectly in order to escape the attack unscathed. Instead, the battle laid bare numerous weaknesses in the fleet's system of radar employment and fighter direction. Equipment deficiencies topped the list of problems. Only about half the planes were carrying the Naval Research Laboratory's new aircraft recognition device, and at one point during the battle *Yorktown*'s radar stopped functioning.[122] Uncertainty over aircraft identity and the limited number of available fighters led Gill to delay sending out the first wave of interceptors, and when he did, he positioned them too low in the sky.[123]

Other issues also led to confusion within *Lexington*'s Radar Plot. To begin, the pilots did a poor job keeping Gill informed about what they were seeing.

This error stemmed both from a lack of understanding on the part of the pilots as to the overall importance of such information and from congestion on the fighter radio net. Net overcrowding also slowed down communications between Gill and his counterpart on *Yorktown*; perhaps even more problematic was that personnel on the two carriers did not always use the officially approved fighter director vocabulary when talking to each other or to the pilots. On neither carrier were plotting symbols used uniformly, which meant that, when a fighter director or one of his assistants looked at the plot, they might see identical symbols for friendly and enemy aircraft.[124] Exacerbating the general confusion was the size of the radar plots on both ships, which were simply "too small for the complicated job of fighter direction." Writing shortly after the battle, *Yorktown*'s skipper Elliott Buckmaster underscored the problem: "The makeshift Radar Plot in this vessel . . . showed itself, during the air attack on May 8, to be woefully inadequate to enable complete use to be made of all the information which the combined radars of own and other ships are capable of furnishing." Buckmaster went on to recommend that the size and scope of the Radar Plot be expanded significantly.[125]

Somewhat unfairly, Gill received considerable blame for the loss of *Lexington*. Sherman objected to this criticism, arguing that the difficulties encountered at Coral Sea with respect to radar employment and fighter direction had been systemic in nature.[126] Such problems notwithstanding, the fleet achieved a historic victory just four weeks later at the Battle of Midway (4–7 June 1942).[127] Participants in the battle noted some improvements in the employment of radar, but the biggest reason two of three U.S. carriers survived was because American dive bombers hit three of four Japanese carriers before they could execute attacks against the U.S. carriers. This state of affairs meant Japanese forces could muster just eighteen dive bombers and ten torpedo planes for a counterstrike.[128] As events unfolded, these aircraft found *Yorktown*, whose fighter director officer, Lieutenant Commander Oscar Pederson, had witnessed firsthand Gill's performance at Coral Sea. Pederson did his best, but he too was hampered by recognition problems, a congested radio net, inconsistent use of the service's official fighter director vocabulary, and an overcrowded Radar Plot.[129] He nevertheless succeeded in directing many of his fighters into favorable intercept positions, and just seven of eighteen dive bombers in the first wave made it through to *Yorktown*. Some quick thinking by the fighter director on board *Enterprise* nearly saved the day, but in the end a simple communications failure led to disaster.

While Pederson and his team worked frenetically to intercept the incoming raid, *Enterprise*'s fighter director officer, Lieutenant Commander Leonard Dow, was in his own Radar Plot monitoring circumstances. Dow had a division of four Wildcats orbiting in reserve over *Enterprise*, but when he perceived that *Yorktown*'s situation had become critical, he sent them into the fray. He vectored the four Wildcats into position with superb timing; they intercepted four Japanese dive bombers just as those planes were about to push over against *Yorktown*. The division leader, Lieutenant Roger Mehle, led his subordinates in for the kill. Upon reaching the optimal firing position, he pulled the trigger on his Browning machine guns. Nothing happened. Regrettably, an electrical failure had disabled Mehle's gunnery system. He peeled off and radioed the other three planes to continue the attack, but for some reason none of them copied the order. Instead, they followed their leader and broke off the attack. *Yorktown* paid for the miscommunication. One of the Japanese planes scored a devastating hit with a 250-kilogram bomb that penetrated the flight deck and exploded deep inside the ship. *Yorktown* billowed heavy black smoke and slowed to a veritable crawl.

Two other bombs also hit home, but Pederson regrouped as the ship's damage control parties put out fires and conducted repairs. Two hours after the attack *Yorktown*'s speed was up to twenty knots and the ship's crew had extinguished most of the fires. Shortly thereafter, the radar on one of *Yorktown*'s escorting cruisers detected a contact forty-five miles out. The cruiser began to relay continuous reports to Pederson, who had six Wildcats circling overhead. Even before his own radar picked up anything, Pederson sent the fighters down a bearing to intercept the unidentified aircraft. Or so he thought. Employing the navy's official fighter director vocabulary, he ordered the pilots down a true bearing. Unfortunately, they appear to have followed a magnetic bearing, which was about ten degrees off. The Wildcats thus missed the intercept. Pederson successfully vectored in the other two Wildcats, but they encountered fierce opposition from the escorting Japanese fighters. As *Yorktown*'s crew scrambled to launch fighters that still had nearly empty gas tanks, ten Japanese torpedo bombers swooped in for an attack. Two of them found their mark, and Buckmaster eventually had to abandon ship. Three days later, *Yorktown* joined *Lexington* on the bottom of the ocean.[130]

The lessons learned at Coral Sea and Midway led to many reforms. Some reforms focused on new and better equipment, such as better recognition devices and more reliable radios that could operate at super frequencies.

Others involved modifications to existing procedures for fighter protection. For example, the fighter director vocabulary was changed to eliminate confusion over true and magnetic compass bearings, and new tactical doctrine specified that protecting fighters should be "stacked" in altitude (e.g., 5,000, 10,000, 15,000 feet). Such a tactic would require more fighters, of course, so the Pacific Fleet reconfigured its squadrons to increase the number of fighters per carrier by one-third (from twenty-seven to thirty-six).[131]

The value of these reforms was offset somewhat by an erroneous conclusion derived from Midway. In that battle *Enterprise* and *Hornet* escaped attack altogether, leading many naval officers to conclude the best tactical disposition for carriers was to have them operate with their own screening formations. In practice, this arrangement often fragmented fighter protection and made fighter direction more difficult. This truth became evident during the Solomons campaign. At the Battle of the Eastern Solomons (24–25 August 1942) enemy dive bombers inflicted heavy damage on *Enterprise*, and at the Battle of Santa Cruz (26 October 1942) *Hornet* was sunk. In both instances the tactical dispersal of participating U.S. carriers created problems.[132] At Santa Cruz, for example, *Enterprise*'s fighter director became overwhelmed directing planes from both carriers in efforts to protect *Hornet*. After the battle, one squadron commander remarked: "I do not believe that one fighter director can efficiently handle the large number of [fighter] aircraft that were available in the air at the time in a manner that obtains anywhere near the maximum amount of protective value possible from them."[133]

In a nutshell, the first year of war in the Pacific repeatedly demonstrated that the fleet's system of air defense worked well against limited raids but that it broke down during large, coordinated attacks. Writing after the loss of *Hornet*, Pacific Fleet commander-in-chief Chester Nimitz candidly observed that "our fighter direction in both practice and in action against small groups has been good but fighter direction against a [large] number of enemy groups . . . is a problem not yet solved."[134] Basic reforms had led to some improvement, but the nation was running out of carriers. Of the seven available at the start of the war, by the end of October four had been sunk and one was undergoing major repairs. Of the remaining two, *Ranger* was desperately needed in the Atlantic for the U-boat war. Only *Enterprise*—which had suffered heavy damage at the Battle of the Eastern Solomons and had been twice hit by Japanese dive bombers at Santa Cruz—was in operational service.

Unfortunately, combat experience also revealed that difficulties in the employment of radar were not limited to aircraft carriers. The Battle of Savo Island (8–9 August 1942)—the first major Pacific naval battle in which carriers played no material role—was an unmitigated tactical disaster for the Allies. Japanese naval forces successfully slipped past two radar-equipped American destroyers screening against just such an approach, and although several ships' radars picked up unidentified aircraft around Savo Island, this information never reached the officer in tactical command (OTC). During a nighttime engagement in which only one side possessed radar-equipped vessels, Japanese forces managed to sink four cruisers and severely damage three other warships. More than one thousand Australian and American sailors lost their lives.[135]

Other early battles in the Guadalcanal campaign revealed similar problems. At the Battle of Cape Esperance (11–12 October 1942), for example, the first U.S. warship to pick up a suspicious radar contact delayed notifying the OTC, and then it accidentally reported unidentified aircraft instead of unidentified ships. Even worse, after the fighting had commenced, difficulties with radar plotting forced the OTC to order a temporary cease-fire because he feared his forces might be firing at one another.[136] Notwithstanding these issues, the battle ended in a tactical victory for the United States. In a frequently quoted passage, one of the naval officers present at Cape Esperance called it "a three-sided battle in which Chance was the winner," but this quote obscures the fact that radar information, however poorly employed, still had provided American naval forces with a tactical advantage.[137]

Of course, this tactical advantage could easily be ceded to an opponent who had spent years developing special tactics for nighttime engagements.[138] At the First Naval Battle of Guadalcanal (12–13 November 1942), the American commander, Rear Admiral Daniel J. Callaghan, initially aligned his five cruisers and eight destroyers in a single line ahead. There was nothing inherently wrong with this disposition, but Callaghan made two critical mistakes with respect to radar. First, three of his ships were carrying a new type of radar, the model SG, which not only possessed better range than earlier models but also had a new type of display called the plan position indicator (PPI). The PPI provided a bird's-eye view of the surrounding area, making it vastly superior to the A-scope for searches. Somewhat inexplicably, Callaghan put the three vessels carrying SG radars fourth, eighth, and last in line. Second, Callaghan chose not to station himself on USS *Helena*, the only cruiser carrying an SG radar. As the admiral should have anticipated and as events unfolded, *Helena*'s SG radar was the first to detect the enemy. *Helena* consistently fed radar information to Callaghan's flagship via TBS radio, but that net soon became congested. Callaghan appears to have gotten confused and delayed opening fire, which permitted Japanese forces to get off the first shot. Fighting quickly devolved into a melee. Although U.S. forces garnered a temporary strategic victory by preventing Japanese vessels from bombarding the Allied airfield on Guadalcanal, the costs were high. Six American warships went to the bottom and hundreds perished, including Callaghan himself.[139]

Two nights later, the senior American naval commander in the region dispatched a new task force under the command of Rear Admiral Willis "Ching" Lee to repel another Japanese naval movement toward Guadalcanal. Lee had a more thorough understanding of radar than Callaghan did, and although Lee's force was numerically smaller (six ships total), it contained two fast battleships, *Washington* and *South Dakota*. Lee, who sailed on the former, skillfully employed his SG radar to get off the first shots of the battle.[140] *South Dakota* quickly followed suit, but sixteen minutes into the action a faulty circuit breaker tripped and power was lost. The crew worked frantically to restore electricity but the battleship's radar was out of commission for thirteen agonizing minutes. During that time, *South Dakota* inadvertently sailed to within five thousand yards of the enemy. Once the crew had restored power the ship's radar operators and plotters scrambled to regain a coherent tactical picture, but the available time was too short. Enemy shells rained down on *South Dakota*, including three that took out the battleship's radar plot.[141] The loss of radar information severely handicapped *South Dakota*, whose commanding officer later stated: "After this ship lost [its radar] equipment, the psychological effect on the officers and crew was depressing. The absence of this gear gave all hands a feeling of being blindfolded."[142]

In a cruelly ironic twist, the vicious attack on *South Dakota* actually helped clarify the tactical picture for Lee. For more than twenty minutes *Washington*'s radar plotting team had been tracking a large unidentified surface contact, possibly a Japanese battleship. Yet, if the large blip on *Washington*'s radar display was an enemy battleship, why was there not a second large blip representing *South Dakota*?[143] Efforts to clarify the tactical picture via TBS radio failed to convince Lee that the radar contact was not *South Dakota*, so he held fire. Suddenly, a Japanese vessel illuminated *South Dakota* with searchlights. Lee had the final piece of information he needed; seconds later his flagship opened fire on Japanese battleship *Kirishima*.[144] According to one recent analysis, *Washington* hit *Kirishima* with as many as twenty sixteen-inch shells before breaking off the action.[145] The badly battered *South Dakota* escaped further harm, and *Kirishima* went to the bottom. Lee's tactical employment of radar, while not perfect, nevertheless had been close to exemplary.[146]

Most accounts of this Second Naval Battle of Guadalcanal (14–15 November 1942) praise Lee's actions, and deservedly so. Often overlooked yet equally important, though, was how well the *Washington*'s personnel processed, interpreted, and disseminated available radar information. The radar plotting team correctly interpreted what they saw, a hard task given the prevalence of "land echoes," and the Radar Plot officer consistently provided accurate information to the main battery plotting room through the ship's sound-powered phone circuits.[147] And while Lee familiarized himself with the tactical situation by examining the SG radar's PPI before heading to the bridge, he also made sure one of the ship's radar experts was there with him during the battle.[148] A few weeks after the battle, *Washington*'s commanding officer, who had stood on the bridge with Lee that night as well, made a telling observation regarding the changed nature of command at sea: "Radar has forced the Captain or O.T.C. to base a greater part of his actions . . . on what he is told rather than what he can see."[149] Many other American naval officers agreed with this sentiment, but what could be done to aid captains and OTCs in their decision making? One possible solution was to provide them with a second brain, a human-machine brain that could collect, organize, process, evaluate, and disseminate relevant combat information. In 1943, that idea would become a reality.

Like most inventions, the Combat Information Center did not stem from any single eureka moment. Indeed, for decades the U.S. Navy had built and experimented with different kinds of shipboard stations for processing vital information. In the late nineteenth century, North Atlantic Station commander Francis Bunce modified bridges and signal houses to improve the efficacy of his ships' signal flags and Ardois lights. At the start of the twentieth century, John Hudgins, Daniel Wurtsbaugh, and others tried to figure out where best to locate shipboard radio rooms. The navy began to install plotting rooms for its gunnery systems in the mid-1910s, and after the First World War senior commanders like Louis Nulton and Samuel Robison pushed for better command facilities on board flagships. During the late 1930s, the service experimented with seaborne control and tracking centers for radio direction-finding intercepts, and the adoption of radar led innovative officers such as Henry Bernstein and Elliott Buckmaster to create the Radar Plot. Yet shipboard personnel utilized each of these facilities to manage just one type of information. Combat experience in the Pacific led many American naval officers to conclude that there was an urgent need for a shipboard facility that could process *all types* of tactical information.

The formal decision to create such facilities dates to late 1942 and originated within the headquarters of Pacific Fleet commander-in-chief Chester Nimitz. The admiral's staff was still surprisingly small, numbering just fifty or so officers.[150] Yet several individuals had been working diligently on a major revision to the fleet's existing radar doctrine. Issued on Thanksgiving Day, Pacific Fleet Tactical Bulletin 4TB-42 stated that every warship needed a facility to process "all available sources of combat intelligence." Heeding a recommendation put forward by his staff, Nimitz designated the new facility a Combat Operations Center. Tactical Bulletin 4TB-42 specified that the Combat Operations Center would be a place where "information from all available sources can be received, assimilated, and evaluated with a minimum delay." It went on to state that the evaluation "of all available information by trained personnel" would allow for the quickest dissemination of such information to "flag and commanding officers, to other control stations concerned over interior communications circuits, and to other ships and aircraft via external communications facilities."[151]

According to one postwar report, Tactical Bulletin 4TB-42 "made a profound impression" on senior officers back in Washington, including U.S. Fleet commander-in-chief (COMINCH) Ernest J. King.[152] Throughout the Guadalcanal campaign, King's readiness division had been receiving action reports from that theater, so it was well informed regarding the performance of ships' radars and radar plots. Because there were clearly some problems, King asked the Bureau of Ships and the Bureau of Ordnance to study "such inadequacies of radar plot as might exist and to obtain information upon which better designs might be based." To discuss both the bureaus' findings and Tactical Bulletin 4TB-42, King's readiness division also organized a conference on the Combat Operations Center concept.[153]

The conference convened in Washington on 8 January 1943; representatives from COMINCH, OPNAV, and the relevant bureaus attended.[154] The attendees embraced the Pacific Fleet's idea of a Combat Operations Center and concluded that a minimum of four plots should always be present: one for navigation, one for air contacts, one for surface contacts, and a spare that might potentially be used for fighter direction. Conference attendees also identified other critical equipment, including interior communications circuits, bearing indicators, remote TBS controllers, clocks with sweep second hands, and PPIs.[155] King was in Casablanca at the time this conference was held, but upon returning to Washington he became involved and directed the vice chief of naval operations to take action on the matter. King's directive, drafted by the COMINCH staff, laid out all of the requisite features for the new shipboard station. It also made one significant revision that King himself must have approved, changing the facility's name from Combat Operations Center to Combat Information Center.[156]

In addition to renaming the new shipboard facility, King's directive authorized the installation of CICs on board all surface combatants. COMINCH delegated many of the details for these installations to the Pacific Fleet, which in turn worked closely with the Bureau of Ships and the Pearl Harbor Navy Yard to develop and construct CICs. COMINCH also laid down some explicit parameters, two of which were particularly important. The first dealt with size. COMINCH specified that whenever possible CIC plans should "provide sufficient space to permit as much as 30% expansion in functions and equipment."[157] The second stricture involved location. On most vessels radar plots were located near the bridge so that relevant information would be "readily and visibly available to the Commanding Officer."[158] The larger relative size of CICs would have made a continuation of this practice difficult, but the issue became moot after the Second Naval Battle of Guadalcanal. The loss of *South Dakota*'s Radar Plot during that battle had blinded the ship and led both the Pacific Fleet and the Bureau of Ships to recommend placing CICs in safer locations.[159] King codified this sentiment when he directed that CICs be placed in protected locations on all newly constructed ships. For battleships and heavy cruisers, this meant behind the armor; for aircraft carriers and destroyers, it meant somewhere below decks. The service retrofitted many older warships and even redesigned the layout of the *Essex*-class carriers to move the CIC from the island structure to a more protected location.[160]

Once COMINCH had established a basic framework for CIC development, the navy's locus of effort shifted largely to the Pacific. And as had been the case for many previous innovations in shipboard command and control, a combination of senior and junior personnel led the way. Throughout 1943, the senior officer who most influenced CIC development was Rear Admiral Mahlon S. Tisdale. An Illinois native, Tisdale was a 1912 graduate of the Naval Academy. He served in a variety of billets over the course of his career, but two in particular gave him a strong background in shipboard command and control systems. In the early 1920s, Tisdale had served as a battleship division radio officer. More critically, in 1940 Tisdale had been in command of *Chester* when that cruiser received one of the first CXAM radars. When the Japanese attacked Pearl Harbor, Tisdale was on shore duty, but six months into the war the Navy Department awarded him command of a Pacific Fleet cruiser division. He saw action at the Eastern Solomons, the Santa Cruz Islands, and Tassafaronga; during the last of these battles, on 30 November 1942, Tisdale actually took over as the OTC after two torpedoes severely damaged the task force flagship.[161]

Tassafaronga was a devastating tactical defeat for the U.S. Navy. The service soon thereafter sent the task force commander to a desk job, but Tisdale emerged from the battle with his reputation essentially intact.[162] Nevertheless, the ghosts of Tassafaronga surely lingered in Tisdale's mind when he assumed command of the Pacific Fleet's destroyers in early 1943. Fortuitously for the admiral, his new staff included an officer who for several months had been investigating the fleet's use of radar information in combat. That individual was Caleb B. Laning, who had some innovative ideas about how best to integrate the CIC into wartime operations. Within months, Tisdale, Laning, and several others would put into place the basic design of the World War II Combat Information Center.

Born on 21 March 1906 in Kansas City, Missouri, Laning was a gifted athlete who enlisted in the navy at the age of eighteen. He garnered an appointment to the Naval Academy after one year of service, earning a commission in the spring of 1929. A technically astute officer, Laning spent some time during the early 1930s on special assignment at the Ford Instrument Company, where he studied fire control and anti-aircraft systems. In 1933 Laning qualified in submarines, and from 1936 to 1938 he studied applied communications at the navy's postgraduate school in Annapolis, Maryland. Like Tisdale, Laning served as a divisional radio officer in the Pacific and possessed combat experience, participating in the Battle of Midway as the executive officer of a destroyer. In October 1942, the Navy Department detailed Laning to the staff of the commander of the Pacific Fleet's destroyers, mainly so he could aid in the development of improved shipboard communications systems.[163]

During the late winter and early spring of 1943, Laning worked diligently to draft plans for optimal CIC arrangements on the various different classes of U.S. destroyers. Laning knew he needed more manpower to accomplish this task, so he combed through action reports and identified personnel who had performed especially well as radar plot officers during the Guadalcanal campaign. Tisdale made sure the Navy Department sent these handpicked individuals to his staff, where Laning quickly put them to work. One junior officer who greatly aided Laning in solidifying his design philosophy for the CIC was J. (Joseph) C. Wylie.[164] Laning and Wylie conceived of the CIC as a square divided into four parts. The aft section of the square was for current information, the forward portion for historical data. Radar and sonar scopes were thus positioned abaft, while plots and status boards went forward. Laning also divided the CIC into port and starboard halves, with one side devoted to aircraft-related activities and the other dedicated to surface and submarine contacts.[165]

In April Tisdale formally submitted to COMINCH detailed specifications for destroyer CICs, calling for prompt action by emphasizing the recent "unprecedented rate of change of combat tactics."[166] Tisdale's counterpart in the Atlantic sought minor modifications related to anti-submarine warfare operations but agreed with the basic principles contained in Laning's proposal. The Bureau of Ships reviewed the proposal as well and recommended to COMINCH the installation of Combat Information Centers on most destroyers in the Pacific theater. In early June, COMINCH approved the bureau's recommendation, directing that every American destroyer of the *Farragut*-class and later be equipped with a CIC.[167]

One week after COMINCH issued this directive, Nimitz promulgated operational guidance for the CIC, telling his subordinates to push forward in organizing teams that would be proficient in maintaining a continuous summary of the tactical information from all sources and in disseminating quickly evaluated data. The Pacific Fleet commander-in-chief's guidance was general in nature, in part because he and his staff recognized that different types of warships (i.e., carriers, battleships, cruisers, and destroyers) had different missions. In addition, they understood that space constraints on smaller vessels precluded a one-size-fits-all approach. Nimitz's staff thus concentrated on the functions of the CIC—that is, on its role as an information station. Absent from the Pacific Fleet commander-in-chief's guidance was a discussion of the CIC's potential role as a cognitive aid to naval commanders.[168]

Tisdale and Laning believed such a discussion was essential, so much so that they took matters into their own hands. In early to mid-May they tasked

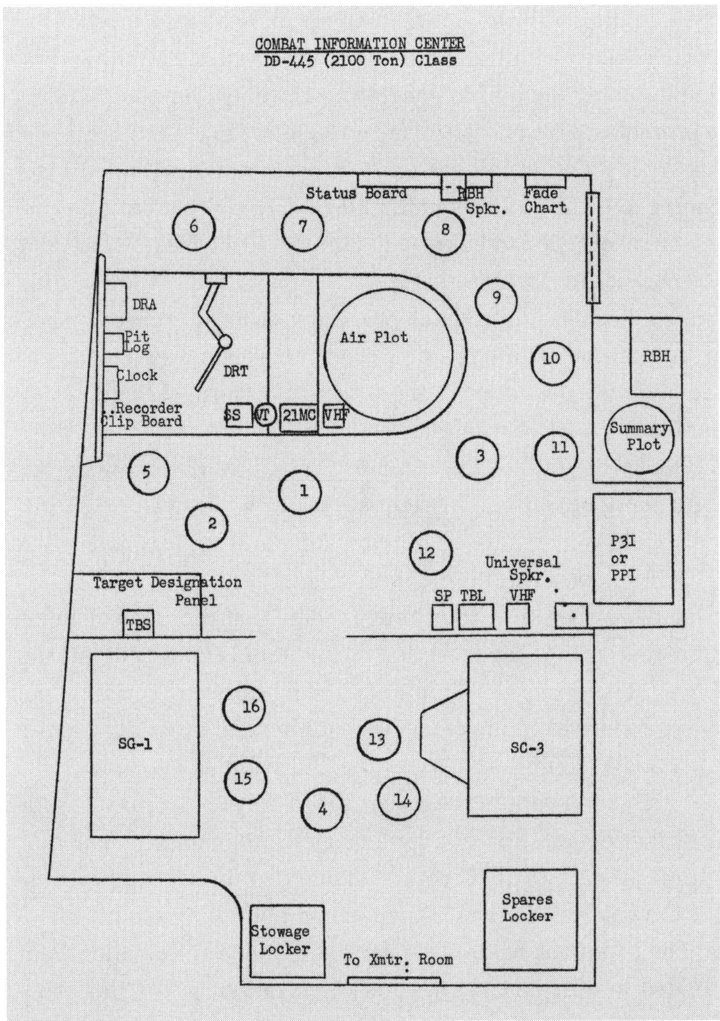

Schematic of a *Fletcher*-class Combat Information Center. The numbered circles represent people. Nos. 1–4 are officers, nos. 5–16 are enlisted personnel. The aft part of the CIC contained both surface search (SG) and early warning (SC) radars, while the forward section had plots for air and surface contacts. A dead-reckoning tracer (DRT) automatically marked own-ship's movement on the surface plot. Despite the best-laid plans of Caleb Laning and J. C. Wylie, space constraints on most *Fletchers* forced the placement of sonar equipment just outside the CIC. *Reprinted from Commander, Fleet Operational Training Command, Pacific to All CIC Activities, 2 May 1945, filing designator A7-3, Confidential Correspondence, 1945, Headquarters of the Commander-in-Chief, U.S. Fleet, Records of the Office of the Chief of Naval Operations (Record Group 38), U.S. National Archives, College Park, Maryland.*

J. C. Wylie to draft a document that would serve as a comprehensive statement of CIC doctrine for the Pacific Fleet's destroyers.[169] Wylie embraced the assignment, taking the lead in generating a twenty-four-page handbook that Tisdale promulgated to his destroyers on 24 June 1943. The admiral also made sure that Wylie's "CIC Handbook for Destroyers" was distributed widely, with copies going to Nimitz and his staff, COMINCH/CNO, other type commanders, the Bureau of Ships and Bureau of Ordnance, the Fleet Radar Center, and select commands in the Atlantic Fleet.[170]

Although Tisdale's "CIC Handbook" focused on destroyer operations, it was the first official document to provide a thorough overview of what CICs could do and how personnel in them should perform their duties. As such, the handbook quickly became an essential publication for all operational commands; by the end of the summer thousands of copies had been produced and distributed throughout the service.[171] Wylie, who later authored an influential Cold War treatise on military and naval strategy, jokingly referred to the "CIC Handbook" as his first best seller.[172]

The handbook began by informing readers that the Combat Information Center created "one of the most drastic and rapid changes in our shipboard experience." Its purpose was simple and straightforward: to assist commanding officers and naval commanders in planning and executing courses of action. For Tisdale, Laning, and Wylie, the CIC had two main functions. First, it was an evaluation center, the place "to which the captain turns for a specific tactical or operational fact, for a general summary, or for an opinion or suggestion." Second, the CIC could assume certain control functions, such as target designation, tactical maneuvering, fighter direction, and weapons control. The handbook made clear that the Combat Information Center was not intended to usurp a captain's authority; rather, it was there to provide "clarification and simplification of work for the Command."[173]

One important postwar analysis of the CIC program argues that some naval officers objected to the idea that the control function of command should be delegated.[174] Yet such objections were based more on philosophical debates about the nature of command at sea than upon real-world experiences, a point stressed by the "CIC Handbook" when it stated that the CIC was not "a child of theory" but rather "a proven result of combat experience."[175] Indeed, the final two years of the Second World War would demonstrate that the CIC offered an invaluable tool for dealing with the complexities of warfare at sea. For both commanding officers and naval commanders, it clarified the tactical situation and facilitated quick decision making. For example, during the Battle of Vella Gulf (6–7 August 1943) and later the Battle of Cape Saint George (25 November 1943), destroyer division commanders Frederick Moosbrugger and Arleigh Burke skillfully employed their CICs to achieve important American victories.[176]

Meanwhile, in the Atlantic the navy moved swiftly to incorporate the CIC into anti-submarine warfare operations. Throughout 1942 convoy escorts had used radar to help combat the U-boat, but during 1943 the service gained a host of new tools to help fight the Battle of the Atlantic: good high-frequency direction finders finally became available after years of development, better airborne radars allowed Allied forces rapidly to search huge swaths of ocean, new weapons like the acoustic homing torpedo made Allied attacks more lethal, a reallocation of resources provided long-range B-24 aircraft to patrol the mid-Atlantic gap, and Admiral King created the Tenth Fleet to coordinate anti-submarine warfare activities and conduct statistics-based research that might help win the U-boat war. Of no less importance, in the spring of 1943 the navy began to deploy escort carriers along with destroyers and destroyer escorts specifically configured for anti-submarine warfare. These hunter-killer groups, as they were called, created a huge problem for U-boats because Allied successes at breaking encrypted German messages frequently permitted hunter-killer groups to search specific ocean areas.[177]

With six possible sources of information—own-ship's radar, lines-of-bearing from direction finders, own-ship's sonar, radio reports from patrol aircraft or other vessels, Ultra intelligence, and old-fashioned visual sightings—the CIC was a natural fit for the Battle of the Atlantic, something that did not go unnoticed by American naval officers. Early in 1943, Atlantic Fleet commander-in-chief Royal Ingersoll arranged a series of conferences with the British Admiralty Delegation and the relevant U.S. Navy bureaus to discuss the issue of CICs on escort carriers. The conferees soon arrived at a consensus that allowed the Bureau of Ships to issue standard CIC plans for the *Bogue*-class escort carriers then entering service. Several months later, COMINCH authorized installation of a standard CIC on the navy's new *Casablanca*-class escort carriers.[178]

The Allies already had turned the tide in the Battle of the Atlantic before the navy could complete installing CICs on its escort carriers, destroyers, and destroyer escorts, but from late 1943 until the end of the war the CIC was an important part of the Allied anti-submarine warfare effort. The escort carrier's

CIC functioned as the hunter-killer group's nerve center, receiving, processing, and distributing important information both to accompanying destroyers and destroyer escorts and to shore-based operations centers located on both sides of the Atlantic. For example, two days before Christmas 1943 a hunter-killer group led by escort carrier USS *Card* received Ultra intelligence indicating the presence of U-boats some eighty-five miles away. Task group commander Arnold J. "Buster" Isbell recognized the risks but nevertheless decided to proceed into what he knew might be a wolves' den.

A little after nine o'clock in the evening, *Card*'s high-frequency direction finder picked up an enemy transmission; fifteen minutes later, *Card*'s radar operator gained a corresponding contact. Isbell sent one of his destroyers, USS *Schenck*, to investigate, but the blips on both vessels' radar screens disappeared. The U-boat had submerged. *Schenck* then got a good sonar contact and dropped a pattern of depth charges. No luck, but the attack prompted Isbell to commence evasive actions with his carrier. A good thing, too, because postwar analysis revealed that around that time a U-boat fired three torpedoes at *Card*. They all missed. Over TBS Isbell ordered another escort, USS *Leary*, to aid *Schenck*. In the ensuing battle *Schenck* destroyed one U-boat but another one torpedoed *Leary*, which sank despite the crew's valiant damage control efforts. *Card*'s CIC had enabled Isbell to monitor the entire battle, and he promptly ordered his carrier back to the scene to aid in rescue efforts. More than one-third of *Leary*'s crew was saved from the frigid Atlantic waters.[179]

While the CIC was instrumental to Allied successes during the last two years of the Battle of the Atlantic, probably the most impressive display of the CIC's capabilities came during the Battle of the Philippine Sea (19–20 June 1944). Historical accounts of the battle, in which the United States inflicted a crushing defeat on the Imperial Japanese Navy, tend to highlight the quantitative and qualitative material advantages of American naval forces. Without question these advantages were substantial. The U.S. task force commander, Marc A. Mitscher, was in charge of fifteen carriers, ninety-seven surface combatants, and more than nine hundred aircraft, while his Japanese counterpart, Jisaburo Ozawa, had command of nine carriers, forty-six surface combatants, and fewer than five hundred aircraft.[180] American forces also consisted of planes, ships, ordnance, and radars technologically superior to anything the Japanese could bring to bear. Although rarely mentioned by historians, another critical U.S. advantage was the CIC, which enabled Mitscher's task force to defend itself against air attack with an effectiveness that simply would have been unattainable earlier in the war.[181] Of more than 370 aircraft launched by Ozawa, roughly two-thirds never returned, and very few succeeded in getting near the American carriers.[182] One exuberant aviator likened the action to "an old-time turkey shoot,"[183] but Mitscher's less colorful analysis was more revealing: "It [the battle] was the first time that a major enemy air blow has been made on our forces without loss or serious damage to one of our carriers. It proved that the long and costly efforts in research, training, and the practical application of radar have not been in vain."[184]

Mitscher's mention of training is particularly telling. The development of the CIC had created a pressing need for skilled personnel to man the new facility, which held roughly fifteen men on a destroyer and more than double that number on a battleship or an aircraft carrier. One of the most important members of any CIC team was the fighter director officer, and in the spring of 1942 the Navy Department had moved Jack Griffin's Fighter Director School from San Diego to Pearl Harbor so it would be closer to the Pacific Fleet. Nevertheless, wartime manning issues limited the number of people the Bureau of Naval Personnel elected to send to the school.[185] While some warships sent personnel for refresher training, over the course of the summer only thirty-two new officers received instruction.[186] Griffin himself detached from the school in early fall to serve on the staff of Rear Admiral Thomas Kinkaid, who made Griffin the fighter director officer on his flagship, USS *Enterprise*. In that capacity Griffin participated in the Battle of the Santa Cruz Islands before returning to Pearl Harbor in early 1943.[187]

Griffin returned to a school that was, in his opinion, underutilized by the service. Senior officers agreed, in part because feedback from the fleet indicated a need to integrate seamlessly fighter director officers into the new CIC.[188] In March 1943 the vice chief of naval operations noted the "urgent and ever increasing demand for officers with fighter director training," and in April Chester Nimitz established a single facility to handle all radar training in the Pacific. Nimitz named the new facility the Pacific Fleet Radar Center and selected Griffin to serve as the officer in charge.[189] Located just north of Pearl Harbor at Camp Catlin, the Pacific Fleet Radar Center consolidated all major Pacific Fleet radar training programs under one command. Owing to the "obvious" interdependence between fighter direction and the functions of the CIC, Nimitz changed the name of Griffin's old school to the Fighter Director and Combat Information Center School.[190]

For the Pacific Fleet commander-in-chief and his staff, the purpose of the Fighter Director and Combat Information Center School was to instruct naval officers "in all phases of radar utilization and control, but specifically in the organization and operation of Combat Information Centers, and in the technique of fighter direction and evaluation of radar information in surface tactics." Initially, the new school offered three regular courses, ranging from two to six weeks in duration. The longest course was CIC Indoctrination, which provided "a thorough understanding of all duties and responsibilities of members of Combat Information Center teams." Instructors covered a wide range of subject matter, including plotting, navigation, radar theory and operations, communications, and tactics. The school also offered courses in Advanced Fighter Direction and Advanced Evaluation, as well as an ad hoc Special Intensive Course for officers of ships in port at Pearl Harbor.[191]

The advanced courses were available only to "superior" students who demonstrated "outstanding promise."[192] Indeed, course policies from mid-1943 reveal clearly that American naval leaders understood CICs had to be manned by personnel who could succeed in a stress-filled world where even small mistakes or minor delays could lead to disaster. Pointing to the necessary attributes of fighter directors and CIC watch officers, the school sought self-confident individuals who possessed "inherent natural ability," could routinely work "at peak efficiency," and who had "an ability to sift a vast amount of information and quickly form a correct opinion and then give well defined orders."[193] Personnel at the Pacific Fleet Radar Center also established optimal criteria for radar operators, giving potential students both a basic intelligence test and a special "Radar Information Test."[194]

Despite the navy's efforts to screen out weak candidates before they arrived at Camp Catlin, attrition was high. Less than three-quarters of officers sent to the Fighter Director and Combat Information Center School graduated; attrition at the Radar Operators School likely was similar.[195] To address the attrition problem and other CIC-related matters, in 1944 the Pacific Fleet held three special conferences on the topic.[196] Even before the first of these convened, however, Jack Griffin obtained approval from Nimitz's staff to modify the Fighter Director and Combat Information Center School's curriculum. Griffin felt greater flexibility was essential if more fleet personnel were to attend his school. Under the original curriculum classes convened once per month, but Griffin and his staff rearranged things so that the school could receive a new class every week. To facilitate this change, Griffin took the six-week CIC Indoctrination course and broke it into three separate courses, which fleet personnel could take in any order. The school retained its three advanced courses, modifying the Special Intensive Course into one in which commanding, executive, and other senior officers could study recent action reports and new literature on CIC design and developments.[197]

The new curriculum at the Fighter Director and Combat Information Center School also included three courses for team training. These courses utilized specially built mock CICs and gave senior officers a way to exercise their CIC teams under simulated combat conditions.[198] This development was noteworthy, for it allowed naval personnel to maintain operational proficiency even when their ships were in port. Indeed, the idea was so well received that the service took steps to establish CIC team trainers for precommissioning crews throughout the United States. The first of these facilities opened on San Clemente Island in October 1943; others followed in places ranging from Astoria, Oregon, to Hollywood, Florida.[199] In the spring of 1944, COMINCH began requiring all prospective commanding and executive officers to receive special instruction "in the functions and use of the Combat Information Center," a policy later extended to include navigators and gunnery officers.[200] By mid-1944, the CIC team trainer had become an integral part of nearly every ship's predeployment workup.[201]

American victory at the Battle of the Philippine Sea showcased the extent to which the U.S. Navy had created a successful system of fleet air defense. The problem of carrier vulnerability, which had been of grave concern to American naval officers for some two decades, seemed like it had been solved. The apparent solution derived not only from the technological superiority of Allied ships, planes, and weapons but also from a new kind of human-machine artifact, the Combat Information Center. Yet success was fleeting. At the Battle of Leyte Gulf (23–26 October 1944), Japanese forces began employing a new tactic, the kamikaze attack.[202] For various reasons, kamikazes placed renewed strain on the fleet's system of air defense. To begin, kamikaze aircraft usually approached singly or in small groups at extremely low altitudes, which made them harder to detect at long ranges. As human-guided missiles, they were more accurate than either torpedoes or bombs. Finally, whereas a traditional air raid typically approached from one or two threat axes, a well-coordinated kamikaze raid came in from multiple directions.[203] As one fighter director officer later recalled, the kamikaze really pressured CIC personnel to make "split-second decisions" in an effort "to save a precious moment or two."[204]

Nowhere were the service's shipboard command and control systems more severely tested than in the spring of 1945 off the island of Okinawa. During that three-month campaign, Japanese forces sank thirty-two American warships, damaged over a hundred more, and killed almost five thousand sailors.[205] Such heavy losses might suggest that the fleet's system of air defense nearly failed, but this was not the view of most on-scene commanders. Flag officers Joseph J. "Jocko" Clark, Ted Sherman, and Allan E. Smith all praised the work of their CIC and fighter director teams, with Smith concluding that the "timely information" provided by screening vessels had been "a most important factor in the security of the entire operation."[206]

Some commanders were less satisfied with the fleet's performance against the kamikaze threat. While Smith felt that losses were "reasonable" given the circumstances,[207] Marc Mitscher, who led the fast carrier task force supporting the Okinawa campaign, voiced displeasure over several equipment problems related to shipboard command and control. According to Mitscher, existing radars too often provided inadequate warning against low-flying planes, the fleet's standard aircraft identification devices worked poorly in congested environments, and there were too many cases of interference between different types of electronic equipment.[208] Mitscher acknowledged as well that a large part of the problem derived from the skill and tenacity with which Japanese attackers exploited these deficiencies. Richmond K. Turner, who commanded the amphibious landing force assaulting Okinawa, expressed similar sentiments.[209]

Although Mitscher and Turner criticized the performance of some pieces of equipment, they lauded the work of others. Mitscher praised the service's new Mark 51 gun directors and the efforts of his radar-equipped night

CIC chart from USS *Hornet* illustrating some of the chaos of 6 April 1945, the date of the first and largest of ten massive kamikaze raids against the U.S. fleet during the Okinawa campaign. The chart reveals that, while American combat air patrols shot down most attackers, a few had to be felled by anti-aircraft fire, and one leaked through to score a hit on destroyer USS *Haynsworth*. *Reprinted from enclosure (e) to Commander Task Group 58.1 to COMINCH, 5 May 1945, box 218, World War II Action and Operations Reports, Records Relating to Naval Activity during World War II, Records of the Office of the Chief of Naval Operations (Record Group 38), U.S. National Archives, College Park, Maryland.*

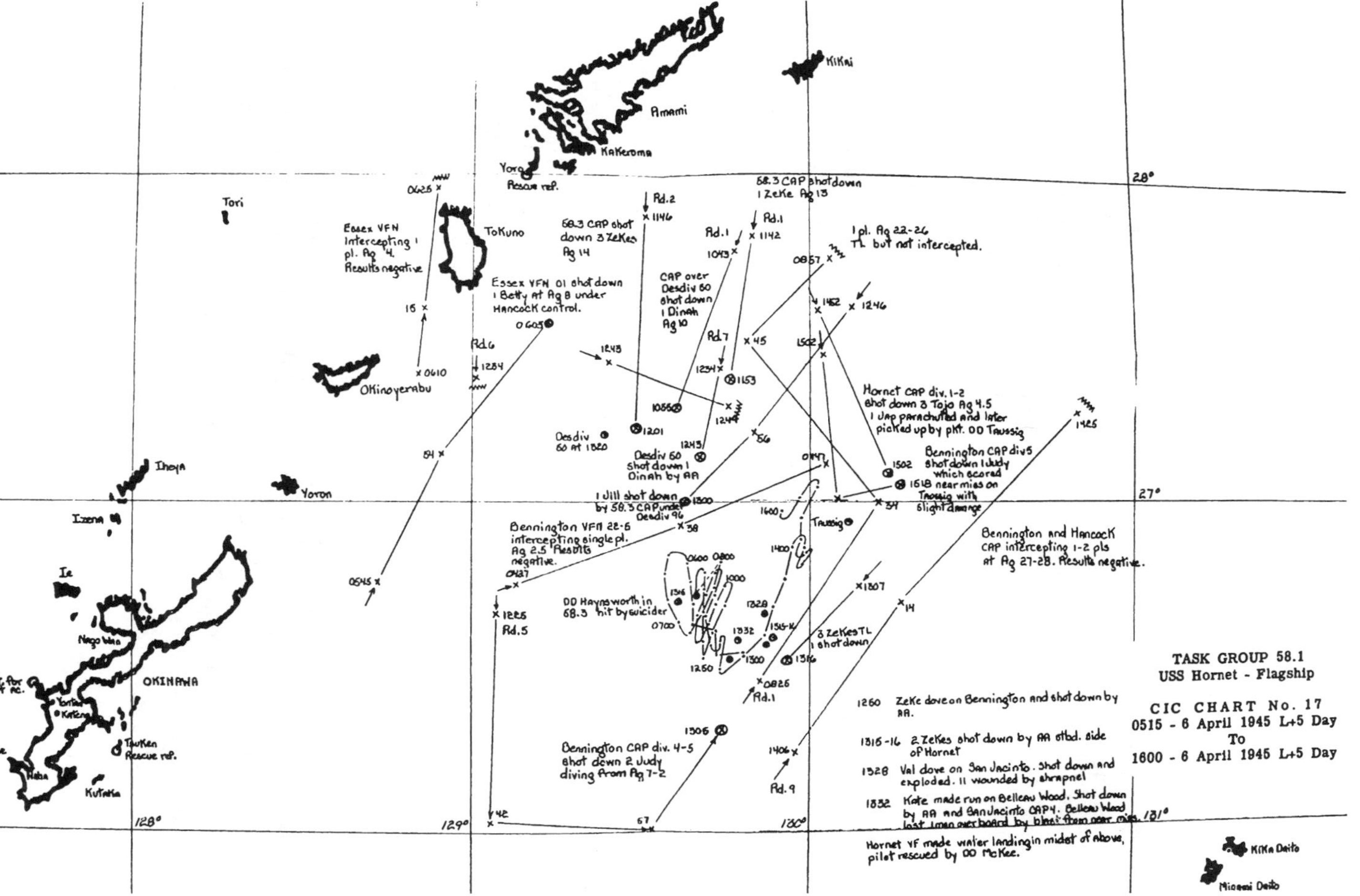
TASK GROUP 58.1
USS Hornet - Flagship
CIC CHART No. 17
0515 - 6 April 1945 L+5 Day
To
1600 - 6 April 1945 L+5 Day
1260 Zeke dove on Bennington and shot down by AA.
1315-16 2 Zekes shot down by AA stbd. side of Hornet
1328 Val dove on San Jacinto. Shot down and exploded. 11 wounded by shrapnel
1332 Kate made run on Belleau Wood. Shot down by AA and San Jacinto CAP4. Belleau Wood lost 1 man overboard by blast from near miss.
Hornet VF made water landing in midst of above, pilot rescued by DD McKee.
Amami
Kakeroma
Kikai
Tokuno
Okinoyerabu
Yoron
Tori
Iheya
Izena
Ie
OKINAWA
Nago Wan
Kutaka
Naha
Kikai Daito
Minami Daito
28°
27°
128°
129°
130°
131°
Essex VFN intercepting 1 pl. Ag 4. Results negative
Essex VFN 01 shot down 1 Betty at Ag 8 under Hancock control.
58.3 CAP shot down 3 Zekes Ag 14
58.3 CAP shot down 1 Zeke Ag 13
1 pl. Ag 22-26 TL but not intercepted.
CAP over Desdiv 60 shot down 1 Dinah Ag 10
Desdiv 60 at 1320
Desdiv 60 shot down 1 Dinah by AA
1 Jill shot down by 58.3 CAP under Desdiv 96
Bennington VFN 22-6 intercepting single pl. Ag 2.5 Results negative.
Hornet CAP div. 1-2 shot down 3 Tojo Ag 4.5 1 Jap parachuted and later picked up by pkt. DD Taussig
Bennington CAP div 5 shot down 1 Judy which scored 1518 near miss on Taussig with slight damage
Bennington and Hancock CAP intercepting 1-2 pls at Ag 27-28. Results negative.
DD Haynsworth in 58.3 hit by suicider
3 Zekes TL 1 shot down
Bennington CAP div. 4-5 shot down 2 Judy diving from Ag 7-2
Taussig
Rd.1
Rd.2
Rd.5
Rd.6
Rd.7
Rd.9

fighters, described by one subordinate commander as "outstanding and highly successful."[210] Turner extolled his task force's communications plan, pointing out that recently added TBS channels were invaluable and characterizing overall circuit discipline as "excellent."[211] Other key elements in defending against kamikaze attacks included reliable anti-aircraft guns like 40mm Bofors, the widespread availability of proximity fuses, and Nimitz's decision to force the Twentieth Air Force to shift its focus from strategic bombing to airfields in southern Japan.[212] As one on-scene naval commander pithily noted, "The only sure way to prevent successful suicide attacks is to keep the planes from taking off."[213]

Yet probably the most critical anti-kamikaze technique employed by the navy during the Okinawa campaign was its use of radar picket ships, which were networked with surrounding naval forces through CICs and other shipboard command and control systems. Commanders positioned such vessels around a force's perimeter in order to give high-value units (e.g., carriers and troop transports) advanced warning of approaching enemy craft and to direct fighters into intercept position. The use of destroyers for radar picket duty appears to have originated in the mind of Caleb Laning, who shared it with his boss, Mahlon Tisdale, around the spring of 1943.[214] Tisdale liked the idea and discussed it with Jack Griffin, by then in charge of the Pacific Fleet Radar Center. Griffin loaned Tisdale two of his best officer students, who went to sea on destroyers and directed fighter intercepts during a task group training exercise.[215] Although far from flawless in execution, these intercepts demonstrated the feasibility of destroyer-based fighter direction and led to the promulgation of a fighter direction manual "for the use of destroyer, cruiser, and other surface vessel officers who may be designated as Fighter Director Officers."[216]

The emergence of the kamikaze threat in the fall of 1944 dictated a renewed focus on the fleet's system of air defense, and the service rapidly organized a series of exercises to identify weak spots the Japanese might find and exploit. Carried out in the South Pacific under the code name Moosetrap, American pilots simulated kamikaze attacks while the fleet tested various intercept procedures. Moosetrap revealed that radar picket ships, mainly destroyers but also some cruisers, should be an essential part of any air defense plan.[217]

At Okinawa, Turner established a ring of sixteen radar picket stations around his landing forces.[218] Praise for the performance of the ships on those stations, especially their CIC and fighter director teams, was uniformly positive.

One on-scene commander commented that the radar picket ships had "destroyed an impressive number of enemy planes," while another noted that "the destroyer pickets were outstanding in picket history, both because of the advance information furnished on enemy planes and because of the numerous successful actions against enemy dive bombers, torpedo planes, buzz bombers, and suiciders."[219] Mitscher told his superiors that the "value of the picket group cannot be too strongly emphasized," although a junior officer on one his carriers chose more heartfelt words: "Thank God for the Pick Line."[220]

Unfortunately, the pickets themselves paid a heavy price for guarding the fleet's carriers, transports, and amphibious assault ships. The most reliable figures available indicate that of the 206 warships that served as radar pickets around Okinawa, 15 were sunk and 45 were damaged. More than thirteen hundred sailors died and nearly another sixteen hundred were wounded.[221] In hindsight, the planners of the Okinawa operation made a huge mistake when they rejected the idea of capturing several outlying islands and installing early-warning radar stations on them. Doing so almost certainly would have reduced the number of warships needed for radar picket duty.[222]

In spite of such tragic losses, one is hard pressed to conclude anything other than that the shipboard command and control systems employed by the navy at Okinawa saved both lives and materiel. Fortunately for countless American servicemen and their families, soon after Okinawa two atomic bombs and a Soviet declaration of war rendered the planned invasion of Japan moot. Yet, even as humankind's most deadly conflict ended, American naval leaders recognized that fleet air defense would be considerably more difficult in a postwar world of jet aircraft and unmanned guided missiles. Indeed, the evolution of shipboard command and control during the Cold War would make for a fascinating study of its own.

A few months after World War II ended, the Office of the Chief of Naval Operations decided to publish a confidential synopsis of the CIC "for the uninitiated who want information on the what and why" of the navy's relatively new shipboard facility. OPNAV described the Combat Information Center as "a weird and eerie jungle of electronic gear, illuminated tables, shining dials and gadgets," with "officers and enlisted personnel . . . wearing earphones and talking strange jargon into microphones, telephones, and squawk boxes that seem to be constantly flashing red lights." Employing a touch of hyperbole, OPNAV claimed the CIC made "Flash Gordon look like a piker and Buck

Rogers an anachronism."[223] Indeed, to many individuals the CIC must have seemed like something from the future. Perhaps even Caleb Laning thought so; at one point during the war, he suggested that the science fiction of his good friend and Naval Academy classmate Robert Heinlein had influenced his thinking, writing in a letter to Heinlein that the "basic ideas" of the CIC were "very similar to some of [your *Astounding Science Fiction*] 'brain-machine' ideas."[224]

At the end of the day, though, there was nothing fictitious about a capable and determined enemy. Recognizing that the international situation might soon present one or more such enemies, in the late 1930s American naval personnel like Arthur J. Hepburn, A. Hoyt Taylor, and Harold G. Bowen pushed for seaborne trials of a 200-megahertz radar recently developed by the Naval Research Laboratory. These men knew that war might be around the corner, and they believed radar could provide American naval forces with a decisive advantage in combat. Thanks to the ingenuity and persistence of Robert M. Page and others, operational tests conducted in 1939 exceeded most observers' expectations. Yet these same tests also revealed the difficulties inherent to the management of radar information.

While personnel at the Naval Research Lab and several other shore commands worked to solve the information management problems created by radar, the most critical solutions came from the fleet itself. In the fall of 1940, USS *California*'s communications officer Henry E. Bernstein designed and oversaw construction of a makeshift room where shipboard personnel could plot radar information. The new facility quickly became known as the Radar Plot. Senior operational officers embraced the concept and pushed the shore establishment to provide both improved facilities and better-trained personnel. Four months before the attack on Pearl Harbor, CNO Harold Stark authorized the permanent installation of radar plots on all U.S. carriers, a step he and others saw as akin to providing a shipboard brain for fleet air defense.

The events of 1942 would reveal that the navy's prewar efforts to integrate radar into the practice of warfare at sea were only partially successful. The fleet's system of air defense worked well against limited raids but became overloaded during large, coordinated attacks. Nor were the challenges of radar employment limited to carrier battles. Of equal concern to American naval leaders was the fact that during the Guadalcanal campaign U.S. warships equipped with radar achieved limited success against Japanese warships without it. Ching Lee's victory at the Second Naval Battle of Guadalcanal was in many ways the exception that proved the rule. Radar technology was not giving American naval forces the tactical advantage it should have been providing.

In an effort to remedy this shortcoming, Pacific Fleet commander-in-chief Chester W. Nimitz promulgated new tactical doctrine. In November 1942 he directed all major surface combatants under his command to establish facilities "in which information from all available sources can be received, assimilated, and evaluated with a minimum of delay."[225] Back in Washington Nimitz's directive generated a flurry of activity and led the Navy Department to authorize installation of such facilities on board all major warships in both the Atlantic and the Pacific. Thus was born the Combat Information Center.

As had been the case for many previous innovations in shipboard command and control, a combination of junior and senior personnel molded the CIC into an effective instrument of war. Two of the most important individuals in this regard were Mahlon S. Tisdale and Caleb B. Laning, who formulated plans for CIC arrangements on the various different classes of American destroyers. Tisdale and Laning believed that the Combat Information Center could serve as a cognitive aid to commanding officers and naval commanders, and they tasked thirty-two-year-old naval officer J. C. Wylie to generate a handbook that would provide an overview of what the CIC could do and how it could assist naval personnel in planning and executing courses of action. American naval officers readily grasped the significance of Wylie's handiwork, and it quickly became essential reading throughout the navy.

The final two years of the Second World War demonstrated that the CIC was invaluable in helping American naval forces deal with the chaos and confusion of anti-submarine warfare, large fleet actions, and massive amphibious assaults. For both commanding officers and OTCs, it clarified tactical situations and facilitated quick decision making. And while CICs contained an array of critical equipment, ranging from plotting tables and sound-powered telephones to sonar scopes and radar repeaters, at the heart of every CIC was a team of well-trained personnel. These individuals held enormous responsibility, sometimes making split-second, life or death decisions for themselves, their superiors, and their shipmates. More broadly, they were some of the first individuals to employ real-time systems for the management and processing of information. Their legacy extends far and wide today.

Conclusion

> We describe ourselves as living in an information age as if this were something completely new.
>
> Ann M. Blair, 2010

One name conspicuously absent from this book is that of Alfred Thayer Mahan. The son of a professor at the U.S. Military Academy, Mahan achieved international fame for *The Influence of Sea Power upon History, 1660–1783* (1890), a book that examined maritime policy, naval warfare, and international politics.[1] Mahan published numerous other works before his death in 1914 and, although he sometimes modified his views, the body of his work retained generally consistent arguments. Most scholars have focused on Mahan's strategic arguments, often interpreting them as deterministic or doctrinaire. They view the post-1890 American navy as "Mahanian"—that is, a service devoted to *guerre d'escadre* (squadron/fleet warfare), command of the sea, and the pursuit of decisive battle.[2] Rarely discussed is Mahan's naval professional argument, which held that dedicated study could help officers make better decisions when faced with the uncertainty created by changing conditions and incomplete information.[3] While Mahan's strategic arguments may have been a principal concern of senior military leaders and civilian policy makers, his naval professional arguments were meant for those who went to sea. On the whole, these individuals had little time to worry about the composition of the fleet. They needed to get the job done with the tools they had.

This is not to suggest that Mahan's strategic arguments were insignificant, just that the naval officer–historian had more than one audience for his work and that he was concerned deeply about decision making under arduous conditions. And while Mahan saw the study of history as one way to minimize uncertainty in the face of rapidly changing conditions and incomplete information, others in the service viewed improved shipboard command and control systems as a means to the same end. To be sure, there were setbacks and some failures along the way; innovation is never as neat and seamless as it appears after the fact. Yet hindsight does not just obscure; it also reveals things not necessarily obvious at the time. Historians speak of interpreting the past within a particular historiographic framework. Naval officers refer simply to lessons learned.

Either way, the history of shipboard command and control systems offers valuable insights into the U.S. Navy's approach to technological innovation following the Civil War. Perhaps most significantly, even after 1890 no single artifact dominated the thinking of American naval officers. Indeed, to the extent that there was a dominant paradigm governing the Navy Department up to and through the Second World War, it was the overriding belief of individuals on shore that their raison d'être was to provide better systems for ships at sea, coupled with the conviction of their seaborne counterparts that they should test, modify, and provide useful feedback on those systems. These groups sometimes disagreed over specifics, but neither waivered in its belief that operations at sea were the sine qua non of the naval profession. No doubt this faith stemmed in part from the fact that the officer on shore one year might well have command at sea the next.

More broadly, this shared institutional mind-set calls into question a basic premise of rational choice theory, the idea that individuals make choices on the basis of self-interest. While this may be true in many instances, the naval officers studied here showed a willingness to put the perceived needs of the service ahead of their own self-interest. From the 1890s to the 1930s, conventional wisdom held that the surest way to promotion was as a gunnery expert, yet many individuals rejected this career path to embrace other technologies. In 1909, for example, Captain Washington Irving Chambers relinquished a battleship command to head the navy's fledgling aviation branch, a decision that almost certainly cost him promotion to flag rank.[4]

Much like aviation, command and control technologies attracted numerous advocates within the service. Not only does rational choice theory fail to explain this state of affairs, but it discounts the importance of organizational norms and practices. This is problematic because the institutional role played by the navy was essential for the successful development of shipboard command and control systems.

Perhaps the most important institutional factor of all was the Navy Department's remarkable consistency in assigning exceptional personnel to work with new command and control technologies. To wit, the department tasked Foxhall Parker to revise and improve ship-to-ship signaling; repeatedly detailed John Hudgins to the North Atlantic Fleet to aid in the adoption of wireless; hired former naval personnel Benjamin Miessner and A. Hoyt Taylor to work in newly created laboratories; and sent Henry Bernstein, an expert in radio and electrical engineering who also had a strong operational background, to oversee installation of the first CXAM radar. Innovations came from unexpected quarters as well. Edward Very's proposals arrived at the signal office unsolicited. Morris Smellow developed the strip cipher, even though he was a newcomer to cryptography. And Caleb Laning was clearly the right person in the right place at the right time for designing the Combat Information Center (CIC). The navy as an institution thus showed itself to be proactive in seeking, and receptive in listening to, new ideas related to shipboard command and control. The service also was willing to promote those with strong expertise in command and control. Both Samuel Robison and Arthur Hepburn eventually rose to become the navy's highest-ranking operational officer, commander-in-chief of the United States Fleet.

The navy's organizational ethos was especially important because the service operated under tight budgets for most of the eight decades covered in this book. New technologies were costly, of course, but naval personnel frequently sought and implemented inexpensive ways to enhance operational efficiency through better command and control systems. In the 1880s, for example, the navy adopted new methods of training and analysis to improve ship-to-ship signaling, and after World War I the fleet achieved notable efficiencies by modifying its communications policies and procedures. Senior naval leaders also addressed fiscal constraints by postponing fleetwide adoption of new command and control technologies when they believed the geostrategic environment allowed them to do so with a minimum of risk. Thus, the service gave the signal office plenty of time to put finishing touches on Very's flares before issuing them to all ships, leveraged European expertise in wireless telegraphy to reduce research and development costs, and waited patiently as Edward Hebern tried to perfect his electromechanical encryption machine. On the other hand, when the international situation warranted, the navy could act with alacrity. Such was the case for radar and TBS radio, two key technologies of the Second World War.

That conflict saw the navy bring together decades of innovations in shipboard command and control to create the CIC, an artifact in which personnel employed a variety of techniques to collect, organize, evaluate, and disseminate information. When operating as designed, CICs gave naval commanders a metaphorical second brain. Inside this human-machine mind, relatively junior personnel made time-critical decisions, sometimes so quickly that the information reached an officer in tactical command or commanding officer only after the fact. In social-scientific terms, cognition at sea was socially distributed.

At a cognitive level, then, the experience of command at sea was considerably different for Marc Mitscher than it was for David Farragut. The same holds true for those who served under them at places like Okinawa and Mobile Bay. Yet one should be careful not to go too far in highlighting this discontinuity, because American naval personnel's attempts to manage the sea of information surrounding them were not all that different in kind from efforts by Renaissance scholars to compile textual information or by Victorian telegraphers to gain information advantages in the business world.[5] Modern pundits would be wise to keep this in mind when they proclaim that we live in an information age unprecedented in human history. Every age is one of information, each in its own way.

Abbreviations

Service Abbreviations

ARSN	*Annual Report of the Secretary of the Navy*
ARSO	*Annual Report of the Army's Chief Signal Officer*
BuAer	Bureau of Aeronautics
BuCR	Bureau of Construction and Repair
BuEng	Bureau of Engineering
BuEq	Bureau of Equipment
BuNav	Bureau of Navigation
BuOrd	Bureau of Ordnance
BuSA	Bureau of Supplies and Accounts
BuSE	Bureau of Steam Engineering
BuShips	Bureau of Ships
CIC	Combat Information Center
C-in-C	commander-in-chief
CinCLant	commander-in-chief, U.S. Atlantic Fleet
CinCPac	commander-in-chief, U.S. Pacific Fleet
CNO	chief of naval operations
COMINCH	commander-in-chief, U.S. Fleet (World War II)
CSO USN	chief signal officer, U.S. Navy
CTF	Commander Task Force
CTG	Commander Task Group
ECM	electric cipher machine
NRL	Naval Research Laboratory
OPNAV	Office of the Chief of Naval Operations
ORN	*Official Records of the Union and Confederate Navies in the War of Rebellion*
OTC	officer in tactical command

SecNav	Secretary of the Navy
SOE	Squadron of Evolution
USMC	United States Marine Corps
USNIP	*United States Naval Institute Proceedings*

Archival Abbreviations

ADM 116/523	Wireless Telegraphy—Experiments, Admiralty: Record Office: Cases, Records of the Navy Board and the Board of Admiralty, National Archives, Kew, U.K.
ADM 116/567	Marconi Wireless Telegraphy System, Admiralty: Record Office: Cases, Records of the Navy Board and the Board of Admiralty, National Archives, Kew, U.K.
AIR 2/2625	Flying Arrangements for Biggin Hill Experiments, Air Ministry and Ministry of Defense: Registered Files, General Records of the Air Boards and Air Ministry, National Archives, Kew, U.K.
CC/100	History of Naval Radio (Series 100), George H. Clark Radioana Collection, Archives Center, National Museum of American History, Smithsonian Institution, Washington, D.C.
IWM/CC	Robert Church Collection, Department of Documents, Imperial War Museum Collections, London.
LOC/HP	Papers of Stanford C. Hooper, Naval Historical Foundation Collection, Manuscript Division, Library of Congress, Washington, D.C.
M89/SL	National Archives Microfilm M89, Letters Received by the Secretary of the Navy from Commanding Officers of Squadrons ("Squadron Letters"), 1841–1886
M964/FP	National Archives Microfilm M964, Records Relating to United States Navy Fleet Problems I to XXII, 1923–1941
M971/FR	National Archives Microfilm M971, Annual Reports of Fleets and Task Forces of the U.S. Navy, 1920–1941
M984/GO	National Archives Microfilm M984, Navy Department General Orders, 1863–1948
M1140/SCC	National Archives Microfilm M1140, Secret and Confidential Correspondence of the Office of the Chief of Naval Operations and the Office of the Secretary of the Navy, 1919–1927
NA/VF	Vertical File, Special Collections and Archives, Nimitz Library, United States Naval Academy, Annapolis, Md.
NDL/HR	World War II Histories and Historical Reports, Special Collections, Navy Department Library, Washington, D.C.

NDL/ZB	Biographical (ZB) Files, Special Collections, Navy Department Library, Washington, D.C.
NOA/OB	Officer Biographies, U.S. Navy Operational Archives, Naval History and Heritage Command, Washington, D.C.
NOA/WWII	World War II Command File, U.S. Navy Operational Archives, Naval History and Heritage Command, Washington, D.C.
NWC/LP	Papers of Caleb B. Laning, Manuscript Collection 116, Naval Historical Collection, Naval War College Archives, Newport, R.I.
NWC/OH	Oral Histories, Naval Historical Collection, Naval War College Archives, Newport, R.I.
PU/MP	Benjamin F. Miessner Papers, Purdue University Archives and Special Collections, West Lafayette, Ind.
RG19/BEQ	General Correspondence, 1899–1910, Records of the Bureau of Equipment, Records of the Bureau of Ships, Record Group 19, National Archives, Washington, D.C.
RG19/NRLC	Naval Research Laboratory General Files, 1923–1940, Confidential, Records of the Bureau of Ships, Record Group 19, National Archives, Washington, D.C.
RG19/NRLS	Naval Research Laboratory General Files, 1923–1940, Secret, Records of the Bureau of Ships, Record Group 19, National Archives, Washington, D.C.
RG19/OM	Orders and Memoranda, 1906–1930, Records of the Bureau of Equipment, Records of the Bureau of Ships, Record Group 19, National Archives, Washington, D.C.
RG19/WT	Records Relating to Wireless Stations and Tests of Wireless Equipment, 1904–1910, Records of the Bureau of Equipment, Records of the Bureau of Ships, Record Group 19, National Archives, Washington, D.C.
RG24/DL	Logs of U.S. Naval Ships, 1801–1915, Logs of Ships and Stations, 1801–1946, Records of the Bureau of Naval Personnel, Record Group 24, National Archives, Washington, D.C.
RG24/FC	Fair Copies of Letters Sent, Records of the Signal Office, 1869–1886, Records of the Bureau of Naval Personnel, Record Group 24, National Archives, Washington, D.C.
RG24/LR	Letters Received From the Chief Signal Officer, Letters Received, 1862–1889, Records of the Bureau of Naval Personnel, Record Group 24, National Archives, Washington, D.C.
RG24/LS	Letters Sent to the Signal Office, Letters Sent to the President, Congressmen, and Executive Departments, Records of the

	Bureau of Naval Personnel, Record Group 24, National Archives, Washington, D.C.
RG24/SS	Letters Received Relating to Signaling at Sea, Letters Received, 1862–1889, Records of the Bureau of Naval Personnel, Record Group 24, National Archives, Washington, D.C.
RG38/AR	World War II Action and Operations Reports, Records Relating to Naval Activity During World War II, Records of the Office of the Chief of Naval Operations, Record Group 38, National Archives, College Park, Md.
RG45/SF1	Subject File, 1775–1910, Records of the Office of Naval Records and Library, Record Group 45, National Archives, Washington, D.C.
RG45/SF2	Subject File, 1911–1927, Records of the Office of Naval Records and Library, Record Group 45, National Archives, Washington, D.C.
RG80/GB	Subject File, 1900–1947, Records of the General Board, General Records of the Navy Department, Record Group 80, National Archives, Washington, D.C.
RG80/FCC	Formerly Confidential Correspondence, 1927–1939, General Records of the Navy Department, Record Group 80, National Archives, Washington, D.C.
RG80/FSC	Formerly Secret Correspondence, 1927–1939, General Records of the Navy Department, Record Group 80, National Archives, Washington, D.C.
RG80/TF	General Correspondence, 1897–1915, General Records of the Navy Department, Record Group 80, National Archives, Washington, D.C.
RG125/EB	Proceedings of Naval and Marine Examining Boards, c. 1890–1941, Records of the Bureau of the Office of the Judge Advocate General (Navy), Record Group 125, National Archives, Washington, D.C.
RG313/CO	Letters Sent to Commanding Officers of Vessels, Squadron of Evolution, 1889–1892, Records of Naval Operating Forces, Record Group 313, National Archives, Washington, D.C.
RG313/ND	Letters Sent to the Navy Department, Squadron of Evolution, 1889–1892, Records of Naval Operating Forces, Record Group 313, National Archives, Washington, D.C.
RG313/R179	Confidential Correspondence, 1942–1945, General Administrative Files, Commander, South Pacific (Red 179), Records of Naval Operating Forces, Record Group 313, National Archives, College Park, Md.

Notes

Introduction

Epigraph. William F. Halsey Jr., foreword to Frederick C. Sherman, *Combat Command: The American Aircraft Carriers in the Pacific War* (New York: E. P. Dutton, 1950), 8.

1. Humankind's use of technology to manage information is centuries old, of course, but only in the 1950s did the term "information technology" become part of the English lexicon. For the earliest-known definition of the term, see Harold J. Leavitt and Thomas L. Whisler, "Management in the 1980's," *Harvard Business Review* 36, no. 6 (November–December 1958): 41–48.

2. David E. Nye, *American Technological Sublime* (Cambridge, Mass.: MIT Press, 1994).

3. See esp. Clifford L. Lord and Archibald Turnbull, *History of United States Naval Aviation* (New Haven: Yale University Press, 1949); Vincent Davis, *The Admirals Lobby* (Chapel Hill: University of North Carolina Press, 1967); Waldo H. Heinrichs, "The Role of the United States Navy," in *Pearl Harbor as History: Japanese-American Relations, 1931–1941*, ed. Dorothy Borg and Shumpei Okamoto (New York: Columbia University Press, 1973), 197–223; and Robert L. O'Connell, *Sacred Vessels: The Cult of the Battleship and the Rise of the U.S. Navy* (New York: Oxford University Press, 1991).

4. O'Connell, *Sacred Vessels*, 4; Davis, *Admirals Lobby*, 75.

5. Thomas C. Hone, Norman Friedman, and Mark D. Mandeles, *American and British Aircraft Carrier Development, 1919–1941* (Annapolis: Naval Institute Press, 1999); William M. McBride, *Technological Change and the United States Navy, 1865–1945* (Baltimore: Johns Hopkins University Press, 2000); Thomas C. Hone and Trent Hone, *Battle Line: The United States Navy, 1919–1939* (Annapolis: Naval Institute Press, 2006); and Craig C. Felker, *Testing American Sea Power: U.S. Navy Strategic Exercises, 1923–1940* (College Station: Texas A&M University Press, 2007).

6. Recent syntheses by distinguished naval historians Michael Palmer and Norman Friedman are notable exceptions to this rule. See Michael A. Palmer, *Command at Sea: Naval Command and Control since the Sixteenth Century* (Cambridge, Mass.: Harvard University Press, 2005), and Norman Friedman, *Network-Centric Warfare: How Navies Learned to Fight Smarter through Three World Wars* (Annapolis: Naval Institute Press, 2009).

7. Although *command and control* entered the American military lexicon after World War II, I use it throughout the book for two reasons: contemporary understandings of

the phrase are consistent with *command*, the word early to mid-twentieth century naval officers would have used to describe what is now known as command and control; and the term is employed widely today and therefore sounds better to the modern reader. Specifically, *command and control* captures succinctly the following idea: the exercise of authority and direction through an arrangement of personnel, equipment, facilities, and procedures employed by a commander to plan, direct, coordinate, and control forces and operations. For a discussion of how and why command and control became a subset of command, see Thomas P. Coakley, *Command and Control for War and Peace* (Washington: National Defense University Press, 1992), 34–38.

8. This expression is from McBride, *Technological Change*, 7. Elsewhere, McBride argues that the battleship was an "obdurate exemplary artifact," one that served as the "technological basis of the navy" from the early 1890s to the early 1940s. William M. McBride, "The Unstable Dynamics of a Strategic Technology: Disarmament, Unemployment, and the Interwar Battleship," *Technology and Culture* 38, no. 2 (April 1997): 388.

9. I use the phrase "warfare at sea" in a broad sense, to include not only combat operations but also preparations for combat at sea.

10. David Kirsch and Paul Maglio, "On Distinguishing Epistemic from Pragmatic Action," *Cognitive Science* 18, no. 4 (October–December 1994): 513–49; Andy Clark, "Embodied, Situated, and Distributed Cognition," in *A Companion to Cognitive Science*, ed. William Bechtel and George Graham (Malden, Mass.: Blackwell, 1998), 506–17; and Richard Menary, "Dimensions of Mind," *Phenomenology and the Cognitive Sciences* 9, no. 4 (December 2010): 561–78.

11. For a useful discussion about the social distribution of cognitive labor in a naval context, see Edwin Hutchins, *Cognition in the Wild* (Cambridge, Mass.: MIT Press, 1995).

12. Samuel Eliot Morison, *New Guinea and the Marianas, March 1944–August 1944*, vol. 8 of *History of United States Naval Operations in World War II* (New York: Little, Brown, 1953), 260. Even when historians acknowledge the importance of command and control systems, they usually make no effort to explain how these systems actually functioned. See, e.g., Nathan Miller, *War at Sea: A Naval History of World War II* (New York: Oxford University Press, 1995), 393, 441.

13. Edward L. Beach, *The United States Navy: 200 Years* (New York: Henry Holt, 1986), 484.

14. A common school of thought holds that military innovation generally comes from the top down. See Stephen P. Rosen, *Winning the Next War: Innovation and the Modern Military* (Ithaca, N.Y.: Cornell University Press, 1991), and Williamson Murray, "Innovation: Past and Future," in *Military Innovation in the Interwar Period*, ed. Williamson Murray and Allan R. Millet (New York: Cambridge University Press, 1996), 300–328. For an alternative interpretation, see Barry R. Posen, *The Sources of Military Doctrine: France, Britain, and Germany between the World Wars* (Ithaca, N.Y.: Cornell University Press, 1984).

CHAPTER 1. Flags, Flares, and Lights: A World before Wireless

Epigraph. "United States Naval Rendezvous, Key West, Fla.," *Frank Leslie's Illustrated Newspaper*, 14 February 1874, p. 379.

1. Under Burriel's orders, Spanish authorities executed twelve more men the following morning.

2. "What Can Be Done with Cuba," *New York Times*, 15 November 1873, p. 6.

3. My recounting of events draws from the following sources: Jeanie Mort Walker, *Life of Captain Joseph Fry: The Cuban Martyr* (Hartford: J. B. Burr, 1875); Jim Dan Hill, "Captain Joseph Fry, of S.S. *Virginius*," *American Neptune* 36, no. 2 (April 1976): 88–100; and Richard H. Bradford, *The* Virginius *Affair* (Boulder: Colorado Associated University Press, 1980).

4. Fry to Grant, [5 November 1873?], reproduced in Walker, *Life of Captain Joseph Fry*, 448–49.

5. Timothy S. Wolters, "Recapitalizing the Fleet: A Material Analysis of Late-Nineteenth-Century U.S. Naval Power," *Technology and Culture* 52 (January 2011): 103–26.

6. Commander, U.S. Naval Forces, North Atlantic Station (G. H. Scott) to SecNav, 22 December 1873, including enclosures; and telegrams sent by Scott to SecNav on 17, 19, 22 December 1873, all on roll 282, M89/SL.

7. Bradford, *The* Virginius *Affair*, 112–14.

8. Case to SecNav, 3, 22 January 1874, roll 283, M89/SL; and Foxhall A. Parker, "Our Fleet Maneuvers in the Bay of Florida, and the Navy of the Future," *USNIP* 1, no. 8 (1874): 163–78. In "Our Fleet Maneuvers," Parker mistakenly claimed that he reported to Key West on 16 January.

9. Case to SecNav, 22 January 1874, roll 283, M89/SL.

10. Case to Henry Hall (U.S. Consul General in Havana), 3 February 1874; Case to SecNav, 22 January 1874, both on roll 283, M89/SL; and Parker, "Our Fleet Maneuvers," 163.

11. Lance C. Buhl, "Maintaining an 'American Navy,' 1865–1889," in *In Peace and War: Interpretations of American Naval History, 1775–1984*, ed. Kenneth J. Hagan, 2d ed. (Westport, Conn.: Greenwood Press, 1984), 145–73. See also William S. Peterson, "Congressional Politics: Building the New Navy, 1876–86," *Armed Forces and Society* 14, no. 4 (December 1988): 489–508.

12. Michael A. Palmer, *Command at Sea: Naval Command and Control since the Sixteenth Century* (Cambridge, Mass.: Harvard University Press, 2005), 125–68.

13. Samuel S. Robison and Mary Robison, *A History of Naval Tactics from 1530 to 1930* (Annapolis: Naval Institute Press, 1942), 447–49.

14. U.S. Navy Department, *Signals for the Use of the United States Navy* (Washington: n.p., 1813).

15. U.S. Navy Department, *Signals for the Use of the United States Navy* (Washington: William A. Harris, 1858).

16. James H. Ward, *A Manual of Naval Tactics* (New York: D. Appleton, 1859), 5, 29.

17. Thornton A. Jenkins, *Code of Flotilla and Boat Squadron Signals for the United*

States Navy (Washington: U.S. Government Printing Office, 1861); and Bureau of Ordnance and Hydrography Circular, approved by SecNav, 22 January 1862, box 161, filing designator DS, RG45/SF1.

18. Linwood S. Howeth, *History of Communications-Electronics in the United States Navy* (Washington: U.S. Government Printing Office, 1963), 8–9. Throughout the period under study in this book, the navy was organized into bureaus. The bureaus were semiautonomous divisions, organized by function, responsible for a majority of the day-to-day activities of the Navy Department.

19. Albert J. Myer, *Manual of Signals* (Washington: n.p., 1864), 30–31.

20. This example is based on the codes in Thornton A. Jenkins, *Code of Flotilla and Boat Squadron Signals for the United States Navy*, 2d ed. (Washington: U.S. Government Printing Office, 1869). Interestingly, this signal book reverses Myer's code, with a left wave signifying "two" and a right wave signifying "one."

21. Rebecca Robbins Raines, *Getting the Message Through: A Branch History of the U.S. Army Signal Corps* (Washington: U.S. Government Printing Office, 1996), 5–29, 143.

22. SecNav to BuNav, 11 February, 29 May 1863, roll 2, U.S. National Archives Microfilm M480, Letters Sent by the Secretary of the Navy to Chiefs of Navy Bureaus, 1842–1886. The quoted phrase is from the earlier letter.

23. Charles H. Davis, Alexander D. Bache, and Joseph Henry to SecNav, 7 July 1863, Minutes of the Permanent Scientific Commission, Records of Boards and Commissions, 1812–1890, Records Collection of the Office of Naval Records and Library (Record Group 45), U.S. National Archives, Washington, D.C.

24. In chronosemic signaling, the interval between successive signals indicates the meaning of the signal. So an interval of five seconds between signals might represent a "one," an interval of ten seconds a "two," and an interval of fifteen seconds a "three." An advantage of such signals was that they could be either audible or visual. B. Franklin Greene, *Chronosemic Signals: A System of Fog Signals, for the Signal Code of the United States Navy, and Adapted to General Signal Communication* (Washington: U.S. Government Printing Office, 1864).

25. David G. Farragut to Charles H. Davis, 20 March 1864, in *ORN*, ser. I, vol. 21, 146.

26. BuNav Circular to Commanders of Squadrons, signed by BuNav, 4 April 1864, box 161, filing designator DS, RG45/SF1. Farragut was not the first officer to propose such a system but his stature may well have influenced BuNav. Extant records indicate that signal expert Thornton Jenkins had proposed a slightly more sophisticated enciphering system as early as March 1863. Jenkins, "Extract from Report of Captain T. A. Jenkins to accompany Naval General Signal Book and Naval Telegraphic Dictionary," March 1863, box 161, filing designator DS, RG45/SF1.

27. Unless otherwise noted, my recounting of events at Mobile Bay comes from John Friend, *West Wind, Flood Tide: The Battle of Mobile Bay* (Annapolis: Naval Institute Press, 2004), 123–202.

28. John C. Kinney, "An August Morning with Farragut," *Scribner's Monthly* 22, no. 2 (June 1881): 203.

29. U.S.S. BROOKLYN (James Alden, Commanding Officer) to Admiral Farragut, 7:25 a.m., 5 August 1864, *ORN*, ser. I, vol. 21, 508.

30. Farragut to Alden, 7:30 a.m., 5 August 1864, *ORN*, ser. I, vol. 21, 508.

31. Farragut to Alden, 7:40 a.m., 5 August 1864, *ORN*, ser. I, vol. 21, 508.

32. Foxhall A. Parker, *The Battle of Mobile Bay and the Capture of Forts Powell, Gaines, and Morgan* (Boston: A. Williams, 1878), 29, and John C. Kinney, "Farragut at Mobile Bay," in *Battles and Leaders of the Civil War* (New York: Century Company, 1884), 4:391.

33. Luce to Rear Admiral John A. Dahlgren, 16 November 1863, box 161, filing designator DS, RG45/SF1.

34. Porter to SecNav, 19 October 1866, box 161, filing designator DS, RG45/SF1.

35. Thornton A. Jenkins, *The United States Naval Signal Code* (Washington: U.S. Government Printing Office, 1867), 6, 26–28, and Jenkins, *Flotilla and Boat Squadron Signals*, 2d ed.

36. BuNav Circular, approved by SecNav, 19 July 1869, RG24/FC.

37. SecNav to Secretary of War, 28 July 1869, RG24/FC. The school was located at Fort Whipple (later renamed Fort Myer).

38. U.S. War Department, *ARSO*, 1869, 198–99; and Chief Signal Officer, U.S. Army, to CSO USN, 1 November 1869, referenced in CSO USN to Chief Signal Officer, U.S. Army, 1 November 1869, RG24/FC.

39. CSO USN (Samuel P. Lee) to SecNav, 20 November 1869, RG24/FC.

40. *ARSO*, 1869, 199; and *ARSO*, 1870, tables I and II, 112.

41. CSO USN to D. Van Nostrand, 8 November 1869; and CSO USN to J. Elliot Condict, 8, 12 November 1869, RG24/FC.

42. CSO USN to Chief Signal Officer, U.S. Army, 7 January 1870, RG24/FC.

43. Form of Inspection for Standard Signal Equipment, 1869, vol. 1, RG24/SS.

44. BuNav Circular, 19 July 1869; CSO USN to Signal Officer of the Pacific Fleet, 18 March 1870; and CSO USN (John J. Almy) to Officer Commanding U.S. Pacific Fleet, 25 February 1871, all in RG24/FC.

45. CSO USN to BuNav (Daniel Ammen), 17, 18 October 1871; CSO USN to Navigation Officer New York Navy Yard, 20 June 1872; CSO USN to Navigation Officer Boston Navy Yard, 20 June 1972, all in RG24/FC; and CSO USN to BuNav, 4 June 1872, vol. 1, RG24/LR.

46. *ARSO*, 1869, 199; *ARSO*, 1870, tables I and II, 112; and *ARSO*, 1871, table I, 308.

47. BuNav Circular, 19 July 1869; and CSO USN to BuNav, 26 June 1871, 6 April 1872, all in RG24/FC. Examples of these quarterly reports may be found in vol. 1, RG24/LR.

48. CSO USN to BuNav, 20 October 1870, *ARSN*, 1870, 51–52; and CSO USN to BuNav, 3 October 1872, *ARSN*, 1872, 96–97. The quoted phrase is from the later letter.

49. Signal Officer North Atlantic Fleet to CSO USN, 13 April 1871, referenced in CSO USN to Signal Officer North Atlantic Fleet, 29 April 1871, RG24/FC.

50. CSO USN to Commander Mare Island Navy Yard, 17 October 1871, RG24/FC; and CSO USN to BuNav, 22 January 1872, vol. 1, RG24/LR. Although the signal office distributed this book in early 1872, it appears to have been printed in late 1871. John J.

Almy, George B. Batch, and A. W. Johnson (hereafter cited as Tactical Signal Book Board) to BuNav, 6 December 1872, RG24/FC.

51. Martha J. Coston, *A Signal Success: The Work and Travels of Mrs. Martha J. Coston; An Autobiography* (Philadelphia: J. B. Lippincott, 1886), 35–56.

52. Charles S. McCauley, John Rodgers, and Henry H. Lewis to SecNav, 27 January 1859, in U.S. House, *Coston's Telegraphic Night Signals, Etc.*, 36th Cong., 2d sess., 1861, Ex. Doc. No. 32.

53. U.S. House, *Claim of Mrs. Martha J. Coston*, 43rd Cong., 2d sess., 1875, Rept. No. 334.

54. Ibid.; CSO USN to BuNav, 21 April 1871; CSO USN to Navigation Officer New York Navy Yard, 20 June 1872; and CSO USN to Navigation Officer Boston Navy Yard, 20 June 1972, all in RG24/FC.

55. Albert J. Myer, *A Manual of Signals* (New York: D. Van Nostrand, 1868), 216; and Jenkins, *Flotilla and Boat Squadron Signals*, 2d ed.

56. Martha J. Coston, "Improvement in Pyrotechnic Night Signals," U.S. Letters Patent No. 115,935, 13 June 1871.

57. BuNav to CSO USN, 11 April 1871, RG24/LS.

58. CSO USN to BuNav, 21 April 1871, RG24/FC.

59. CSO USN to BuNav, 19 January 1872, RG24/FC.

60. Excerpt of a report by George A. Norris on William H. Ward's proposed signaling system, in CSO USN to BuNav, 14 May 1872, RG24/FC.

61. Tactical Signal Book Board to BuNav, 8 October 1872, RG24/FC.

62. Foxhall A. Parker, *Squadron Tactics under Steam* (New York: D. Van Nostrand, 1864).

63. Foxhall A. Parker, *Fleet Tactics under Steam* (New York: D. Van Nostrand, 1870).

64. Clark G. Reynolds, "Parker, Foxhall Alexander, Jr.," in *Famous American Admirals* (Annapolis: Naval Institute Press, 2002), 245–46.

65. Parker, "Omissions, Corrections, Additions, Etc. to U.S. Naval Signal Book," n.d., vol. 1, RG24/LR; and Tactical Signal Book Board to BuNav, 8 October 1872, RG24/FC.

66. Tactical Signal Book Board to BuNav, 8 October 1872, RG24/FC.

67. Tactical Signal Book Board to BuNav, 6 December 1872, RG24/FC; and Tactical Signal Book Board to BuNav, 10 January 1873, vol. 1, RG24/LR.

68. Tactical Signal Book Board to BuNav, 10 January 1873, vol. 1, RG24/LR.

69. BuNav to CSO USN, 1 July 1873, RG24/LS; and Parker to BuNav, 30 October 1873, *ARSN* 1873, 98.

70. BuNav to CSO USN, 9 October, 20 November 1873, RG24/LS; and Parker to BuNav, 30 October 1873, in *ARSN*, 1873, 98.

71. Parker to BuNav, 23 October 1874, in *ARSN*, 1874, 71–72.

72. Palmer, *Command at Sea*, 218–22.

73. Parker, *Fleet Tactics*, 219–20.

74. Case to SecNav, 3, 22 January 1874, roll 283, M89/SL.

75. Case to SecNav, 22 January 1874, roll 283, M89/SL; and *ARSN*, 1874, 6–13.

76. Parker, "Journal of the Movements of the U.S.N.A. Fleet," vol. I; and Case to SecNav, 7 February 1874, both on roll 283, M89/SL.

77. Case to SecNav, 6 March 1874, roll 283, M89/SL.

78. Memorandum written by Nelson off Cadiz, 9 October 1805, reproduced in Nicholas Harris Nicolas, *The Dispatches and Letters of Vice Admiral Lord Nelson* (London: Henry Colburn, 1846), 7:89–92.

79. Philip H. Colomb, "Modern Naval Tactics," *Journal of the Royal United Service Institution* 9, no. 34 (1865): 1–26. The quoted passages are from p. 16.

80. Jenkins, *United States Naval Signal Code*, 66–67.

81. General Order No. 6, issued by Case on 31 January 1874; Case to Henry Hall (U.S. Consul General in Havana), 3 February 1874; and Case to SecNav, 7 February 1874, all on roll 283, M89/SL.

82. Parker, "Journal of the Movements of the U.S.N.A. Fleet," vol. I; and Parker, "Journal of Movements, North Atlantic Fleet, from Feb. 21 to Feb. 28, 1874, Inclusive," both on roll 283, M89/SL.

83. Case to SecNav, 7 February 1874, roll 283, M89/SL.

84. Parker, "Our Fleet Maneuvers," 166.

85. David D. Porter to SecNav, 7 November 1874, in *ARSN*, 1874, 199.

86. Case to SecNav, 11 February 1874; and Parker, "Journal of the Movements of the U.S.N.A. Fleet," vol. II, both on roll 283, M89/SL.

87. Parker, "Journal of the Movements of the U.S.N.A. Fleet," vol. II; and Case to SecNav, 15 February 1874, both on roll 283, M89/SL.

88. Parker, "Journal of Exercises, North Atlantic Fleet, Feb. 15 to Feb. 20 Inclusive;" and Parker, "Journal of Movements, North Atlantic Fleet, from Feb. 21 to Feb. 28, 1874, Inclusive," both on roll 283, M89/SL.

89. Case to SecNav, 6 March 1874, roll 283, M89/SL; and Parker, "Our Fleet Maneuvers," 176 (emphasis in original).

90. Parker, "Our Fleet Maneuvers," 175.

91. Case, "General Order No. 9," 28 March 1874, enclosure to Case to SecNav, 1 April 1874, roll 283, M89/SL.

92. Bradley A. Fiske, *From Midshipman to Rear Admiral* (New York: Century, 1919), 55; Fiske's duty station at the time of this incident is given in Summary of Service, Bradley Allen Fiske folder, NA/VF.

93. Albert P. Niblack, "Proposed Day, Night, and Fog Signals for the Navy, with Brief Description of the Ardois Night System," *USNIP* 17, no. 2 (1891): 254.

94. William H. Gardner, Henry A. Adams, Thornton A. Jenkins, and Robert D. Minor to William J. McCluney, 24 November 1859; and Louis M. Goldsborough to Joshua R. Sands, 18 August 1860, all in U.S. House, *Coston's Telegraphic Night Signals*, 36th Cong., 2d sess., 1861, Ex. Doc. No. 32.

95. Parker to BuNav, 23 October 1874, in *ARSN*, 1874, 71–72; and Parker to BuNav, 12 October 1875, in *ARSN*, 1875, 82–83.

96. Very to CSO USN, 17 August 1874, vol. 1, RG24/LR.

97. Ibid.

98. Summary of Service, Edward Wilson Very folder, NA/VF; and Paul J. Scheips, "Edward Wilson Very (1847–1910)," dated 1962, in Edward Wilson Very folder, NDL/ZB.

99. CSO USN to BuNav, 23 May 1874, vol. 1, RG24/LR.

100. CSO USN to BuNav, n.d., vol. 1, RG24/LR. Parker did not date this letter but BuNav received it on 25 July 1874.

101. CSO USN to BuNav, 28 January 1875, vol. 1, RG24/LR; BuNav to CSO, 29 January, 27 May, 13 July, 15 October 1875, RG24/LS; and CSO USN to BuNav, 30 June 1875, vol. 2, RG24/LR.

102. CSO USN (by direction) to E. F. Linton and three other firms, 22 December 1875, RG24/FC.

103. CSO USN to BuNav, 16 February 1876, vol. 2, RG24/LR.

104. CSO USN (John C. Beaumont) to BuNav, 21 February 1877, RG24/FC.

105. CSO USN to BuNav, 2 May 1878, vol. 2, RG24/LR; and CSO USN to BuNav, 26 September 1876, in *ARSN*, 1876, 100–101.

106. Very to CSO USN, 17 August 1874, vol. 1, RG24/LR.

107. Edward W. Very, "Improvement in Signal-Cartridges," U.S. Letters Patent No. 190,263, 1 May 1877.

108. CSO USN to BuNav, 10 October 1877, RG24/FC.

109. CSO USN to BuNav, 28 October 1877, in *ARSN*, 1877, 151.

110. CSO USN to BuNav, 23 May 1874, vol. 1, RG24/LR.

111. William B. Cogar, "Edwin Longnecker," in *Dictionary of Admirals of the U.S. Navy* (Annapolis: Naval Institute Press, 1991), 2:167–68.

112. "Experiments on the Very Night Signals," enclosure to Edwin Longnecker to BuNav, 5 February 1878, vol. 2, RG24/LR.

113. W. H. Turner to CSO USN, 1, 3 May 1878, vol. 2, RG24/LR; and J. H. Moore to CSO USN, 1, 3 May 1878, RG24/FC.

114. CSO USN to BuNav, 2, 4 May 1878, vol. 2, RG24/LR; and BuNav to CSO USN, 4 May 1878, RG24/LS.

115. Moore to BuNav (William D. Whiting), 17 August 1878, vol. 2, RG24/LR.

116. Although Very's original code used red and white, by the late 1870s the navy had replaced white with green, apparently because the latter color offered a starker contrast to red. Very to CSO USN, 17 August 1874, vol. 1, RG24/LR; and "Experiments on the Very Night Signals," enclosure to Edwin Longnecker to BuNav, 5 February 1878, vol. 2, RG24/LR.

117. Moore to BuNav, 17 August 1878, vol. 2, RG24/LR.

118. CSO USN to BuNav, 18, 25 September 1878, vol. 2, RG24/LR; and CSO USN to BuNav, 15 October 1878, RG24/FC.

119. CSO USN to BuNav, 18 December 1878, vol. 2, RG24/LR.

120. CSO USN (Clark H. Wells) to BuNav, 23 October 1879, RG24/FC.

121. Summary of Service, Edward Wilson Very folder, NA/VF; and Very to BuNav, 22 August 1879, vol. 2, RG24/LR.

122. William C. Babcock (for CSO USN) to Very, 22 March 1877; and W. H. Turner to Very, 23 February 1878, both in RG24/FC.

123. Very to BuNav, 22 August 1879, vol. 2, RG 24/LR; and Very to BuNav, 19 September 1879, vol. 3, RG 24/LR. The quoted attributes are from the latter document.

124. Very to CSO USN, 7 January 1880, vol. 1, RG24/SS.

125. CSO USN to BuNav, 2, 19 December 1874, vol. 1, RG24/LR; and CSO USN to BuNav, 12 October 1875, in *ARSN*, 1875, 82.

126. CSO USN to BuNav, 26 September 1876, in *ARSN*, 1876, 100.

127. Very to CSO USN, 7 January 1880, vol. 1, RG24/SS.

128. Very to BuNav, 5 January 1880, vol. 3, RG 24/LR.

129. Ibid.

130. Ibid. Very excluded from this requirement signals sent by Myer's wigwag.

131. Very to CSO USN, 26 October 1880, vol. 1, RG24/FC.

132. CSO USN to BuNav, 1 November 1880, RG24/FC.

133. Coston, *A Signal Success*, 291–305; and William C. Babcock (for CSO USN) to Very, 22 March 1877, RG24/FC.

134. BuNav to Very, 6 July 1881, RG24/LS.

135. BuNav Circular to Commanders of Squadrons, signed by BuNav, 31 January 1882, box 161, filing designator DS, RG45/SF1; C-in-C North Atlantic Station to BuNav, 5 February 1882; USS *Nipsic* to BuNav, 25 May 1882; and USS *Adams* to BuNav, 7 June 1882, vol. 1, RG24/SS.

136. USS *Pensacola* to F. Hanford, Frank R. Heath, and F. H. Hunicke, 21 March 1882; and *Pensacola* to C-in-C Pacific Station, 28 April 1882, both in vol. 1, RG24/SS. *Pensacola* was flagship of the Pacific Station.

137. Hanford, Heath, and Hunicke to USS *Pensacola*, 26 April 1882, vol. 1, RG24/SS.

138. BuNav to CSO USN (P. C. Johnson), 11 August 1882, RG24/LS; and CSO USN to BuNav, 18 August 1882, vol. 3, RG24/LR.

139. General Order No. 301, 21 October 1882, roll 1, M984/GO.

140. Regrettably, Parker did not live to see the navy's official adoption of Very's system. He died of heart failure on 10 June 1879 at the age of just fifty-seven.

141. Hanford, Heath, and Hunicke to USS *Pensacola*, 26 April 1882, vol. 1, RG24/SS.

142. CSO USN to BuNav, 18 August, 14 December 1883, vol. 3, RG24/LR.

143. S. Dana Greene, "Notes on Electrical Testing and Measuring Apparatus for Ships," *USNIP* 12, no. 4 (1886): 483.

144. Joseph B. Murdock, "The Naval Use of the Dynamo Machine and Electric Light," *USNIP* 8, no. 3 (1882): 343–85.

145. C-in-C SOE to SecNav, 18 November 1889, RG313/ND.

146. Daniel H. Wicks, "New Navy and New Empire: The Life and Times of John Grimes Walker" (Ph.D. diss., University of California at Berkeley, 1979), 214–15.

147. C-in-C SOE to SecNav, 18 November 1889, RG313/ND.

148. Wicks, "New Navy and New Empire," 220–27.

149. C-in-C SOE to SecNav, 16 March 1890; and C-in-C SOE to BuEq, 22 March 1890, both in RG313/ND.

150. C-in-C SOE to BuEq, 22 March 1890, RG313/ND.

151. Ibid.

152. C-in-C SOE to BuEq, 22 March 1890, RG313/ND; and Sidney A. Staunton to C-in-C SOE, 15 May 1891, box 162, filing designator DS, RG45/SF1.

153. Lewis R. Hamersly, *The Records of Living Officers of the U.S. Navy and Marine Corps*, 7th ed. (New York: L. R. Hamersly, 1902), 216.

154. Staunton to C-in-C SOE, 15 May 1891, box 162, filing designator DS, RG45/SF1.

155. C-in-C SOE to BuEq, 20 May 1891, RG313/ND.

156. C-in-C SOE to SecNav, 19 October 1891, box 464, filing designator OO, RG45/SF1.

157. Niblack, "Brief Description of the Ardois Night System," 258–60. A year later Niblack reported the price per outfit had dropped to $1,180. Niblack, "The Signal Question Up to Date," *USNIP* 18, no. 1 (1892): 59.

158. Sidney A. Staunton and [illegible] to C-in-C SOE, 11 December 1890, vol. 1, RG313/CO. Niblack reported that under favorable conditions Ardois lights could be read at distances of up to three miles. Niblack, "Brief Description of the Ardois Night System," 259.

159. Biographical sketch of Albert Parker Niblack, 1859–1929, 11 November 1937, Albert Parker Niblack folder, NDL/ZB; and Report on the Fitness of Officers, Captain A. P. Niblack, 25 January 1912 to 31 March 1912, sections completed by Niblack, box 809, RG125/EB.

160. Niblack to SecNav, 7 August 1889, in Albert Parker Niblack folder, NDL/ZB.

161. Report on the Fitness of Officers, Ensign A. P. Niblack, 17 April 1889 to 10 May 1891, signed by Captain H. B. Robeson, box 809, RG125/EB.

162. Evaluating Niblack's professional aptitude, one superior later wrote: "I regard Lt. Niblack an unusually fine officer and would be willing to trust him for any position including that of command." Report on the Fitness of Officers, Lieutenant A. P. Niblack, 4 August 1894 to 31 December 1894, signed by Rear Admiral R. W. Meade, box 809, RG125/EB. Many of Niblack's fitness reports contain similar comments.

163. C-in-C SOE to S. A. Staunton, T. E. D. W. Veeder, and A. P. Niblack, 28 March 1891, vol. 1, RG313/CO; Niblack, "Brief Description of the Ardois Night System," 257–63; Niblack, "Signal Question Up to Date," 57–65; Albert P. Niblack, "Naval Signaling," *USNIP* 18, no. 4 (1892): 471–78; and C-in-C SOE to Commanding Officer, USS *Atlanta*, 11 November 1891, vol. 2, RG313/CO.

164. Niblack, "Signal Question Up to Date," 59–63.

165. Wolters, "Recapitalizing the Fleet," 118.

166. Sidney A. Staunton, "Naval Signaling: Discussion," *USNIP* 19, no. 1 (1893): 97.

167. Niblack, "Signal Question Up to Date," 61–65; and General Order No. 345, 3 April 1886, roll 1, M984/GO. In 1890, the navy shifted from English Morse code to American Morse code. General Order No. 380, 2 January 1890, roll 1, M984/GO.

168. R. T. Mulligan et al., "Naval Signaling: Discussion," *USNIP* 18, no. 4 (1892): 490–505.

169. General Order No. 407, 26 January 1893, box 161, filing designator DS, RG45/

SF1; and Albert P. Niblack, "The Signal Question Once More," *USNIP* 28, no. 3 (1902): 553–56.

170. Staunton, "Naval Signaling: Discussion," 97.

171. Note the striking similarities in the following passages: "Rapidity is, therefore, limited by the consideration of simplicity, and increased by mechanical perfection and reliability, which last restricts too rapid methods of signaling. Tactics demand rapidity, but the crowning virtue is reliability. Therefore, *that method is best which is most reliable and which is as simple and rapid as is consistent with absolute reliability*" (Niblack), and "As a general rule, it diminishes the labor and shortens the time of signaling, and increases the scope and value of the system, to increase the number of elements; but this should never reach a point which produces confusion; *i.e.*, precision and certainty should never be sacrificed" (Staunton). Niblack, "Naval Signaling," 482; and Staunton, "Naval Signaling: Discussion," 106.

172. Hamersly, *Records of Living Officers*, 287; and Record of Service of Lieutenant (j.g.) A. P. Niblack, U.S. Navy, 3 August 1896, box 809, RG125/EB.

173. Report on the Fitness of Officers, Lieutenant A. P. Niblack, 4 August 1894 to 31 December 1894, 1 January 1895 to 9 May 1895, both signed by Rear Admiral R. W. Meade, box 809, RG125/EB. The quoted passage is from the latter report.

174. Records of Service of A. P. Niblack, U.S. Navy, 16 June 1902, 14 June 1907, 10 March 1910, and 28 September 1918, all in box 809, RG125/EB. Niblack eventually made flag rank, retiring from naval service in 1923. Biographical sketch of Albert Parker Niblack, 1859–1929, Albert Parker Niblack folder, NDL/ZB.

175. Yeoman 1st Class Hugo L. R. Lehmann to SecNav via Commanding Officer, USS *Bancroft*, 10 January 1901, and corresponding endorsements, all in box 161, filing designator DS, RG45/SF1. *Bancroft*'s commanding officer forwarded Lehmann's suggestions to the Bureau of Navigation, which in turn tasked the North Atlantic Station to investigate and report upon their utility.

176. C-in-C Asiatic Station to SecNav, 17 August 1893, 18 August 1896, box 469, filing designator OO, RG45/SF1.

177. "A Surprise for the Navy: Commodore Bunce in Command of the Atlantic Station," *New York Times*, 20 June 1895, p. 5.

178. C-in-C North Atlantic Station to SecNav, 15 September 1896, box 472, filing designator OO, RG45/SF1.

179. Rather ironically, shipboard officers' increased attention to naval signaling marginalized the role of the signal office, whose functions Secretary Tracy moved into the Bureau of Equipment as part of a broader reorganization that occurred during the summer of 1889. General Order No. 372, 25 June 1889, roll 1, M984/GO.

180. USS *Iowa* Signal Record Book, 22 February 1898 to 26 November 1898, entries for 21, 22, 23 April 1898, Signal-Records Books from Vessels, Sept. 1897–Nov. 1898, Communications "Logs" and Other Records, 1897–1922, Records of the Bureau of Naval Personnel (Record Group 24), U.S. National Archives, Washington, D.C.

CHAPTER 2. Sparks and Arcs: The Navy Adopts Radio

Epigraph. Albert M. Beecher, "Wireless Telegraphy—United States Navy," *Transactions of the American Institute of Electrical Engineers* 19, no. 8 (1902): 578.

1. Jack London, "The Story of an Eye-Witness," *Collier's*, 5 May 1906, 22–23. My recounting of events also draws from Simon Winchester, *A Crack in the Edge of the World: America and the Great California Earthquake of 1906* (New York: Harper Collins, 2005).

2. Bureau of Navigation, Movements of Vessels, 19 April 1906, Daily Reports of Movements of Vessels, 1 September 1897 to 31 December 1915, Records of the Office of the Chief of Naval Operations, 1887–1945, Records Collection of the Office of Naval Records and Library (Record Group 45), U.S. National Archives, Washington, D.C.; and "Log of the United States Flagship *Chicago*," 18 April 1906, *Chicago*, vol. 26, 16 December 1905 to 6 August 1906 (hereafter cited as *Chicago* Deck Log with date, RG24/DL).

3. Typescript of the recollections of Stanford C. Hooper, pp. 2–3, boxes 24–25, LOC/HP.

4. Amalgamation of the U.S. Navy's line officers and engineers is well covered in the literature. See esp. William M. McBride, *Technological Change and the United States Navy, 1865–1945* (Baltimore: Johns Hopkins University Press, 2000), 21–37; and Donald Chisholm, *Waiting for Dead Men's Shoes: Origins and Development of the U.S. Navy's Officer Personnel System, 1793–1941* (Stanford, Calif.: Stanford University Press, 2001), 419–66. A critical assessment of amalgamation may be found in Mark R. Hagerott, "Commanding Men and Machines: Admiralship, Technology, and Ideology in the Twentieth Century" (Ph.D. diss., University of Maryland, 2008), 41–67.

5. Hooper Recollections, 3–4, 8–9, LOC/HP.

6. *Chicago* Deck Log, 18 April 1906, RG24/DL.

7. *Chicago* Deck Logs, 18, 19, 20 April 1906, RG24/DL.

8. *Chicago* Deck Logs, 22 April to 2 May 1906, RG24/DL.

9. Hooper Recollections, 10, LOC/HP.

10. Report on the Fitness of Officers, Commander Charles J. Badger, 31 December 1905 to 30 June 1906, signed by Rear Admiral Caspar F. Goodrich, box 36, RG125/EB.

11. Susan J. Douglas, "Technological Innovation and Organizational Change: The Navy's Adoption of Radio, 1899–1919," in *Military Enterprise and Technological Change: Perspectives on the American Experience*, ed. Merritt Roe Smith (Cambridge, Mass.: MIT Press, 1985), 117–73.

12. Ibid., 126.

13. Bradley A. Fiske, "Discussion," *USNIP* 28, no. 4 (1902): 939, 941.

14. Douglas, "The Navy's Adoption of Radio," 119. See also Linwood S. Howeth, *History of Communications-Electronics in the United States Navy* (Washington: U.S. Government Printing Office, 1963), 65. Fiske's biographer never really addresses his subject's opinions about wireless telegraphy. Paolo E. Coletta, *Admiral Bradley A. Fiske and the American Navy* (Lawrence: University Press of Kansas, 1979).

15. Bradley A. Fiske, "War Signals," *USNIP* 29, no. 4 (1903): 931–34.

16. Ibid., 932.

17. To cite one such anecdote, Douglas recounts the story of a lieutenant who "didn't give a damn about wireless" and ordered the antennas on his warship moved because he objected to their "unsymmetrical appearance." Douglas, "The Navy's Adoption of Radio," 148–49. She fails to mention that this action quickly was countermanded and the antennas moved back to their original (and functioning) location. George H. Clark, "Radio in War and Peace," 43, unpublished manuscript, box 289, CC/100.

18. In 1882–83 Fiske took leave from the navy to study electricity and gain experience in the emerging field of electrical engineering. Before returning to active service, he patented a new insulator and authored a successful textbook on electricity. Bradley A. Fiske, *Electricity in Theory and Practice; or, The Elements of Electrical Engineering* (New York: D. Van Nostrand, 1883). For more on the induction telegraph developed by Woods, see Rayvon Fouché, *Black Inventors in the Age of Segregation: Granville T. Woods, Lewis H. Latimer, and Shelby J. Davidson* (Baltimore: Johns Hopkins University Press, 2003), 28–55.

19. Bradley A. Fiske, *From Midshipman to Rear Admiral* (New York: Century, 1919), 72–76, 99–101.

20. Fiske to Stephen B. Luce, 6 January 1888, Bradley Allen Fiske folder, NDL/ZB.

21. James G. O'Hara and Willibald Pricha, *Hertz and the Maxwellians: A Study and Documentation of the Discovery of Electromagnetic Wave Radiation, 1873–1894* (London: Peter Peregrinus, 1987), 1–21, and Jed Z. Buchwald, *The Creation of Scientific Effects: Heinrich Hertz and Electric Waves* (Chicago: University of Chicago Press, 1994), 262–98.

22. Sungook Hong, *Wireless: From Marconi's Black-Box to the Audion* (Cambridge, Mass.: MIT Press, 2001), 2–16.

23. A. J. L. Blond, "Technology and Tradition: Wireless Telegraphy and the Royal Navy, 1895–1920" (Ph.D. diss., University of Lancaster, 1993), 11–12, 25–31.

24. Throughout the latter nineteenth century, most officers considered the electrically detonated torpedo a formidable means of harbor defense. Timothy S. Wolters, "Electric Torpedoes in the Confederacy: Reconciling Conflicting Histories," *Journal of Military History* 72, no. 3 (July 2008): 760–64, 781 n. 107.

25. Buchwald, *Scientific Effects*, 233–61.

26. Hong, *Wireless*, 2–6.

27. Report of Captain Jackson, 22 May 1897, submitted to C-in-C Devonport, in "Admiralty Correspondence Relating to Signor Marconi's System of Wireless Telegraphy," ADM 116/523.

28. Secretary of the Admiralty (Evan MacGregor) to C-in-C Devonport, 21 August 1896, ADM 116/523; and Blond, "Technology and Tradition," 34–35.

29. Report of Captain Jackson, 16 September 1896, ADM 116/523.

30. Hugh G. J. Aitken, *Syntony and Spark: The Origins of Radio* (New York: John Wiley and Sons, 1976), 216–24.

31. Report of Captain Jackson, 22 May 1897, ADM 116/523.

32. W. P. Jolly, *Marconi* (London: Constable, 1972), 47–48, and Aitken, *Syntony and Spark*, 224–33.

33. Jackson to Frederick T. Hamilton (Commanding Officer, HMS *Defiance*), 19 January 1899, ADM 116/523.

34. Secretary of the Admiralty to Jackson, 1 July 1899, ADM 116/523; and Managing Director, The Wireless and Telegraph Signal Company, Ltd. (H. Jameson Davis) to Director of Navy Contracts, 3 July 1899, in "Marconi W/T System—Agreement between Company and Admiralty," ADM 116/567.

35. Reports of Captain Jackson, 25 July, 10 August 1899, ADM 116/523. A thorough account of the Royal Navy's employment of wireless telegraphy during these maneuvers is available in Rowland F. Pocock and Gerald R. M. Garratt, *The Origins of Maritime Radio: The Story of the Introduction of Wireless Telegraphy in the Royal Navy between 1896 and 1900* (London: Her Majesty's Stationary Office, 1972), 24–32.

36. Report of Captain Jackson, 10 August 1899, ADM 116/523.

37. Covering remarks by Vice Admiral Sir Compton Domville to "Report of Captain Jackson, 10 August 1899," 11 August 1899, ADM 116/523.

38. Secretary of the Admiralty to The Wireless and Telegraph Signal Company, 7 December 1899; Managing Director, The Wireless and Telegraph Signal Company (Major Flood-Page) to Secretary of the Admiralty, 11, 13, 29 December 1899; and minutes of a meeting between Flood-Page and the Controller of the Navy and the Director of Navy Contracts held on 14 December 1899, dated 15 December 1899, ADM 116/567.

39. Secretary of the Admiralty to Postmaster-General, 26 February 1900, ADM 116/567. Britain began fighting the Boer War in October 1899.

40. Secretary of the Admiralty to Marconi's Wireless Telegraph Company, 8 May 1900; and Marconi's Wireless to the Admiralty, 12, 31 May 1900, ADM 116/567. The two parties signed the final contract nine months later. "Contract for Supply of Apparatus and Payment of Royalty," between Marconi's Wireless Telegraph Company, Ltd., and the Admiralty, 20 February 1901, ADM 116/567.

41. Pocock and Garratt, *Origins of Maritime Radio*, 43–44.

42. The word *radio* appeared in the English language around the turn of the century, coming into common usage in the United States sometime between 1906 and 1912. Susan J. Douglas, *Inventing American Broadcasting, 1899–1922* (Baltimore: Johns Hopkins University Press, 1987), xxviii–xxix. My research indicates the American navy made the shift from "wireless" to "radio" in 1912. For evidence of this, see the "Description of Radio Installation for Ships Only" forms, boxes 1–11, Semiannual Reports of Naval Radio Stations and Ships, 1910–17, Records of the Radio Division, Records of the Bureau of Ships (Record Group 19), U.S. National Archives, Washington, D.C.

43. Howeth, *History of Communications-Electronics*, 25–28.

44. Report on the Fitness of Officers, Commander Royal B. Bradford, 27 July 1891 to 31 December 1891 and 1 January 1892 to 5 May 1892, both signed by Rear Admiral John G. Walker; and Report on the Fitness of Officers, Commander Royal B. Bradford, November 1893 to January 1894, signed by Rear Admiral George E. Belknap, all in box 111, RG125/EB.

45. George W. Denfeld to BuEq, 1 November 1899; Edward F. Qualtrough to BuEq, 3 November 1899; John T. Newton to BuEq, 9 November 1899; and John B. Blish to BuEq, 13 November 1899, box 83, RG19/BEQ.

46. Denfeld to BuEq, 1 November 1899; and Blish to BuEq, 13 November 1899, box 83, RG19/BEQ.

47. Newton to BuEq, 9 November 1899, box 83, RG19/BEQ.

48. Denfeld to BuEq, 1 November 1899, box 83, RG19/BEQ.

49. Blish to BuEq, 13 November 1899, box 83, RG19/BEQ.

50. Qualtrough to BuEq, 3 November 1899, box 83, RG19/BEQ.

51. SecNav to John T. Newton, 23 October 1899, in "Report of Board on Marconi System of Wireless Telegraphy," 13 November 1899 (hereafter cited as Marconi Board Report), box 83, RG19/BEQ.

52. Marconi to Newton, Blish, and Frank K. Hill (hereafter cited as Marconi Board Members), 29 October 1899, in Marconi Board Report, box 83, RG19/BEQ.

53. Ibid.

54. "Experiments with three stations, two on vessels underway and one on shore," and "Data for experiments," in Marconi Board Report, box 83, RG19/BEQ.

55. BuEq to SecNav, 1 December 1899, box 83, RG19/BEQ.

56. John B. Blish, "Notes on the Marconi Wireless Telegraph," *USNIP* 25, no. 4 (1899): 857–64.

57. Marconi Board Members to BuEq, Marconi Board Report, box 83, RG19/BEQ.

58. Ibid. Hong's work shows that, at the time of these tests, Marconi's apparatus could not have prevented interference. Marconi and his assistants had been working on the interference problem for more than a year, but as revealed by the infamous Maskelyne affair, even as late as 1903 they still had not solved that problem. Hong, *Wireless*, 89–118.

59. SecNav to The Wireless and Telegraph Signal Company, 11 November 1899, box 459, RG80/TF (the abbreviation "TF" derives from the fact that officials folded these documents with a trifold before filing).

60. BuEq to SecNav, 1 December 1899 (No. 10191-6); and 1st endorsement to No. 10191-6 by BuNav, 4 December 1899, box 83, RG19/BEQ; and 3rd endorsement to No. 10191-6 by Assistant SecNav, 9 December 1899, box 459, RG80/TF.

61. Fourth endorsement to No. 10191-6 by SecNav, 29 December 1899, box 459, RG80/TF.

62. BuEq to SecNav, 5 January 1900, box 459, RG80/TF.

63. BuEq to SecNav, 1 October 1900, in *ARSN*, 1900, 320.

64. Ibid., 265–66.

65. Marconi's Wireless Telegraph Company to the Admiralty, 12, 31 May 1900, ADM 116/567.

66. Aitken, *Syntony and Spark*, 247–54; and Hong, *Wireless*, 62–64, 90–100, 117–18.

67. BuEq to SecNav, 1 October 1901, in *ARSN*, 1901, 380.

68. Office of Naval Intelligence, *Notes on Naval Progress: July, 1900* (Washington: U.S. Government Printing Office, 1900), 284–94.

69. Hong, *Wireless*, 92.

70. BuEq to SecNav, 1 October 1901, in *ARSN*, 1901, 381.

71. BuEq to SecNav, 1 December 1899, box 83, RG19/BEQ.

72. SecNav to Marconi Wireless Telegraph Company, 25 September 1903, box 85, RG19/BEQ. During the period under study in this book, federal fiscal years ran from 1 July to 30 June.

73. Regarding relations with Japan, see esp. Edward S. Miller, *War Plan Orange: The U.S. Strategy to Defeat Japan, 1897–1945* (Annapolis: Naval Institute Press, 1991), 19–30.

74. Paul M. Kennedy, *The Samoan Tangle: A Study in Anglo-German-American Relations, 1878–1900* (New York: Harper & Row, 1974), 189–306.

75. Acting BuEq to SecNav, 9 August 1901, box 1, RG19/WT; and Acting BuEq to BuNav, 13 September 1901, box 83, RG19/BEQ.

76. First endorsement to No. 12479-5 by Assistant SecNav, 3 October 1901, box 531, RG80/TF.

77. Third endorsement to No. 12479-5 by SecNav, 2 November 1901, box 531, RG80/TF; Barber to William H. Beehler (American naval attaché in Berlin), 12 January 1902, box 83, RG19/BEQ; and 3rd endorsement to No. 12479-18 by BuEq, 15 April 1902, box 84, RG19/BEQ.

78. Third endorsement to No. 12479-18 by BuEq, 15 April 1902, box 84, RG19/BEQ; Barber to BuEq, 7 May 1902, box 88, RG19/BEQ; Barber to BuEq, 17 June, 29 July 1902, box 85, RG19/BEQ; and Barber to BuEq, 15 April 1907, box 89, RG19/BEQ.

79. Barber to Beehler, 12, 14 June 1902; Beehler to Barber, 12, 13, 18 June 1902; Barber to BuEq, 11 July 1902, box 85, RG19/BEQ; and Barber to BuEq, 11 February 1907, box 89, RG19/BEQ.

80. Barber to BuEq, 4, 6 December 1901, box 85, RG19/BEQ.

81. BuEq to SecNav, 9 December 1901, box 83, RG19/BEQ.

82. BuEq to Barber, 14 December 1901, box 83, RG19/BEQ; and BuEq to Barber, 13 January 1902, box 84, RG19/BEQ. Barber later received permission to purchase two sets from Braun-Siemens-Halske. Barber to BuEq, 29 July 1902, box 85, RG19/BEQ.

83. BuEq to SecNav, 12 February 1902, box 89, RG19/BEQ.

84. William C. Bean, "When the Navy Went Window Shopping for Wireless, No. 2," as told to George H. Clark, January 1939, box 296, CC/100.

85. Summary of Service, 4 March 1935, John Milton Hudgins folder, NDL/ZB.

86. Frank M. Bennett, *The Steam Navy of the United States: A History of the Growth of the Steam Vessel of War in the U.S. Navy, and of the Naval Engineer Corps* (Pittsburgh: W. T. Nicholson, 1896), 677.

87. Summary of Service, 4 March 1935, John Milton Hudgins folder, NDL/ZB.

88. BuEq to SecNav, 1 October 1902, in *ARSN*, 1902, 373–74.

89. BuEq to SecNav, 15 March 1902, box 89, RG19/BEQ; and Bradford to SecNav, 14 April 1902, box 531, RG80/TF.

90. Barber to BuEq, 9 May 1902, box 88, RG19/BEQ.

91. Hudgins to BuEq, 24 May 1902, box 83, RG19/BEQ. Hudgins noted that *Nashville* held "indications of waves" out to a distance of about fifteen miles.

92. Barber to BuEq, 17 June 1902, box 85, RG19/BEQ.

93. Barber to BuEq, 11 July 1902, box 85, RG19/BEQ; and Hudgins to BuEq, 7 August 1902, box 62, RG19/BEQ.

94. Hudgins to BuEq, 7 August 1902, box 62, RG19/BEQ.

95. Ibid.

96. Acting SecNav to Commander Conway H. Arnold, 14 August 1902; and BuEq to Arnold, 18 August 1902, box 289A, CC/100.

97. BuEq to Arnold, 18 August 1902, box 289A, CC/100; and Wireless Telegraph Board (Conway H. Arnold, George L. Dyer, Charles J. Badger, Albert M. Beecher, and John M. Hudgins) to BuEq, 3 December 1902, box 85, RG19/BEQ.

98. "Marconi to Participate," *New York Herald*, 21 June 1902, a clipping attached to Marconi Wireless Company of America to Barber, 21 June 1902, box 85, RG19/BEQ; and BuEq to SecNav, 1 October 1902, in *ARSN*, 1902, 376.

99. Wireless Telegraph Board to BuEq, 3 December 1902, box 85, RG19/BEQ; BuEq to BuSA, 28 February 1903; and BuEq to Barber (telegram), 17 March 1903, box 87, RG19/BEQ.

100. Barber to BuEq, 30 March 1903, box 87, RG19/BEQ.

101. BuEq to Barber, 13 April 1903, box 87, RG19/BEQ.

102. Allgemeine Elektricitäts-Gesellchaft to BuEq, 10 July 1903, box 85, RG19/BEQ.

103. BuEq to BuCR, 21 March 1903, box 19, RG19/BEQ; BuEq to SecNav, 10 June 1903, box 18, RG19/BEQ; Wireless Telegraph Board to BuEq, 10 July 1903, box 85, RG19/BEQ; BuEq to C-in-C North Atlantic Fleet, 27 July 1903, box 19, RG19/BEQ; and BuEq to SecNav, 17 September 1903, in *ARSN*, 1903, 374.

104. BuEq to C-in-C North Atlantic Fleet, 27 July 1903, box 19, RG19/BEQ.

105. C-in-C European Squadron to SecNav, 20 April 1904; and C-in-C North Atlantic Fleet to SecNav, 1 July 1904, box 473, filing designator OO, RG45/SF1.

106. Douglas, "The Navy's Adoption of Radio," 149.

107. C-in-C North Atlantic Fleet to SecNav, 1 July 1904, box 473, filing designator OO, RG45/SF1.

108. Ibid.

109. Wireless Telegraph Board to BuEq, 28 August 1903, box 293, CC/100.

110. BuEq to Barber (telegram), 9 September 1903; Barber to BuEq, 10 September 1903, box 88, RG19/BEQ; and BuEq to SecNav, 17 September 1903, in *ARSN*, 1903, 374.

111. BuEq (George A. Converse) to SecNav, 1 October 1904, in *ARSN*, 1904, 381; and BuEq to SecNav, 1 October 1905, in *ARSN*, 1905, 319. Converse relieved Bradford on 21 October 1903.

112. Second endorsement to No. 12479 by BuSA, 6 December 1909, box 534, RG80/TF; and undated memorandum by George H. Clark, "List Showing Purchases of Wireless Telegraph Material by the Navy Department from 1903 to August 1, 1907 [and] from August 1, 1907, to June 15, 1910," box 292, CC/100.

113. Fessenden to SecNav, 2 May 1903, box 586, RG80/TF.

114. Assistant SecNav to Fessenden, 7 May 1903, box 586, RG80/TF.

115. SecNav to Secretary of Agriculture, 15 September 1903, box 586, RG80/TF;

Fessenden to SecNav, 23 February 1904, box 293, CC/100; and BuEq to SecNav, 16 December 1904, box 32, RG19/BEQ.

116. BuEq to BuSA, 21 October 1904, box 12, RG19/BEQ. For several years after the formation on Telefunken official navy correspondence continued to refer to "Slaby-Arco" equipment.

117. BuEq to SecNav, 1 October 1906, in *ARSN*, 1906, 384; and SecNav to William S. Greene, 17 December 1908, box 534, RG80/TF.

118. George H. Clark, "Analysis of Companies from whom the U.S. Navy Purchased Wireless Sets in 1906, 1907, and 1908," 2 November 1939, box 289A, CC/100.

119. SecNav to Burton L. French, 30 April 1909, box 534, RG80/TF.

120. Betts, Betts, Sheffield, and Betts to SecNav, 17 July 1903, box 459, RG80/TF; and Fessenden to SecNav, 23 February 1904, box 293, CC/100.

121. Julian Corbett, *Maritime Operations in the Russo-Japanese War, 1904–1905* (Annapolis and Newport: Naval Institute Press and Naval War College Press, 1994), 2:332–33.

122. Ibid., 217–39. Criticism of the Russian navy often focused on the fleet commander's decision not to interfere with Japanese radio transmissions. Seaton Schroeder, "Gleanings from the Sea of Japan," *USNIP* 32, no. 1 (1906): 60–65; and President Naval War College to President General Board, 21 February 1912, box 49, subject file 419, RG80/GB.

123. John M. Hudgins, *Instructions for the Use of Wireless-Telegraph Apparatus* (Washington: U.S. Government Printing Office, 1903), box 292, CC/100; and Navy Department, *Instructions for the Transmission of Messages by Wireless Telegraphy, U.S. Navy* (Washington: U.S. Government Printing Office, 1904), box 290, CC/100.

124. President Naval War College via BuNav to BuEq, 22 March 1904; BuEq via BuNav to President Naval War College, 8 April 1904, box 83, RG19/BEQ; and BuEq to C-in-C North Atlantic Fleet, 18 January 1905, box 32, RG19/BEQ.

125. Third endorsement to No. 118,540 by Lieutenant Louis A. Kaiser, 24 August 1905, box 36, RG19/BEQ.

126. BuEq (Henry N. Manney) to C-in-C North Atlantic Fleet, 8 January 1906, box 32, RG19/BEQ. Manney relieved Converse on 15 March 1904.

127. Third endorsement to No. 118,540 by Lieutenant Louis A. Kaiser, 24 August 1905, box 36, RG19/BEQ.

128. Bradley A. Fiske and R. C. Smith, "Discussion," *USNIP* 28, no. 4 (1902): 939–44; and Fiske, "War Signals," *USNIP* 29, no. 4 (1903): 931–34.

129. SecNav to C-in-C North Atlantic Fleet, 28 December 1905; and C-in-C Atlantic Fleet to SecNav, 9 February 1906, box 473, filing designator OO, RG45/SF1.

130. C-in-C Atlantic Fleet to SecNav, 2 February 1906, summarized in Howeth, *History of Communications-Electronics*, 111–12.

131. Commander, Third Squadron, Atlantic Fleet to his commanding officers, 15 February 1906, Letters Sent to Commanding Officers of Vessels, Squadrons and Fleets, 1865–1940, Records of Naval Operating Forces (Record Group 313), U.S. National Archives, Washington, D.C.

132. BuEq to Bruce Cornwall, 17 May 1905, box 34, RG19/BEQ.

133. Henry T. Wade, "Wireless Telephony in the United States Navy," *Scientific American*, 28 September 1907, 221–22.

134. "The Acquisition of the Wireless Telephone by the U.S. Navy," dictated to George H. Clark by Frank Butler, 10 December 1946, box 292, CC/100.

135. Description #18383-A, Wireless Telephone, De Forest System, Bu. Eq. Req. #74, 22 August 1907, as listed on a cover sheet dated 20 November 1907, box 292, CC/100.

136. "Notes on the De Forest Wireless Telephone in the U.S. Navy," dictated to George H. Clark by William C. Bean, January 1939, box 296, CC/100; and Lee De Forest, *Father of Radio: The Autobiography of Lee De Forest* (Chicago: Wilcox & Follett, 1950), 233. Bean was one of the two enlisted assistants who accompanied John Hudgins to Europe in 1902.

137. Memorandum for BuCR, written by BuEq, 23 November 1907, box 1, RG19/WT.

138. For more on the Great White Fleet, see esp. James R. Reckner, *Teddy Roosevelt's Great White Fleet* (Annapolis: Naval Institute Press, 1988).

139. De Forest, *Father of Radio*, 233.

140. "Acquisition of the Wireless Telephone," by Frank Butler, box 292, CC/100.

141. "The Story of the De Forest Wireless Telephones in the U.S. Navy during the World Cruise, 1907–1908," told to George H. Clark by Henry J. Meneratti, June 1938, box 293, CC/100.

142. "Notes on the De Forest Wireless Telephone," by William C. Bean, box 296, CC/100; and Howeth, *History of Communications-Electronics*, 171.

143. "Story of the De Forest Wireless Telephones," by Henry J. Meneratti, box 293, CC/100.

144. "Acquisition of the Wireless Telephone," by Frank Butler, box 292, CC/100.

145. "Story of the De Forest Wireless Telephones," by Henry J. Meneratti, box 293, CC/100; Recollections of Henry Meneratti, recorded in Hooper Recollections, 293–94, LOC/HP; and "Notes on the De Forest Wireless Telephone," by William C. Bean, box 296, CC/100.

146. Wireless Specialty Apparatus Company to Commander Cleland Davis, 28 March 1908; and Commanding Officer, USS *Rhode Island* to BuEq (William S. Cowles), 27 May 1908, box 1, RG19/WT. Cowles relieved Manney on 22 March 1906.

147. Acting BuEq to Annie May Simpson, 18 May 1909, box 289A, CC/100.

148. Cleland Davis to C-in-C Atlantic Fleet, 8 July 1909, box 88, RG19/BEQ.

149. C-in-C Atlantic Fleet to BuEq, 18 July 1909, box 88, RG19/BEQ.

150. Second endorsement to No. 196537 by Acting BuEq, 27 May 1909, box 1, RG19/WT; and BuEq to SecNav, 14 October 1909, in *ARSN*, 1909, 279–80.

151. Louis W. Austin, "The Work of the U.S. Naval Radio-Telegraphic Laboratory," *Journal of the American Society of Naval Engineers* 24, no. 1 (February 1912): 124.

152. Chen-Pang Yeang, "Scientific Fact or Engineering Specification? The U.S. Navy's Experiments on Wireless Telegraphy circa 1910," *Technology and Culture* 45, no. 1 (January 2004): 1–29.

153. Austin, "Radio-Telegraphic Laboratory," 124–41; and Louis W. Austin, "Report

of the U.S. Naval Radiotelegraphic Laboratory," an enclosure to Austin to BuSE, 10 February 1915, box 293, CC/100.

154. Clark, "Radio in War and Peace," 59–61.

155. George V. L. Meyer, "Annual Report of the Secretary of the Navy," 4 December 1909, in *ARSN*, 1909, 5.

156. George C. Sweet to BuEq, 17 August 1903; and Memorandum for BuNav, written by BuEq, 26 August 1903, box 49, RG19/BEQ.

157. Alexander W. Rilling, "The First Fifty Years of Graduate Education in the United States Navy, 1909–1959," (Ph.D. diss., University of Southern California, 1972), 94–107. For student throughput, see table 2, p. 103.

158. Frank F. Fletcher to BuNav, 8 October 1903, box 22, RG19/BEQ; Charles O. Paullin, *Paullin's History of Naval Administration, 1775–1911* (Annapolis: Naval Institute Press, 1968), 469–70; and Frederick S. Harrod, *Manning the New Navy: The Development of a Modern Naval Enlisted Force, 1899–1940* (Westport, Conn.: Greenwood Press, 1978), 89–90.

159. Navigator, USS *Birmingham* to Commanding Officer, USS *Birmingham*, 9 June 1910, box 13, RG19/BEQ.

160. John Hudgins via BuEq to SecNav, 15 February 1904, box 83, RG19/BEQ.

161. BuEq to BuNav, 6 February 1904, box 83, RG19/BEQ; and BuNav to SecNav, 30 September 1905, in *ARSN*, 1905, 378.

162. Samuel S. Robison, *Manual of Wireless Telegraphy for the Use of Naval Electricians* (Washington: U.S. Government Printing Office, 1906); Navy Biography, 25 April 1957, Samuel Shelburne Robison folder, NOA/OB; and Clark, "Radio in War and Peace," 57.

163. Clark, "Radio in War and Peace," 91–92; BuSE to A. Frederick Collins, 20 June 1912, box 289A, CC/100; and Howeth, *History of Communications-Electronics*, 137–42.

164. Navigator, USS *Kentucky* to Commanding Officer, USS *Kentucky*, 18 April 1905, box 294, CC/100.

165. W. J. Baker, *A History of the Marconi Company* (London: Methuen, 1970), 74–76.

166. Reginald A. Fessenden, "Receiver for Electromagnetic Waves," U.S. Letters Patent No. 727,331, 5 May 1903.

167. Clark, "Radio in War and Peace," 46–49.

168. Navigator, USS *Birmingham* to Commanding Officer, USS *Birmingham*, 9 June 1910, box 13, RG19/BEQ.

169. Greenleaf Whittier Pickard, "Means for Receiving Intelligence Communicated by Electric Waves," U.S. Letters Patent No. 836,531, 20 November 1906.

170. Howeth, *History of Communications-Electronics*, 173–74.

171. Biographical information is from two Civil Service employment forms completed by Miessner c. July 1916 and February 1917. See folder 1, box 3, PU/MP.

172. Benjamin F. Miessner to Charles Miessner, 27 May 1908, folder 4, box 10, PU/MP.

173. Benjamin J. [*sic*] Miessner, "Detector for Wireless Apparatus," U.S. Letters Patent No. 1,104,065, 21 July 1914. Miessner filed for this patent in 1910.

174. George H. Clark to BuSE, 25 April 1911, folder 9, box 19, PU/MP; and Howeth, *History of Communications-Electronics*, 148–49.

175. "Explosion in a Turret Kills 6 on *Kearsarge*," *New York Times*, 15 April 1906, p. 1.

176. Record of the Service of Lieutenant Daniel W. Wurtsbaugh, U.S. Navy, 10 August 1908, box 111, RG125/EB.

177. Daniel W. Wurtsbaugh to C-in-C Atlantic Fleet, 7 January 1909, box 75, RG19/BEQ.

178. Memorandum for BuCR, written by BuEq, 4 January 1902, box 23, RG19/BEQ; BuEq to BuCR, 21 March 1903, box 19, RG19/BEQ; and Equipment Officer Boston Navy Yard to BuEq, 25 March 1903, box 85, RG19/BEQ.

179. Acting BuEq to Commandant Boston Navy Yard et al., 16 June 1903, box 85, RG19/BEQ.

180. Sixth endorsement to No. 78134 by Equipment Officer New York Navy Yard, 28 October 1903, box 20, RG19/BEQ.

181. Hudgins to Commanding Officer, USS *Kearsarge*, 30 May 1904, box 26, RG19/BEQ.

182. Navigator, USS *Kentucky* to Commanding Officer, USS *Kentucky*, 18 April 1905, box 294, CC/100; and Fifth endorsement to No. 145207 by BuEq, 22 December 1906, box 59, RG19/BEQ.

183. Wurtsbaugh to C-in-C Atlantic Fleet, 7 January 1909, box 75, RG19/BEQ.

184. C-in-C Atlantic Fleet to SecNav, 10 January 1909, box 75, RG19/BEQ. Atlantic Fleet commander-in-chief Charles Sperry's comments are worth quoting at length, for they reveal clearly his enthusiasm for the new arrangement: "For long distance work the present rig of aerial in an exposed and lofty position is necessary, and perhaps the present location of wireless rooms above decks is also necessary; but for short distance work during an action it would be a great tactical advantage to be able to shift to a battle wireless safe from an enemy's fire . . . The importance of a well-protected battle wireless installation can hardly be overestimated."

185. General Board endorsement, 24 February 1909, subject file 419, box 49, RG80/GB; and Memorandum for BuCR, written by BuEq, 2 March 1909, box 75, RG19/BEQ.

186. Record of Service of Lieutenant-Commander Daniel W. Wurtsbaugh, U.S. Navy, 25 November 1914, box 111, RG125/EB.

187. Lillian C. White, *Pioneer and Patriot: George Cook Sweet, Commander U.S.N., 1877–1953* (Delray Beach, Fla.: Southern Publishing Company, 1963), 61–62.

188. Sweet to Commander, Fifth Division, U.S. Atlantic Fleet, 10 August 1910, summarized in Howeth, *History of Communications-Electronics*, 175–76.

189. Clark, "Radio in War and Peace," 60–61.

190. Sweet to Commander, Fifth Division, U.S. Atlantic Fleet, 10 August 1910, summarized in Howeth, *History of Communications-Electronics*, 175–76.

191. White, *Pioneer and Patriot*, 69–72.

192. Summary of Service, 12 November 1937, Thomas Tingey Craven folder, NA/VF; and Record of Service of Lieutenant Commander Thomas T. Craven, U.S. Navy, 16 July 1914, box 126, RG125/EB.

193. Clark, "Radio in War and Peace," 81.

194. Navy Biography, 27 April 1955, Stanford Caldwell Hooper folder, NOA/OB.

195. Clark, "Radio in War and Peace," 78–80.

196. Hooper Recollections, 24, LOC/HP.

197. Biographical sketch of Charles Hamilton Maddox through 25 July 1941, in Charles Hamilton Maddox folder, NA/VF.

198. BuSE to Head of Engineering Experiment Station, Annapolis, Md., 15 August 1911, box 1, LOC/HP.

199. Maddox as quoted in Clark, "Radio in War and Peace," 86.

200. Recollections of Charles H. Maddox, recorded in Hooper Recollections, 185–86, LOC/HP. Maddox studied mainly under George W. Pierce, an up-and-coming physicist who later would become the director of Harvard's Cruft Laboratory.

201. Hooper Recollections, 33, LOC/HP.

202. Ibid., 33–34.

203. Ibid., 34–36; Clark, "Radio in War and Peace," 84; and Douglas, "The Navy's Adoption of Radio," 158.

204. Hooper to BuSE, 12 April 1912, box 296, CC/100. This document is a handwritten transcription done by George Clark, but there is no reason to doubt its accuracy.

205. Ibid.

206. Clark, "Radio in War and Peace," 117–18; and Hooper Recollections, 36, LOC/HP.

207. Biographical sketch of Charles Hamilton Maddox through 25 July 1941, in Charles Hamilton Maddox folder, NA/VF; and Navy Department, *Register of the Commissioned and Warrant Officers of the United States Navy and Marine Corps, January 1, 1915* (Washington: U.S. Government Printing Office, 1915), 273.

208. William H. H. Southerland, *Fleet Regulations, United States Pacific Fleet*, September 1912, box 296, CC/100.

209. Most armored ships had conning towers; unarmored ships did not.

210. Hooper to BuSE, 12 April 1912, box 296, CC/100.

211. First endorsement to No. 84672-493-W by BuSE, 27 December 1912, box 296, CC/100. This document is another handwritten transcription by George Clark.

212. Hooper Recollections, 51, LOC/HP; "Adm. C. J. Badger Dies at Age of 79," *Washington Evening Star*, 8 September 1932, newspaper clipping found in Charles Johnston Badger folder, NDL/ZB; and Navy Department, *Register of the Commissioned and Warrant Officers of the United States Navy and Marine Corps, January 1, 1914* (Washington: U.S. Government Printing Office, 1914), 243.

213. Hooper as quoted in Clark, "Radio in War and Peace," 123.

214. Hooper to BuSE, 12 April 1912, box 296, CC/100.

215. Clark, "Radio in War and Peace," 125.

216. Charles J. Badger, "Specifications by C in C for Standard Radio Sets on Ships," 7 April 1913, box 295, CC/100.

217. BuSE (Robert S. Griffin) to BuSE Material Department, 7 March 1914; and Badger to SecNav, 10 March 1914, box 290, CC/100.

218. Badger, "Specifications by C in C for Standard Radio Sets on Ships," 7 April 1913, box 295, CC/100. At this time radio transmissions were measured in wavelengths

instead of frequencies. In succeeding chapters, I use the modern terminology (i.e., hertz).

219. Clark, "Radio in War and Peace," 105–6, 133–35, 139–40; and Guy Hill to Lieutenant Commander C. R. Miller, 29 May 1913, box 289A, CC/100.

220. Critical for understanding why an enemy would be unable to jam quickly a U.S. fleet's newly selected transmitting frequency is the fact that each channel on the wave changer was discrete. An enemy either would have needed a variable device (not possible given the existing state of the art) or would have had to possess wave changers nearly identical to those operated by the U.S. fleet. The latter would not have been possible without a security breach and would have required an enemy to know well in advance that it was going to fight against a U.S. fleet.

221. Clark, "Radio in War and Peace," 106, 135, 140.

222. Fleet Radio Officer (Hooper) to C-in-C Atlantic Fleet (Badger), 29 March 1914, box 1, LOC/HP.

223. Austin to BuSE Radio Division (Arthur J. Hepburn), 5, 16 June 1914; Austin to Hepburn, 30 July 1914, box 292, CC/100.

224. Henry J. Meneratti to George H. Clark, 13 April 1915, box 295, CC/100.

225. Ibid.

226. In the U.S. Navy, officers between the ranks of ensign and lieutenant commander, inclusive, traditionally are considered "junior officers."

227. Clark, "Radio in War and Peace," 141F–141G. Clark does not give a specific year for this incident, but it had to have occurred late in Hooper's tenure as Fleet Radio Officer.

CHAPTER 3. War and Peace: Coordinating Naval Forces

Epigraph. Fleet Radio Officer, "Fleet Problem VI—Report of Operations—Communications," in enclosure (b) to C-in-C U.S. Fleet to CNO, 14 March 1926, roll 7, M964/FP.

1. John Jellicoe, *The Grand Fleet, 1914–1916: Its Creation, Development and Work* (New York: George H. Doran, 1919), 341–43.

2. My recounting of events draws from Jellicoe's recollections and several other sources, including Frederick Dreyer, *The Sea Heritage: A Study of Maritime Warfare* (London: Museum Press, 1955), 115–32, and Arthur J. Marder, *Jutland and After (May 1916–December 1916)*, vol. 3, *From the Dreadnought to Scapa Flow: The Royal Navy in the Fisher Era, 1904–1919* (New York: Oxford University Press, 1966), 36–95. Although Marder's work offers a detailed description of events, recent scholarship has overturned many of his assumptions and conclusions. See esp. Nicholas A. Lambert, "'Our Bloody Ships' or 'Our Bloody System'? Jutland and the Loss of the Battle Cruisers, 1916," *Journal of Military History* 62 (January 1998): 29–55, and Jon Tetsuro Sumida, "A Matter of Timing: The Royal Navy and the Tactics of Decisive Battle, 1912–1916," *Journal of Military History* 67 (January 2003): 85–136.

3. Percy Leonard Herbert Burgess to Robert Church, [1972], Burgess folder, box 1, IWM/CC.

4. Marder, *Dreadnought to Scapa Flow*, 3:45–55, 93–110, 151–86, 215–24.

5. Andrew Gordon, *The Rules of the Game: Jutland and British Naval Command* (London: John Murray, 1996; Annapolis: Naval Institute Press, 2000).

6. Arthur John Brister, "W/T Apparatus and Organization," attached to Brister to Church, 23 July 1972, Brister folder, box 1, IWM/CC.

7. Gordon, *Rules of the Game*, 506.

8. Arthur Hezlet, *Electronics and Sea Power* (New York: Stein and Day, 1975), 127.

9. Analyses of Jutland can be found throughout the professional literature of the period. Writings by American naval officers in the decade after the battle include George Dewey, "Lessons of Skagerrack," *Sea Power: The Nation's Defense* 1, no. 2 (July 1916): 7–10, 35; Holloway H. Frost, "The Results and Effects of the Battle of Jutland," *USNIP* 47, no. 9 (1921): 1327–54; Alfred W. Hinds, "Practical Lessons for the American Navy from the Battle of Jutland," *USNIP* 47, no. 12 (1921): 1877–84; and Thomas C. Hart, "What Might Have Happened at Jutland," *USNIP* 51, no. 7 (1925), 1143–51.

10. C-in-C U.S. Fleet, "Report on United States Fleet Problem Number One," 19 June 1923 (hereafter cited as "Report on Fleet Problem One"), p. 3, roll 1, M964/FP.

11. Archibald Owen, comp., *Navy Yearbook*, 65th Cong., 3d sess., 1918–19, S.Doc. 418, 383.

12. Robert W. Love Jr., *History of the U.S. Navy, 1775–1941* (Harrisburg, Pa.: Stackpole Books, 1992), 457–58. Predictably, naval historians have been more critical of Daniels than his biographer. See Joseph L. Morrison, *Josephus Daniels: The Small-d Democrat* (Chapel Hill: University of North Carolina Press, 1966).

13. Josephus Daniels, "Annual Report of the Secretary of the Navy," 1 December 1914, in *ARSN*, 1914, 6–7, 23–26.

14. Ibid., 29.

15. Daniels, "Annual Report of the Secretary of the Navy," 1 December 1913, in *ARSN*, 1913, 18.

16. Daniels to Thomas A. Edison, 7 July 1915, reproduced in Lloyd N. Scott, *Naval Consulting Board of the United States* (Washington: U.S. Government Printing Office, 1920), 286–88.

17. William M. McBride, "The 'Greatest Patron of Science'? The Navy-Academia Alliance and U.S. Naval Research, 1896–1923," *Journal of Military History* 56 (January 1992): 7–33.

18. William M. McBride, *Technological Change and the United States Navy, 1865–1945* (Baltimore: Johns Hopkins University Press, 2000), 92. See also McBride, "The Navy-Academia Alliance," 25–27.

19. Bradley Fiske was arguably the most vocal expansionist. Paolo E. Coletta, *Admiral Bradley A. Fiske and the American Navy* (Lawrence: University Press of Kansas, 1979), 131–86.

20. Historians often credit Congress with authorizing "a navy second to none," but this phrase appears nowhere in the Navy Act of 1916. Owen, *Navy Yearbook*, 399–467. As historian William Braisted points out, the bill was really "a move to recover for the United States second place [behind Britain] among the naval powers." William R. Brai-

sted, *The United States Navy in the Pacific, 1909–1922* (Austin: University of Texas Press, 1971), 202.

21. Typescript of the recollections of Stanford C. Hooper, pp. 232–34, 490–93, boxes 24–25, LOC/HP. The quoted passage is from p. 493.

22. Ibid., 232–34, 493, and Robert S. Griffin, *History of the Bureau of Engineering, Navy Department, during the World War* (Washington: U.S. Government Printing Office, 1922), 11–12.

23. George H. Clark, "Radio in War and Peace," 154–59, unpublished manuscript, box 289, CC/100; and "Report of Board on Organization U.S. Naval Radio Service," 20 February 1915, "Rewritten and corrected to include Supplementary Report of the Board of Organization U.S. Naval Radio Service," 14 July 1916, box 1, LOC/HP.

24. Hooper Recollections, 233X, LOC/HP. I have converted wavelengths to their corresponding frequencies (i.e., 2,000 meters = 750 kilohertz) since this will be more familiar to the modern reader.

25. "Report of Board on Organization of Radio Service," box 1, LOC/HP.

26. Edwin H. Dodd to Stanford C. Hooper, 18 July 1916, box 1, LOC/HP.

27. "Report of Board on Organization of Radio Service," box 1, LOC/HP.

28. Superintendent, Naval Radio Service to BuNav, 20 September 1915, in *ARSN*, 1915, 285.

29. Ibid., 284–86.

30. Recollections of Arthur J. Hepburn, recorded in Hooper Recollections, 244–62, LOC/HP; "Adm. Arthur J. Hepburn Dies," *New York Times*, 1 June 1964, p. 29; and *Annual Register of the United States Naval Academy*, 1 October 1908 (Washington: U.S. Government Printing Office, 1908), 19.

31. George H. Clark, "Report of Tests between Arlington and USS *Salem*," 15 May 1913, box 289A, CC/100; and Linwood S. Howeth, *History of Communications-Electronics in the United States Navy* (Washington: U.S. Government Printing Office, 1963), 143.

32. Heterodyning is the generation of a new frequency by combining two other frequencies. Fessenden's heterodyne receiver had an oscillator circuit that produced a signal close in frequency to the one being received. When the two signals combined, a "beat" frequency in the audible range was created. This process allowed operators to hear the dots and dashes of Morse code as "a pure musical sound." George H. Clark, "Report of Tests between Arlington and USS *Salem*," 15 May 1913, box 289A, CC/100.

33. Recollections of Arthur J. Hepburn, recorded in Hooper Recollections, 260, LOC/HP.

34. Hepburn to BuSE, 3 April 1913, quoted in Howeth, *History of Communications-Electronics*, 147; and George H. Clark, "Arc Apparatus Purchased by U S Navy," compiled in June 1935, box 290, CC/100.

35. Recollections of Arthur J. Hepburn, recorded in Hooper Recollections, 256, LOC/HP; and Superintendent, Naval Radio Service to BuNav, 18 March 1914, in *ARSN*, 1914, 213–16.

36. Clark, "Radio in War and Peace," 107.

37. In this era the bureau used the term *research and design* to refer to what now

would be known as *research and development*. Because the two phrases are so similar, I have retained the original terminology.

38. Hooper later claimed he was solely responsible for this expansion, but radio aide George Clark explicitly gave the credit to Hepburn. The timing of events supports Clark's version. Hooper reported to the radio division the last week of April, yet all six experts had been hired by the end of June. Hooper also wrote a rather self-serving letter in the fall of 1916 in which he listed his accomplishments with respect to naval radio organization, and this document does not mention hiring radio aides. A plausible explanation is that Hooper assisted Hepburn in obtaining these positions after he returned from Europe but before he took over the radio division. Hooper Recollections, 724–28, LOC/HP; Clark, "Radio in War and Peace," 352–53; Hooper to SecNav, 9 October 1916, box 1, LOC/HP; and Superintendent, Naval Radio Service, to BuNav, 20 September 1915, in *ARSN*, 1915, 269–70.

39. Clark, "Radio in War and Peace," 111–12.

40. Ibid., 94.

41. Recollections of Arthur J. Hepburn, recorded in Hooper Recollections, 258–59, LOC/HP. John Firth, Greenleaf Whittier Pickard, and Philip Farnsworth had founded the Wireless Specialty Apparatus Company in 1907.

42. C-in-C Atlantic Fleet to SecNav, 16 May 1914, box 290, CC/100.

43. Superintendent, Naval Radio Service, to BuNav, 20 September 1915, in *ARSN*, 1915, 267, 269–70; and Louis A. Gebhard, *Evolution of Naval Radio-Electronics and Contributions of the Naval Research Laboratory* (Washington: U.S. Government Printing Office, 1979), 18–20. The frequency ranges of these receivers were 60–600 kilohertz, 30–300 kilohertz, and 1,200–3,000 kilohertz for types A, B, and C, respectively. Howeth, *History of Communications-Electronics*, 594.

44. Lee De Forest, "Oscillation-Responsive Device," U.S. Letters Patent No. 824,637, 26 June 1906; De Forest, "Space Telegraphy," U.S. Letters Patent No. 879,532, 18 February 1908; Hugh G. J. Aitken, *The Continuous Wave: Technology and American Radio, 1900–1932* (Princeton, N.J.: Princeton University Press, 1985), 233–49; and Sungook Hong, *Wireless: From Marconi's Black-Box to the Audion* (Cambridge, Mass.: MIT Press, 2001), 181–89.

45. Clark, "Radio in War and Peace," 100–101, 163–64.

46. Bureau of Steam Engineering Semi-Monthly Radio Report, 15 June 1916, box 1, LOC/HP.

47. Aitken, *Continuous Wave*, 216–23.

48. Clark, "Radio in War and Peace," 101, 167–68; and Gebhard, *Evolution of Naval Radio-Electronics*, 19.

49. Reginald Fessenden's original heterodyne receiving system employed a small arc to generate local oscillations in the receiving circuit.

50. Clark, "Radio in War and Peace," 168–69.

51. George H. Clark, "Arc Apparatus Purchased by U S Navy," compiled in June 1935, box 290, CC/100; and Clark, "Radio in War and Peace," 99–100, 165.

52. George H. Clark, "Classes of Receiving Units Used by the U.S. Navy during the War Period," n.d., box 295, CC/100.

53. Clark, "Radio in War and Peace," 162.

54. Recollections of Charles H. Maddox, recorded in Hooper Recollections, 187–91, LOC/HP.

55. Maddox as quoted in Clark, "Radio in War and Peace," 172.

56. Howeth, *History of Communications-Electronics*, 190; and Recollections of Charles H. Maddox, recorded in Hooper Recollections, 190, LOC/HP.

57. SecNav to Maddox, 21 August 1934, Charles Hamilton Maddox folder, NDL/ZB.

58. Clark, "Radio in War and Peace," 174.

59. Frank F. Fletcher to SecNav, 9 July 1916, in U.S. Senate, *Target Practice in the Navy*, 64th Cong., 1st sess., 1916, Doc. No. 496.

60. Roy A. Grossnick, *United States Naval Aviation, 1910–1995* (Washington: U.S. Government Printing Office, 1997), 17.

61. Clark, "Radio in War and Peace," 175.

62. BuSE to Commander of Air Service, USS *North Carolina*, 13 May 1916, box 1, LOC/HP.

63. Clark, "Radio in War and Peace," 175–77, and Howeth, *History of Communications-Electronics*, 268.

64. Hooper Recollections, 238, LOC/HP; and Theodore Johnson Jr., "Naval Aircraft Radio (First Half)," *Proceedings of the Institute of Radio Engineers* 8, no. 1 (February 1920): 5–7.

65. Benjamin F. Miessner, "Brief History of my Birth, Parentage, and Early Life—Statement," [1918], folder 6, box 32, PU/MP.

66. Clark, "Radio in War and Peace," 179–82, and Theodore Johnson Jr., "Naval Aircraft Radio (Second Half)," *Proceedings of the Institute of Radio Engineers* 8, no. 2 (April 1920): 109.

67. Benjamin F. Miessner, "Special Radio Report on Noise Conditions Affecting Radio Reception on Airplanes," 10 January 1917, folder 5, box 33, PU/MP; Commandant [U.S. Navy Aeronautic Station, Pensacola] to BuSE, 8 February 1917, and BuSE to Commandant Navy Aeronautic Station, Pensacola, 14 February 1917, folder 1, box 3, PU/MP.

68. A. Hoyt Taylor, *Radio Reminiscences: A Half Century* (Washington: Naval Research Laboratory, 1948), 113–14. This typescript is available in the Navy Department Library, Washington, D.C.

69. BuSE to Commander of Air Service, USS *North Carolina*, 13 May 1916, box 1, LOC/HP.

70. Johnson, "Naval Aircraft Radio (First Half)," 10, and Johnson, "Naval Aircraft Radio (Second Half)," 109.

71. BuSE to SecNav, 11 October 1918, in *ARSN*, 1918, 530.

72. Norman Friedman, *U.S. Submarines through 1945* (Annapolis: Naval Institute Press, 1995), 19–73.

73. George H. Clark to Lieutenant Commander Cleland Davis, 12 June 1909, 6 July 1909, box 293, CC/100.

74. Chief Clerk, BuEq to SecNav, 1 June 1910, box 436, RG80/TF.

75. "Memorandum on Tests Made and Conclusion," written by Chief Clerk, BuEq, 2 June 1910, box 534, RG80/TF.

76. Friedman, *U.S. Submarines through 1945*, 61.

77. Engineer Officer to Commandant [New York Navy Yard?], 22 May 1912, box 292, CC/100.

78. D. Pratt Manney, "Report on Tests of Two Methods for Radio Communication from a Submerged Submarine," 30 November 1915, box 293, CC/100.

79. For more on these tests, see Timothy S. Wolters, "Early Experiments in Submarine Wireless," *Submarine Review* 29 (July 2011): 119–29.

80. Daniels, "Report of the Secretary of the Navy," 1 December 1915, in *ARSN*, 1915, 42–43; and "Wireless Speech Flies to Pacific," *New York Times*, 30 September 1915, p. 1. Western Electric was the manufacturing arm of AT&T, which helps explain the company's considerable interest in the radiotelephone. The relationship between Western Electric and AT&T before World War I is covered in George David Smith, *The Anatomy of a Business Strategy: Bell, Western Electric, and the Origins of the American Telephone Industry* (Baltimore: Johns Hopkins University Press, 1985).

81. Multiplexing permits the sending of multiple information streams over a single channel. The technique originated with the telegraph.

82. Clark, "Radio in War and Peace," 232–33; Hooper to Lieutenant H. W. McCormack, 3 March 1917, box 1, LOC/HP; and Griffin, *History of the Bureau of Engineering*, 98–100.

83. The most comprehensive account of American naval operations during World War I is William N. Still Jr., *Crisis at Sea: The United States Navy in European Waters in World War I* (Gainesville: University Press of Florida, 2006). Very useful for establishing the broader context of U.S. naval operations is Paul G. Halpern, *A Naval History of World War I* (Annapolis: Naval Institute Press, 1994).

84. Jerry W. Jones, *U.S. Battleship Operations in World War I* (Annapolis: Naval Institute Press, 1998). In August 1918 the United States sent a second division. It operated independently of the Grand Fleet.

85. Mary Klachko with David F. Trask, *Admiral William Shepherd Benson: First Chief of Naval Operations* (Annapolis: Naval Institute Press, 1987), 62–73; and Still, *Crisis at Sea*, 12–13, 307.

86. Quoted in Tracy B. Kittredge, *Naval Lessons of the Great War: A Review of the Senate Investigation of the Criticisms by Admiral Sims of the Policies and Methods of Josephus Daniels* (Toronto: Doubleday, Page, 1921), 285. On the historiographic debates surrounding U.S. naval preparedness before World War I, see Joseph W. Kirschbaum, "The 1916 Naval Expansion Act: Planning for a Navy Second to None" (Ph.D. diss., George Washington University, 2008), 1–18.

87. Joseph K. Taussig, "Destroyer Experiences during the Great War: Convoying Merchant Ships," *USNIP* 49, no. 2 (1923), 221–48. On World War I convoy operations,

see esp. Glenn Ansel Stackhouse, "The Anglo-American Atlantic Convoy System in World War I" (Ph.D. diss., University of South Carolina, 1993), 22–63.

88. For this attribution, see Still, *Crisis at Sea*, 342.

89. Willem Hackmann, *Seek and Strike: Sonar, Anti-Submarine Warfare and the Royal Navy, 1914–1954* (London: Her Majesty's Stationery Office, 1984), 64–71.

90. Jones, *U.S. Battleship Operations*, 22–55, and Joseph K. Taussig, *The Queenstown Patrol, 1917: The Diary of Commander Joseph Knefler Taussig, U.S. Navy*, ed. William N. Still Jr. (Newport, R.I.: Naval War College Press, 1996).

91. BuEng to C-in-C Pacific Fleet, c. May 1922, box 296, CC/100; and Hooper Recollections, 233X–234, LOC/HP.

92. Howeth, *History of Communications-Electronics*, 595.

93. Lieutenant Paul H. Bastedo to Commander U.S. Naval Forces Operating in European Waters, 24 October 1917, box 1,RG19/OM.

94. BuSE to Commander North Atlantic and West Indies Station [Royal Navy], 16 October 1918, box 1, RG19/OM; and Clark, "Radio in War and Peace," 318–25.

95. Lewis Bayly, *Pull Together! The Memoirs of Admiral Sir Lewis Bayly* (London: George G. Harrap, 1939), 225; Still, *Crisis at Sea*, 335; and Howeth, *History of Communications-Electronics*, 292–94.

96. A. J. L. Blond, "Technology and Tradition: Wireless Telegraphy and the Royal Navy, 1895–1920" (Ph.D. diss., University of Lancaster, 1993), 355, and Norman Friedman, *Network-Centric Warfare: How Navies Learned to Fight Smarter through Three World Wars* (Annapolis: Naval Institute Press, 2009), 7.

97. Taussig, *Queenstown Patrol*, 53.

98. Taussig, "Convoying Merchant Ships," 239.

99. Theodore Johnson Jr., "Naval Radio Tube Transmitters," *Proceedings of the Institute of Radio Engineers* 9, no. 5 (October 1921): 393–97; Griffin, *History of the Bureau of Engineering*, 37, 99; and Walter S. Lemmon, "Recent Development of Radio Telephones," *Proceedings of the Radio Club of America* 1, no. 4 (June 1920): 2–7.

100. Griffin, *History of the Bureau of Engineering*, 99.

101. William S. Sims, *The Victory at Sea* (New York: Doubleday, Page, 1920; Annapolis: Naval Institute Press, 1984), 217, and Michael Simpson, *Anglo-American Naval Relations, 1917–1919* (Aldershot: Scolar Press for the Navy Records Society, 1991), 194.

102. Commander USS *Hannibal* (Escort Commander) to All vessels concerned, 21 December 1917, box 387, filing designator OD, RG45/SF2.

103. Sims, *Victory at Sea*, 120–21.

104. Geoffrey L. Rossano, *Stalking the U-Boat: U.S. Naval Aviation in Europe during World War I* (Gainesville: University Press of Florida, 2010), 2, 81, 92–95. Rossano does not specify if this Tellier carried a radio or if its apparatus was inoperative. Although many seaplanes carried radios by 1917, the Telliers operated by the U.S. Navy out of Le Croisic were older, single-engine models that may not have had the lifting capacity for on-board radio equipment. James J. Davilla and Arthur M. Soltan, *French Aircraft of the First World War* (Stratford, Conn.: Flying Machines Press, 1997), 529–31.

105. Rossano, *Stalking the U-Boat*, 4, 71–72.

106. Officer-in-Charge, Development of Radio for Aircraft to Commandant, Aeronautic Station, Pensacola, 7 December 1916, folder 1, box 3, PU/MP.

107. C-in-C Atlantic Fleet to Atlantic Fleet, 29 January 1917, folder 1, box 3, PU/MP. In the same letter, Mayo also called for volunteer radio operators for submarine duty.

108. Commandant [U.S. Navy Aeronautic Station, Pensacola] to BuSE, 1 March 1917, and BuSE to Commandant Navy Aeronautic Station, Pensacola, 9 March 1917, both in folder 1, box 3, PU/MP.

109. "Report of Conference on subject Aircraft Radio, Bureau of Steam Engineering, May 30th, 1917," 2 June 1917, folder 1, box 3, PU/MP; Civil Service announcement No. 1266, issued 7 June 1917, and R. D. Kirkpatrick, "Memo relative to the Radio Laboratory," 6 July 1917, with a hand-written endorsement by Miessner, n.d., folder 5, box 32, PU/MP.

110. Lieutenant Robert A. Lavender to BuSE, 14 September 1917, box 1, RG19/OM.

111. Company brochure entitled "Simon Radio," [1919?], folder 10, box 29, PU/MP; Emil J. Simon to BuSE, 2 October 1917, box 1, RG19/OM; "U.S. Navy 1/2 KW Aircraft Radio Transmitter Supplied under Requisition 580 N. S. A. Contract 33259," 26 November 1917, folder 10, box 1, PU/MP; and BuSE to SecNav, 11 October 1918, in *ARSN*, 1918, 530.

112. Assistant Secretary of the Navy (Franklin D. Roosevelt) to Benjamin F. Miessner (telegram), 10 January 1918; Ralph G. Pennoyer to Miessner (telegram), 11 January 1918; and William E. Cox to Miessner, 5 June 1918, all in folder 6, box 32, PU/MP.

113. BuSE to Miessner, 1, 3 July 1918, and Miessner to BuSE, 2 July 1918, folder 6, box 32, PU/MP.

114. Taylor, *Radio Reminiscences*, 50–72, 102–20.

115. Ibid., 105.

116. Howeth, *History of Communications-Electronics*, 602, and Johnson, "Naval Aircraft Radio (Second Half)," 111–12.

117. William A. Eaton to BuSE, 3 August 1918, box 2, RG19/OM.

118. Griffin, *History of the Bureau of Engineering*, 122–23.

119. Johnson, "Naval Aircraft Radio (First Half)," 48–49; Benjamin F. Miessner, "Telephone Transmitter," U.S. Letters Patent No. 1,507,081, 2 September 1924. Miessner wrote that he sold this patent to Magnavox for $3,300. Hand-written note by Miessner, folder 9, box 19, PU/MP.

120. Johnson, "Naval Aircraft Radio (Second Half)," 139.

121. Griffin, *History of the Bureau of Engineering*, 121–22; Noel Shirley, ed., "Wartime Memoirs of John Jay Schieffelin," *Over the Front* 9, no. 2 (1994): 115; and Rossano, *Stalking the U-Boat*, 240–41.

122. Frost, "Results and Effects of the Battle of Jutland," 1336–41.

123. Hinds, "Practical Lessons for the American Navy," 1881–82.

124. Hart, "What Might Have Happened at Jutland," 1149–50.

125. William M. Blumerkrance [Blumenkranz] to BuSE, 20 November 1918, box 290, CC/100.

126. Lieutenant Paul H. Bastedo to Commander U.S. Naval Forces Operating in Eu-

ropean Waters, 24 October 1917, box 1, RG19/OM; Blumenkranz to BuSE, 20 November 1918, box 290, CC/100; and Blumenkranz to BuSE, 11 January 1919, box 1, RG19/OM. The quoted phrase is from the report dated 20 November 1918.

127. Captain Ernest L. Bennett, "History of Training Division, Bureau of Navigation, Navy Department, during the World War," September 1920, 242–48, 324–27, box 941, filing designator ZNA, RG45/SF2. The number of personnel trained is reported in the undesignated enclosure "List of Training Schools and their Locations Operated by the U.S. Navy between April 6th, 1917 and November 11th, 1918," 10–11.

128. Notes on a lecture delivered by T. A. M. Craven to the Naval Post-Graduate School, 16 February 1924, contained in the Radio Division's "Monthly Radio and Sound Report," April 1924, box 291, CC/100.

129. C-in-C Atlantic Fleet to SecNav, 1 July 1920, roll 1, M971/FR.

130. Thomas C. Hone, "Spending Patterns of the United States Navy, 1921–1941," *Armed Forces and Society* 8 (Spring 1982): 447–48.

131. Emily O. Goldman, *Sunken Treaties: Naval Arms Control between the Wars* (University Park: Pennsylvania State University Press, 1994), and Richard W. Fanning, *Peace and Disarmament: Naval Rivalry and Arms Control, 1922–1933* (Lexington: University Press of Kentucky, 1995). For international events leading to the Washington Naval Conference, see Roger Dingman, *Power in the Pacific: The Origins of Naval Arms Limitation, 1914–1922* (Chicago: University of Chicago Press, 1976).

132. George H. Clark, "Arc Apparatus Purchased by U S Navy," compiled in June 1935, box 290, CC/100; Clark, "Chronological 'History-Service' Chart of Navy Radio," [1928?], box 291, CC/100.

133. Radio Division, "Monthly Sound and Radio Report," December 1921, box 291, CC/100. Only an extract of the full report is located here. The Bureau of Steam Engineering became the Bureau of Engineering in June 1920.

134. C-in-C U.S. Fleet, "Annual Report, 1 July 1923 to 30 June 1924," 28 August 1924, p. 27, roll 4, M971/FR; C-in-C U.S. Fleet, "Annual Report, 1 July 1924 to 30 June 1925," 9 July 1925, p. 24, roll 5, M971/FR; and C-in-C Scouting Fleet, "Annual Report, Fiscal Year 1926," 25 June 1926, p. 14, roll 5, M971/FR.

135. "Extracts from War College Confidential Maneuver Rules, June 1923," roll 63, M1140/SCC.

136. C-in-C U.S. Fleet to CNO, 16 November 1925, roll 60, M1140/SCC.

137. Jon Tetsuro Sumida, *In Defence of Naval Supremacy: Finance, Technology, and British Naval Policy, 1889–1914* (New York: Routledge, 1993), 71–107, 146–58, 196–248; Sumida, "The Quest for Reach: The Development of Long-Range Gunnery in the Royal Navy, 1901–1912," in *Tooling for War: Military Transformation in the Industrial Age*, ed. Stephen D. Chiabotti (Chicago: Imprint Publications, 1996), 49–96; and David A. Mindell, *Between Human and Machine: Feedback, Control, and Computing before Cybernetics* (Baltimore: Johns Hopkins University Press, 2002), 19–68.

138. Frederic Dreyer to Vice Admiral Commanding Second Battle Squadron, 4 May 1914; and Hooper Recollections, 172–73, LOC/HP. I am indebted to Jon Sumida for providing me with a copy of Dreyer's letter.

139. Hannibal Choate Ford, "Battle-Tracer," U.S. Letters Patent No. 1,293,747, 11 February 1919. Ford filed for this patent on 4 August 1914.

140. Mindell, *Between Human and Machine*, 30–33, and Friedman, *Network-Centric Warfare*, 42–43.

141. C-in-C Battle Fleet, "Annual Report, 1 July 1923 to 30 June 1924," 22 July 1924 (hereafter cited as "Battle Fleet Report for FY 1924"), pp. 42–43, roll 4, M971/FR.

142. Commander ZED SECTOR (C.B.D.3) [Nulton] to C-in-C (O.T.C.) [C-in-C Battle Fleet], 14 January 1924, roll 1, M964/FP. I have corrected the date, erroneously typed as "1923." For similar reports, see C-in-C Battle Fleet, "Annual Report, 1 July 1924 to 30 June 1925," c. July 1925 (hereafter cited as "Battle Fleet Report for FY 1925"), p. 5, roll 5, M971/FR; and C-in-C U.S. Fleet, "Annual Report, 8 November 1927 to 30 June 1928," 26 September 1928, p. 13, roll 7, M971/FR.

143. Commander ZED SECTOR (C.B.D.3) to C-in-C (O.T.C.), 14 January 1924, roll 1, M964/FP.

144. C-in-C Atlantic Fleet to Battleship Forces One and Two, 27 May 1918; and C-in-C Atlantic Fleet to Atlantic Fleet, 10 February 1919, both in box 292, CC/100.

145. C-in-C Atlantic Fleet to Atlantic Fleet, 10 February 1919, box 292, CC/100.

146. C-in-C U.S. Fleet, "Annual Report, 1 July 1926 to 30 June 1927," 27 September 1927, p. 48, roll 6, M971/FR.

147. Albert A. Nofi, *To Train the Fleet for War: The U.S. Navy Fleet Problems, 1923–1940* (Washington: U.S. Government Printing Office, 2010), 1–2; and Peter M. Swartz, "Sea Changes: Transforming U.S. Navy Deployment Strategy, 1775–2002," CNA Draft Report D0006679.A1, 31 July 2002, 36–38. I am indebted to Peter Swartz for providing me with a copy of this report.

148. Goldman, *Sunken Treaties*, 320–21; and Thomas H. Buckley, *The United States and the Washington Conference, 1921–1922* (Knoxville: University of Tennessee Press, 1970), 90–103.

149. General Order No. 94, 9 December 1922, roll 2, M984/GO.

150. There also existed a fleet base force to provide logistical support. The relationship and composition of each fleet or force changed over time (e.g., the Scouting Fleet, by then called the Scouting Force, moved to the Pacific in 1932); I provide here a general overview drawn primarily from Craig C. Felker, *Testing American Sea Power: U.S. Navy Strategic Exercises, 1923–1940* (College Station: Texas A&M University Press, 2007), 147 n. 1; Thomas C. Hone and Trent Hone, *Battle Line: The United States Navy, 1919–1939* (Annapolis: Naval Institute Press, 2006), 128–29; and McBride, *Technological Change*, 144.

151. The best work on American war plans against Japan remains Edward S. Miller, *War Plan Orange: The U.S. Strategy to Defeat Japan, 1897–1945* (Annapolis: Naval Institute Press, 1991).

152. "Reorganized Fleet to Hold Big Tests," *New York Times*, 26 December 1922, p. 24. The navy held large fleet maneuvers both in the winter of 1916–17 and in early 1921, but because few records from these maneuvers seem to have survived this claim can be neither confirmed nor refuted.

153. Edwin Denby, "Annual Report of the Secretary of the Navy," 15 November 1923, in *ARSN*, 1923, 5.

154. Nofi, *To Train the Fleet for War*, 3, 51–56.

155. C-in-C U.S. Fleet, "Report on Fleet Problem One," 138.

156. Ibid., 137–38.

157. Ibid., 138.

158. Extracts from the report of the Blue Fleet Commander, in ibid., 115.

159. C-in-C U.S. Fleet, "Report on Fleet Problem One," 138.

160. Extracts from the report of Commanding Officer, USS *Delaware*, in ibid., 116.

161. Extracts from the report of Commander Destroyer Squadron Nine, in ibid., 120.

162. Extracts from the report of Commander Air Patrol, in ibid., 117.

163. Extracts from the reports of commanding officers of USS *Farenholt*, USS *Mervine*, and USS *Sumner*, in ibid., 103–4.

164. Extracts from the report of Commanding Officer, USS *Selfridge*, in ibid., 101.

165. Extracts from the reports of Commander Destroyer Squadron Twelve and Destroyer Division Thirty-One, in ibid., 94, 96.

166. Extracts from the reports of commanding officers of USS *Wood*, USS *Corry*, USS *Sumner*, and USS *Robert Smith*, in ibid., 100, 104, 106.

167. Extracts from the report of Commanding Officer, USS *Melvin*, in ibid., 105.

168. CNO to Captain William H. Standley, 25 July 1923, roll 1, M964/FP.

169. A handwritten note by an unknown officer, attached to "Minutes of Meeting of Special Board to Consider Problems for the Winter Maneuvers of the U.S. Fleet," 31 July 1923, roll 1, M964/FP.

170. Richard W. Turk, "Edward Walter Eberle," in *The Chiefs of Naval Operations*, ed. Robert W. Love Jr. (Annapolis: Naval Institute Press, 1980), 39.

171. Edward W. Eberle, "A Messenger-Pigeon Service for Naval Purposes," *USNIP* 21, no. 4 (1895): 843.

172. Edward W. Eberle, "Homing Pigeons as Messengers of the Fleet," *USNIP* 23, no. 4 (1897): 615–43.

173. Clark G. Reynolds, "Eberle, Edward Walter," in *Famous American Admirals* (Annapolis: Naval Institute Press, 2002), 105–6; Record of Service of Admiral Edward Walter Eberle, U.S. Navy, 13 October 1924, Edward Walter Eberle folder, NDL/ZB; and C-in-C Battle Fleet to SecNav, 12 January 1923, roll 3, M971/FR.

174. Lieutenant Commander William T. Swinburne in the case of Ensign S. S. Robison from October 1890 to July 1892; and Report on the Fitness of Officers, Ensign S. S. Robison, 1 January 1898 to 30 June 1898, signed by Captain Frank Wildes, both in box 92, RG125/EB.

175. Record of the Service of Lieutenant Samuel S. Robison, U.S. Navy, 7 July 1905; Record of the Service of Lieutenant-Commander Samuel S. Robison, U.S. Navy, 5 January 1910; and Report on the Fitness of Officers, 1 July 1906 to 15 July 1906, signed by Rear Admiral William S. Cowles, all in box 92, RG125/EB.

176. Report on the Fitness of Officers, 1 January 1910 to 31 March 1910, signed by Captain William F. Halsey [Senior]; and Report on the Fitness of Officers, 1 April 1911 to

30 September 1911, signed by Engineer-in-Chief Hutch I. Cone, both in box 92, RG125/EB.

177. Navy Biography, 25 April 1957, Samuel Shelburne Robison folder, NOA/OB; and Report on the Fitness of Officers, 1 April 1922 to 30 September 1922, signed by Edwin Denby on 21 May 1923, box 92, RG125/EB.

178. C-in-C Battle Fleet, "Battle Fleet Report for FY 1924," p. 46, roll 4, M971/FR.

179. C-in-C U.S. Fleet to CNO, 16 November 1925, roll 60, M1140/SCC; and C-in-C Battle Fleet, "Battle Fleet Report for FY 1925," p. 86, roll 5, M971/FR.

180. BuEng to CNO, 23 July 1925, roll 63, M1140/SCC.

181. C-in-C Battle Fleet, "Battle Fleet Report for FY 1924," pp. 50, 52, 77, roll 4, M971/FR; and C-in-C Battle Fleet, "Battle Fleet Report for FY 1925," p. 48, roll 5, M971/FR.

182. C-in-C Battle Fleet, "Battle Fleet Report for FY 1924," p. 48, roll 4, M971/FR.

183. C-in-C Battle Fleet, "Battle Fleet Report for FY 1925," pp. 51–53, roll 5, M971/FR. The quoted passage is from p. 53. Transmission time seems to have been measured from a message's arrival in the radio room to the end of transmission for that message.

184. E. B. Potter, *Nimitz* (Annapolis: Naval Institute Press, 1976), 129–31, 136–42. Nimitz credited classmate Roscoe MacFall with conceiving this formation.

185. Mark A. Campbell, "The Influence of Air Power upon the Evolution of Battle Doctrine in the U.S. Navy, 1922–1941" (M.A. thesis, University of Massachusetts at Boston, 1992), 83–86.

186. Remarks of Admiral S. S. Robison, 14 March 1925, roll 4, M964/FP. See also Trent Hone, "The Evolution of Fleet Tactical Doctrine in the U.S. Navy, 1922–1941," *Journal of Military History* 67, no. 4 (October 2003): 1108–11.

187. Commander Battleship Divisions, Battle Fleet to C-in-C Battle Fleet, 15 January 1924, roll 1, M964/FP; and Potter, *Nimitz*, 139–40.

188. Commander Fleet Base Force to C-in-C Battle Fleet, 16 January 1924, roll 1, M964/FP.

189. In 1924 the service divided its annual maneuvers into three parts, giving each a numeric designator (i.e., Two, Three, and Four). Even though fleet problems became longer and more complex over time, after 1930 the navy grouped all parts of the annual maneuvers under a single numeric heading. The only years that saw multinumbered fleet problems were 1924 and 1930.

190. C-in-C Scouting Fleet to C-in-C U.S. Fleet, 12 February 1924, roll 3, M964/FP.

191. Commander Battleship Divisions, Battle Fleet to C-in-C Battle Fleet, 15 January 1924, roll 1, M964/FP.

192. C-in-C Battle Fleet to C-in-C U.S. Fleet, 1 February 1924, roll 1, M964/FP.

193. Remarks of Admiral S. S. Robison, 14 March 1925, roll 4, M964/FP.

194. C-in-C Battle Fleet, "Battle Fleet Report for FY 1924," p. 46, roll 4, M971/FR; and C-in-C Battle Fleet, "Battle Fleet Report for FY 1925," p. 48, roll 5, M971/FR.

195. C-in-C Scouting Fleet, "Communications Instructions, Problem No. 5 (BLUE SCOUTING FLEET) Modification," 20 February 1925, roll 3, M964/FP.

196. C-in-C U.S. Fleet (Charles F. Hughes), "Annual Report, 1 July 1926 to 30 June 1927," 27 September 1927, p. 48, roll 6, M971/FR; and C-in-C U.S. Fleet (Henry A. Wi-

ley), "Annual Report, 8 November 1927 to 30 June 1928," 26 September 1928, p. 11, roll 7, M971/FR.

197. Chester W. Nimitz, introduction to Howeth, *History of Communications-Electronics*, xi.

198. C-in-C U.S. Fleet, "Annual Report, 3 October 1925 to 30 June 1926," 25 August 1926, p. 45, roll 5, M971/FR.

199. C-in-C Battle Fleet, "Annual Report, 4 October 1925 to 30 June 1926," 1 July 1926, p. 101, roll 5, M971/FR.

200. Commander Destroyer Squadrons, Battle Fleet, "Annual Report for Fiscal Year 1926–1927," c. July 1927, p. 5, roll 6, M971/FR.

201. C-in-C U.S. Fleet, "Annual Report, 1 July 1926 to 30 June 1927," 27 September 1927, p. 46, roll 6, M971/FR.

202. BuEng to C-in-C Pacific Fleet, c. May 1922, box 296, CC/100; BuEng to CNO, 23 July 1925, roll 63, M1140/SCC; and BuEng to CNO, 5 April 1928, roll 6, M971/FR.

203. The modern HF band is 3,000 to 30,000 kilohertz, but American naval personnel in the 1920s used "high frequency" to refer to anything over 2,000 kilocycles.

204. The new facility, located in southeastern D.C., initially was known as the Naval Experiment and Research Laboratory. In 1925 the Navy shortened this to Naval Research Laboratory. David K. Allison, *New Eye for the Navy: The Origin of Radar at the Naval Research Laboratory* (Washington: U.S. Government Printing Office, 1981), 49.

205. BuEng to SecNav, 1 September 1923, in *ARSN*, 1923, 321.

206. Gebhard, *Evolution of Naval Radio-Electronics*, 31–37; and Allison, *New Eye for the Navy*, 35–37, 41–46.

207. Submarine communications was only one of several problems the navy hoped to solve with underwater sound signaling. See chapter 4 for a discussion of efforts to employ underwater sound signals for tactical ship-to-ship communications.

208. Taylor, *Radio Reminiscences*, 236–37.

209. C-in-C Battle Fleet, "Battle Fleet Report for FY 1924," p. 62, roll 4, M971/FR.

210. Ibid.

211. BuEng to SecNav, 15 September 1925, in *ARSN*, 1925, 290.

212. C-in-C U.S. Fleet, "Annual Report, 3 October 1925 to 30 June 1926," 25 August 1926, pp. 42–43, roll 5, M971/FR.

213. BuEng to SecNav, 1 September 1926, in *ARSN*, 1926, 249; BuEng to SecNav, 1 September 1927, in *ARSN*, 1927, 259; and BuEng to SecNav, 31 August 1928, in *ARSN*, 1928, 304.

214. Taylor, *Radio Reminiscences*, 181.

215. Ibid., 184–85, 193–99.

216. Taylor, *Radio Reminiscences*, 198–99; and BuEng to SecNav, 1 September 1927, 257, in *ARSN*, 1927.

217. C-in-C U.S. Fleet, "Annual Report, 1 July 1926 to 30 June 1927," 27 September 1927, p. 49, roll 6, M971/FR; C-in-C Scouting Fleet, "Annual Report, 9 November 1927 to 30 June 1928," 1 July 1928, p. 32, roll 7, M971/FR; and C-in-C U.S. Fleet, "Annual Report, 8 November 1927 to 30 June 1928," 26 September 1928, p. 61, roll 7, M971/FR.

218. Lieutenant Commander C. N. Ingraham, "Navy Radio High Frequency Communications," n.d., box 291, CC/100. This document is the final draft of an article published in an unknown venue in December 1927. For a clipping of the published article, see box 294, CC/100.

219. Taylor, *Radio Reminiscences*, 193.

220. Lieutenant Commander C. N. Ingraham, "Navy Radio High Frequency Communications," box 291, CC/100; BuEng to SecNav, 31 August 1928, in *ARSN*, 1928, 302; and Taylor, *Radio Reminiscences*, 237–39.

221. Jon Tetsuro Sumida, "Forging the Trident: British Naval Industrial Logistics, 1914–1918," in *Feeding Mars: Logistics in Western Warfare from the Middle Ages to the Present*, ed. John A. Lynn (Boulder, Colo.: Westview Press, 1993), 217–49.

222. Report of the Paymaster General, 12 September 1929, Statement 8, in *ARSN*, 1929, 862; and Report of the Paymaster General, 21 September 1932, Statement 8, in *ARSN*, 1932, 784.

CHAPTER 4. A Most Complex Problem: Demanding Information

Epigraph. Comment by Chief Observer, in C-in-C U.S. Fleet, "United States Fleet Problem X, 1930," 7 May 1930, p. 65, roll 13, M964/FP.

Epigraph. Arleigh A. Burke, quoted in E. B. Potter, *Admiral Arleigh Burke* (New York: Random House, 1990), 73.

1. "Victory for the Middies," *New York Times*, 27 November 1892, p. 1; "Middies Again Triumphant," *New York Times*, 3 December 1893, p. 3; and Thomas Wildenberg, *All the Factors of Victory: Admiral Joseph Mason Reeves and the Origins of Carrier Airpower* (Washington: Brassey's, 2003), 11–25.

2. Wildenberg, *All the Factors of Victory*, 37–46, 67–78, 90–124.

3. Thomas C. Hone, Norman Friedman, and Mark D. Mandeles, *American and British Aircraft Carrier Development, 1919–1941* (Annapolis: Naval Institute Press, 1999), 30–31, 40–43, and Wildenberg, *All the Factors of Victory*, 125–34.

4. C-in-C Battle Fleet to C-in-C U.S. Fleet, 21 February 1926, roll 7, M964/FP; and Fleet Observer Blue to Chief Observer Blue, 16 March 1927, roll 8, M964/FP. In this chapter I have changed the naming convention for fleet problems from Arabic to Roman numerals. This shift is in accord with practices of the period. Fleet Problem V was the first in which contemporaries used Roman numerals; by the time of Fleet Problem VIII the original convention had nearly disappeared.

5. C-in-C U.S. Fleet to CNO, "Fleet Problem VIII—Report of Commander in Chief, United States Fleet," c. May 1928, pp. 7–8, roll 11, M964/FP; "'Attackers' Subdue Oahu in War Games," *New York Times*, 18 May 1928, p. 2; and Wildenberg, *All the Factors of Victory*, 152–57, 161–62.

6. Joseph M. Reeves to William A. Moffett, 16 February 1928, box 10, Papers of William Adger Moffett, 1920–1948, Manuscript Collection 198, Special Collections and Archives, Nimitz Library, United States Naval Academy, Annapolis, Md.

7. Lewis Freeman, "Saratoga's Raid Left Fleet Behind," *New York Times*, 18 February 1929, p. 10; and Remarks of Battle Fleet Chief of Staff (Arthur J. Hepburn), 5 February 1929, roll 12, M964/FP.

8. Lewis Freeman, "Commander's Story of Saratoga's Raid," *New York Times*, 18 February 1929, p. 14.

9. C-in-C U.S. Fleet, "United States Fleet Problem IX, January, 1929," 18 March 1929 (hereafter cited as "Report on Fleet Problem IX"), part III, p. 21, roll 12, M964/FP. This roll of microfilm contains two identical versions of the commander-in-chief's report, one printed, the other in typescript. All references are to the typed report.

10. Lewis Freeman, "Commander's Story of Saratoga's Raid," *New York Times*, 18 February 1929, p. 14.

11. Ibid.

12. C-in-C U.S. Fleet, "Report on Fleet Problem IX," part III, pp. 21–23.

13. Rear Admiral Joseph M. Reeves, "Statement on Fleet Problem Nine," 2 February 1929, roll 12, M964/FP.

14. Unit Observer, Black Striking Force to Fleet Observer Black, U.S. Fleet Problem IX, 4 February 1929, roll 12, M964/FP.

15. C-in-C U.S. Fleet, "Report on Fleet Problem IX," part III, p. 29.

16. Craig C. Felker, *Testing American Sea Power: U.S. Navy Strategic Exercises, 1923–1940* (College Station: Texas A&M University Press, 2007), 121, and Charles Melhorn, *Two-Block Fox: The Rise of the Aircraft Carrier, 1911–1929* (Annapolis: Naval Institute Press, 1974), xii. See also Archibald D. Turnbull and Clifford L. Lord, *History of United States Naval Aviation* (New Haven, Conn.: Yale University Press, 1949), 270–72; Gerald E. Wheeler, *Admiral William Veazie Pratt, U.S. Navy: A Sailor's Life* (Washington: U.S. Government Printing Office, 1974), 268, 274–75; Wildenberg, *All the Factors of Victory*, 1–10, 188–95; and Albert A. Nofi, *To Train the Fleet for War: The U.S. Navy Fleet Problems, 1923–1940* (Washington: U.S. Government Printing Office, 2010), 109–19.

17. Hone et al., *American and British Carrier Development*, 48–49.

18. Robert L. O'Connell, *Sacred Vessels: The Cult of the Battleship and the Rise of the U.S. Navy* (New York: Oxford University Press, 1991), 285–87; Clark G. Reynolds, *The Fast Carriers: The Forging of an Air Navy* (New York: McGraw-Hill, 1968; Annapolis: Naval Institute Press, 1992), 16–18; and Thomas Wildenberg, *Destined for Glory: Dive Bombing, Midway, and the Evolution of Carrier Air Power* (Annapolis: Naval Institute Press, 1998), 62–64.

19. C-in-C U.S. Fleet, "Report on Fleet Problem IX," part III, p. 28.

20. O'Connell, *Sacred Vessels*, 271, 283; Clark G. Reynolds, *Admiral John H. Towers: The Struggle for Naval Air Supremacy* (Annapolis: Naval Institute Press, 1991), 223–24; and William M. McBride, *Technological Change and the United States Navy, 1865–1945* (Baltimore: Johns Hopkins University Press, 2000), 6–7.

21. O'Connell, *Sacred Vessels*, 3.

22. Thomas C. Hone, "Spending Patterns of the United States Navy, 1921–1941," *Armed Forces and Society* 8 (Spring 1982): 448–51.

23. Felker, *Testing American Sea Power*, 61–87. See also Thomas C. Hone and Trent Hone, *Battle Line: The United States Navy, 1919–1939* (Annapolis: Naval Institute Press, 2006), 110–25.

24. John T. Kuehn, *Agents of Innovation: The General Board and the Design of the Fleet That Defeated the Japanese Navy* (Annapolis: Naval Institute Press, 2008).

25. C-in-C U.S. Fleet, "Report on Fleet Problem IX," part III, pp. 18–28.

26. Ibid., part II, p. 5. The report for Fleet Problem V contains a paragraph labeled "Dissemination of Information," but this paragraph is only two sentences long. C-in-C U.S. Fleet, "Report on Fleet Problem V," 29 May 1925, p. 4, roll 4, M964/FP.

27. C-in-C U.S. Fleet, "Report on Fleet Problem IX," part II, p. 5.

28. Remarks of Battle Fleet Chief of Staff (Arthur J. Hepburn), 5 February 1929, roll 12, M964/FP.

29. C-in-C U.S. Fleet, "Report on Fleet Problem IX," part II, p. 5.

30. Lieutenant (j.g.) Thomas H. Dyer, "Outline of Remarks on Black Communications," n.d., roll 12, M964/FP.

31. Fleet Observer Black (William Glassford), quoted in C-in-C U.S. Fleet, "Report on Fleet Problem IX," part IV, p. 1.

32. C-in-C U.S. Fleet, "Report on Fleet Problem IX," part IV, pp. 1–2. I have found no evidence that the Navy Department ever implemented this suggestion.

33. Rear Admiral Joseph M. Reeves, "Statement on Fleet Problem Nine," 2 February 1929, roll 12, M964/FP.

34. Surface warships hypothetically damaged or sank opposing carriers in ten of the eleven fleet problems for which surviving documentation is reasonably thorough. Unfortunately, extant records for Fleet Problem XVII (1936) shed no light on the subject. Regarding missing records for Fleet Problem XVII, see Timothy K. Nenninger, *Records Relating to United States Navy Fleet Problems I to XXII, 1923–1941* (Washington: National Archives and Records Service, 1975), 3–4, 22; and Nofi, *To Train the Fleet for War*, 215 n. 1.

35. C-in-C U.S. Fleet, "United States Fleet Problem X, 1930," 7 May 1930 (hereafter cited as "Report on Fleet Problem X"), pp. 44–45, roll 13, M964/FP.

36. C-in-C U.S. Fleet, "United States Fleet Problem XI, 1930," 14 July 1930 (hereafter cited as "Report on Fleet Problem XI"), pp. 30, 48–49, 73, roll 13, M964/FP.

37. C-in-C U.S. Fleet, "United States Fleet Problem XII, 1931," 1 April 1931 (hereafter cited as "Report on Fleet Problem XII"), pp. 10–13, 32–33, roll 13, M964/FP.

38. C-in-C U.S. Fleet, "United States Fleet Problem XIII, 1932," 23 May 1932 (hereafter cited as "Report on Fleet Problem XIII"), pp. 6–9, roll 14, M964/FP.

39. Remarks of Vice Admiral William H. Standley, in C-in-C U.S. Fleet, "United States Fleet Problem XIV," 20 April 1933 (hereafter cited as "Report of Fleet Problem XIV"), section three, p. 21, roll 15, M964/FP.

40. Ibid., section two, p. 10.

41. C-in-C U.S. Fleet, "Report of Fleet Problem XV," 1 June 1934 (hereafter cited as "Report of Fleet Problem XV"), p. 19, roll 16, M964/FP. One of the carriers was constructive (i.e., a surrogate ship representing a carrier). The navy sometimes employed

constructive units during fleet problems to test vessels, aircraft, or capabilities that did not exist. See Nofi, *To Train the Fleet for War*, 26–27.

42. C-in-C U.S. Fleet, "Report of Fleet Problem XV," 19.

43. Ibid., 20–22.

44. Remarks of Vice Admiral Harris Laning, 25 May 1934, roll 16, M964/FP.

45. C-in-C U.S. Fleet, "Fleet Problem Sixteen," 15 September 1935 (hereafter cited as "Report of Fleet Problem XVI"), pp. 25–26, 32, roll 18, M964/FP.

46. C-in-C U.S. Fleet, "Operation Plan No. 1–35," 1 March 1935; and Remarks of Admiral Harris Laning, 15 June 1935, both on roll 18, M964/FP. The quoted phrase is from Laning's remarks.

47. Remarks of Vice Admiral Henry V. Butler, 15 June 1935, roll 18, M964/FP.

48. C-in-C U.S. Fleet, "Report of Fleet Problem XVI," 32–33.

49. C-in-C U.S. Fleet, "Annex Affirm to U.S. Fleet Operation Order No. 7-37," 1 March 1937; and Commander Battle Force (Commander BLACK Fleet) to C-in-C U.S. Fleet, "FPXVIII—Narrative of Events and Track Charts," 11 May 1937, both on roll 22, M964/FP. Before Fleet Problem XVIII the navy redesignated *Langley* a seaplane tender, although black employed *Langley* much like a carrier during the exercise.

50. Commander Aircraft, Base Force (Commander White Air Force) to C-in-C U.S. Fleet, 12 May 1937, roll 23, M964/FP.

51. Commander Submarine Division Eight to C-in-C U.S. Fleet, 9 May 1937, roll 23, M964/FP.

52. Commander Battleships, WHITE Fleet to C-in-C U.S. Fleet, 10 May 1937, roll 23, M964/FP.

53. Black Fleet Umpire to C-in-C U.S. Fleet, 9 May 1937, roll 23, M964/FP.

54. On 1 April 1931 the Battle Fleet and the Scouting Fleet became the Battle Force and the Scouting Force, respectively. General Order No. 211, 10 December 1930, roll 3, M984/GO.

55. Ernest J. King and Walter M. Whitehill, *Fleet Admiral King: A Naval Record* (New York: W. W. Norton, 1952), 274; and Commander Aircraft, Battle Force to C-in-C U.S. Fleet, 4 June 1937, roll 23, M964/FP.

56. Nofi, *To Train the Fleet for War*, 221.

57. Commander Battle Force (Commander BLACK Fleet) to C-in-C U.S. Fleet, 23 June 1937, roll 23, M964/FP.

58. My assessment of Bloch's interest in command and control stems from reports and correspondence generated during the navy's 1937, 1938, and 1939 annual maneuvers.

59. Commander Battle Force (Commander BLACK Fleet) to C-in-C U.S. Fleet, 23 June 1937, roll 23, M964/FP.

60. Commander Battle Force (Charles P. Snyder) to C-in-C U.S. Fleet, 15 May 1940, roll 32, M964/FP.

61. The reality of this continuing threat would be painfully demonstrated in June 1940 when German battleships *Scharnhorst* and *Gneisenau* sank the British carrier *Glorious*. Of note, poor command and control played a major role in that disaster, for which

see Vernon W. Howland, "The Loss of HMS *Glorious*: An Analysis of the Action," *Warship International* 31, no. 1 (1994): 47–62; and John Winton, *Carrier* Glorious: *The Life and Death of an Aircraft Carrier* (London: Leo Cooper, 1986), 147–82.

62. "Narrative of Events," enclosure (a) to Commander BLACK Fleet (Commander Scouting Force) to C-in-C U.S. Fleet, 1 March 1939, roll 26, M964/FP; and "Original Narrative of Events," enclosure (a) to Commander Battle Force to C-in-C U.S. Fleet, 19 March 1939, roll 28, M964/FP.

63. Ship Umpire (USS *Ranger*) to C-in-C U.S. Fleet, 27 February 1939, roll 28, M964/FP; Ship's Umpire (USS *Yorktown*) to C-in-C U.S. Fleet, 27 February 1939, Ship Umpire [USS *Lexington*] to C-in-C U.S. Fleet, 28 February 1939, and Ship's Umpire, USS *Enterprise* to C-in-C U.S. Fleet, 28 February 1939, roll 30, M964/FP. See enclosure (a) to each of these for the umpires' reports.

64. Commander Battle Force to C-in-C U.S. Fleet, "Part II, Fleet Problem XXI—Report of Commander Black Fleet," 15 May 1940; and "Summaries of Damage, Ammunition Expended, and Torpedoes Fired," enclosure (c) to Commander Battleships, Battle Force to C-in-C U.S. Fleet, 29 April 1940, both on roll 32, M964/FP.

65. Commanding Officer, USS *Brooklyn* to C-in-C U.S. Fleet, 2 March 1939, roll 27, M964/FP; Commander Battle Force to C-in-C U.S. Fleet, 15 May 1940; and "Narrative of Events," enclosure (b) to Commander Battleships, Battle Force (Commander WHITE Fleet) to C-in-C U.S. Fleet, 29 April 1940, roll 32, M964/FP.

66. Particularly important was the development of more powerful radial engines. Curtis Alan Utz, "Carrier Aviation Policy and Procurement in the U.S. Navy, 1936–1940" (M.A. thesis, University of Maryland, 1989).

67. Mark A. Campbell, "The Influence of Air Power upon the Evolution of Battle Doctrine in the U.S. Navy, 1922–1941" (M.A. thesis, University of Massachusetts at Boston, 1992), 201–5. Using Campbell's methodology, I divide aircraft range by a factor of three to calculate approximate combat radius. All performance figures are averages taken from table A.

68. Wildenberg, *Destined for Glory*, 10–19, 83–98.

69. Ibid., 65–82.

70. These aircraft were the Vought SB2U Vindicator and the Curtiss SBC Helldiver, introduced to the fleet in 1935 and 1937, respectively. Roy A. Grossnick, *United States Naval Aviation, 1910–1995* (Washington: U.S. Government Printing Office, 1997), 493.

71. Remarks of Rear Admiral Clarence S. Kempff, 15 June 1935, roll 18, M964/FP.

72. Timothy S. Wolters, "Managing a Sea of Information: Shipboard Command and Control in the United States Navy, 1899–1945" (Ph.D. diss., Massachusetts Institute of Technology, 2003), 198.

73. Grossnick, *United States Naval Aviation*, 494.

74. Extracts from the report of Commander Battleship Divisions, Battle Fleet, in C-in-C U.S. Fleet, "Report on United States Fleet Problem Number One," 19 June 1923 (hereafter cited as "Report on Fleet Problem One"), p. 3, roll 1, M964/FP.

75. Gary E. Weir, *Building American Submarines, 1914–1940* (Washington: U.S. Government Printing Office, 1991), 38–46; John D. Alden, *The Fleet Submarine in the U.S.*

Navy (Annapolis: Naval Institute Press, 1979), 16–19; and Norman Friedman, *U.S. Submarines through 1945* (Annapolis: Naval Institute Press 1995), 292–93.

76. Enclosure (a) to Ship Umpire (USS *Ranger*) to C-in-C U.S. Fleet, 27 February 1939, roll 28, M964/FP; and Commander Submarine Force to C-in-C U.S. Fleet, 24 March 1939, roll 26, M964/FP. The quoted passage is from the latter document.

77. Commander Submarine Squadron Four to Commander Patrol Wing Two (Commander BLACK Force), 12 April 1940, roll 32, M964/FP. The two submarines that attacked *Lexington* were *Cachalot* and *Pollack*; the submarine that attacked *Saratoga* was *Shark*. They had been commissioned in 1933, 1937, and 1936, respectively.

78. Commander Convoy Guard (Vice Admiral Luke McNamee), quoted in C-in-C U.S. Fleet, "Report on Fleet Problem XIII," 24.

79. Comment of Commander Aircraft Squadrons, Scouting Fleet, in C-in-C U.S. Fleet, "Report on Fleet Problem X," 66.

80. C-in-C U.S. Fleet, "Report on Fleet Problem IX," part III, p. 31.

81. Commander BLACK Fleet to BLACK Fleet, "Information and Intelligence, Fleet Problem XX," 4 February 1939, roll 25, M964/FP.

82. C-in-C Battle Fleet to C-in-C U.S. Fleet, 21 February 1926, roll 7, M964/FP.

83. Comments by Chief Observer [Blue], in C-in-C U.S. Fleet, "Report on Fleet Problem X," 71; and Comments by Commander [Submarine] V–3, in C-in-C U.S. Fleet, "Report on Fleet Problem X," 70.

84. Commander, BLACK Fleet to BLACK Fleet, "Notes on Communication Fleet Problem XI," 29 March 1930, roll 13, M964/FP.

85. Remarks by Commander Patrick N. L. Bellinger, in C-in-C U.S. Fleet, "Report of Fleet Problem XIV," 20 April 1933, section three, p. 13, roll 15, M964/FP.

86. Orange Decrypting Unit (Lieutenant (j.g.) F. H. Bond) to C-in-C U.S. Fleet, 15 May 1928, roll 11, M964/FP.

87. A special decrypting unit also was assigned to the blue fleet, but because black maintained radio silence for much of the exercise, this unit had few opportunities available to it.

88. Lieutenant (j.g.) Thomas H. Dyer, "Outline of Remarks on Black Communications," n.d., roll 12, M964/FP.

89. C-in-C U.S. Fleet, "Report on Fleet Problem IX," part II, p. 6; and Remarks of Commander Black Fleet (William V. Pratt), 5 February 1929, roll 12, M964/FP.

90. C-in-C U.S. Fleet to CNO, 3 February 1929, roll 12, M964/FP.

91. Members of the black decrypting unit included Laurance F. Safford, Joseph J. Rochefort, and Thomas H. Dyer.

92. Fleet Observer Black (William Glassford), quoted in C-in-C U.S. Fleet, "Report on Fleet Problem IX," part IV, p. 1; Lieutenant (j.g.) Thomas H. Dyer, "Outline of Remarks on Black Communications," n.d.; and C-in-C U.S. Fleet to CNO, 3 February 1929, roll 12, M964/FP.

93. Fleet Observer Black (William Glassford), quoted in C-in-C U.S. Fleet, "Report on Fleet Problem IX," part IV, p. 1; and Commander Scouting Fleet to C-in-C U.S. Fleet, 9 February 1929, roll 12, M964/FP.

94. District Communications Officer, Fifteenth Naval District to Commandant Fifteenth Naval District, 30 January 1929, roll 12, M964/FP.

95. Comment of Commander Scouting Fleet (Commander BLACK Fleet), in C-in-C U.S. Fleet, "Report on Fleet Problem X," 70.

96. Commander Battle Force, "Annual Report, 1 July 1933 to 5 May 1934," 5 May 1934, p. 34, roll 9, M971/FR; Commander Battle Force, "Annual Report, 1 July 1934 to 30 June 1935," 12 July 1935, p. 16, roll 10, M971/FR; and Comment by Commanding Officer, USS *Barney*, in C-in-C U.S. Fleet, "Report of Fleet Problem XIV," section three, p. 21.

97. Commander Cruisers, Scouting Force to CNO (Director of Naval Communications), 1 April 1933, filing designator A6-3(7), RG80/FCC.

98. CNO to C-in-C U.S. Fleet, 24 August 1936, filing designator A6-3(7), RG80/FCC.

99. C-in-C U.S. Fleet to CNO, 3 February 1929, roll 12, M964/FP; and Commander Battleship Divisions, Battle Fleet to CNO, 26 April 1930, roll 13, M964/FP. The quoted phrase is from the latter document.

100. Commander BLACK Fleet to BLACK Fleet, "Notes on Communication Fleet Problem XI," 29 March 1930, roll 13, M964/FP.

101. Commander Battleship Divisions, Battle Fleet to CNO, 26 April 1930, roll 13, M964/FP.

102. Comment by Communications Officer, Staff, Chief Observer in C-in-C U.S. Fleet, "Report on Fleet Problem XI," 72–73. Of note, codes and ciphers are two distinct forms of encryption. A code is a system for encrypting messages that replaces each word or phrase of the original text with a different character or characters. The list of replacements is contained in a single source, the codebook. A cipher is a system for encrypting messages that replaces each letter of the original text with a different letter. A "key" is used to determine which letters replace the original letters. Ciphers are potentially more secure than codes, one reason being that a key can be changed more easily than a codebook. To encrypt is either to encipher or to encode. Likewise, to decrypt is either to decipher or to decode. Unfortunately, naval officers of the era did not always make this distinction and sometimes used the term "code" or its derivatives to refer to any type of encryption or decryption. In my own prose, I endeavor to maintain the literal definitions of the two terms but have chosen not to alter the language of contemporaries.

103. Vice Admiral Frank H. Clark (Commander BLACK Raiding Force), "Communication Plan No. 1," in Annex C2 of "U.S. Fleet Problem XIV and Contributory Plans, BLACK," n.d., roll 15, M964/FP.

104. C-in-C U.S. Fleet, "Annual Report, 1 July 1932 to 10 June 1933," 10 June 1933, p. 55, roll 9, M971/FR.

105. Commander, BLACK Fleet to BLACK Fleet, "Notes on Communication Fleet Problem XI," 29 March 1930, roll 13, M964/FP; C-in-C U.S. Fleet, "Annual Report, 1 July 1929 to 30 June 1930," 1 August 1930, p. 61, roll 8, M971/FR; and Commander Black Force, Annex II to "Black Operation Plan No. 1–32," in "United States Fleet Problem XIII, Brief Estimate of the Situation," 3 March 1932, roll 15, M964/FP.

106. CNO to C-in-C U.S. Fleet, c. June 1934, roll 16, M964/FP.

107. Comment by C-in-C Battle Fleet (Commander BLUE Fleet), in C-in-C U.S. Fleet, "Report on Fleet Problem X," 63.

108. Tactical flexibility and individual initiative are consistent themes in records relating to the fleet's annual maneuvers. See, e.g., C-in-C U.S. Fleet, "Report of Fleet Problem XV," 66. Numerous others have pointed to the U.S. Navy's long-standing adherence to the principle of "the initiative of the subordinate," but the only one to do so within the context of naval command and control is Michael Palmer, *Command at Sea: Naval Command and Control since the Sixteenth Century* (Cambridge, Mass.: Harvard University Press, 2005). See esp. chapter 8.

109. C-in-C U.S. Fleet, "Report of Fleet Problem XV," 53.

110. Ibid.

111. Commander Black Fleet, "Appendix '1' to Annex Affirm of Black Operation Plan 1S-39," effective date of 13 February 1939, p. 8(b), roll 25, M964/FP.

112. Memorandum Study, "Analysis of Military Communications of Fleet Problem XIX," compiled by Thomas B. Inglis, 20 March 1939, roll 24, M964/FP.

113. Comment by Commander BLACK Fleet, in C-in-C U.S. Fleet, "Report on Fleet Problem XI," 70–71. The service later designated the broadcast method of radio communication as the "F" or "FOX" method, also sometimes referred to by naval personnel as the "no answer" method. C-in-C U.S. Fleet, "Annual Report, Fiscal Year 1935," 11 October 1936, p. 25; Commander Battle Force, "Annual Report, 1 July 1935 to 24 June 1936," 21 May 1936, p. 22; and Commander Battle Force, "Annual Report, 1 July 1936 to 30 June 1937," c. July 1937, p. 14, all on roll 10, M971/FR.

114. Navy Department, *War Instructions, 1923* (W.P.L. 7) (Washington: U.S. Government Printing Office, 1923), box 5, Records of the Strategic Plans/War Plans Division (W.P.L. Series), Records of the Office of the Chief of Naval Operations (Record Group 38), U.S. National Archives, College Park, Md.

115. Pratt named the new publication "Tentative Fleet Dispositions and Battle Plans." C-in-C U.S. Fleet, "Annual Report, 1 July 1929 to 30 June 1930," 1 August 1930, p. 16, roll 8, M971/FR.

116. The Hones draw an instructive analogy: "Like a quarterback calling a play at the line of scrimmage, the OTC would set the plan in motion. It would be up to his teammates—subordinate ship and task group commanders—to be familiar with the plan and to execute it appropriately." Hone and Hone, *Battle Line*, 84.

117. Comment by Commander BLACK Fleet, in C-in-C U.S. Fleet, "Report on Fleet Problem XI," 65; and Comment by Commander BLUE Fleet, in C-in-C U.S. Fleet, "Report on Fleet Problem XI," 65.

118. C-in-C U.S. Fleet, "Annual Report, 1 July 1930 to 30 June 1931," 28 July 1931, p. 10, roll 8, M971/FR; C-in-C U.S. Fleet, "Annual Report, 1 July 1931 to 30 June 1932," 21 July 1932, p. 23, roll 9, M971/FR; and Navy Department, *General Tactical Instructions, 1934* (F.T.P. 142) (Washington: U.S. Government Printing Office, 1934), box 108, NOA/WWII.

119. C-in-C U.S. Fleet, "Report of Fleet Problem XV," 57.

120. CNO to C-in-C U.S. Fleet, c. June 1934, roll 16, M964/FP.

121. C-in-C U.S. Fleet, "Annual Report, 1 July 1933 to 15 June 1934," 15 June 1934, p. 30, roll 9, M971/FR; and C-in-C U.S. Fleet, "Annual Report, 1 July 1935 to 30 June 1936," 24 June 1936, p. 42, roll 10, M971/FR. The quoted passage is from the latter document.

122. CNO to C-in-C U.S. Fleet, 7 December 1934, roll 9, M971/FR.

123. Commander Battle Force "Annual Report, 1 July 1934 to 30 June 1935," 12 July 1935, p. 17; Commander Battle Force, "Annual Report, 1 July 1936 to 30 June 1937," c. July 1937, p. 21; and C-in-C U.S. Fleet, "Annual Report, 1 July 1937 to 30 June 1938," 3 August 1938, p. 14, all on roll 10, M971/FR.

124. C-in-C U.S. Fleet, "Annual Report, 1 July 1939 to 30 June 1940," 27 July 1940, p. 23, roll 10, M971/FR.

125. Josephus Daniels, "Annual Report of the Secretary of the Navy," appendix H, 1 December 1918, in *ARSN*, 1918, 287.

126. Commander Scouting Fleet, "Annual Report, 1 July 1926 to 30 June 1927," 30 June 1927, p. 33; and C-in-C U.S. Fleet, "Annual Report, 1 July 1926 to 30 June 1927," 27 September 1927, p. 52, both on roll 6, M971/FR.

127. Office of Naval Operations Code and Signal Section, *Instructions for the Cylindrical Cipher Device* (CSP 493) (Washington: U.S. Government Printing Office, 1926), box 33, Records of the Naval Security Group Repository, Crane, Ind., General Records of the Chief of Naval Operations (Record Group 38), U.S. National Archives, College Park, Md.

128. Memorandum for Assistant Chief of Naval Operations, "Use of Decryptors in Fleet Problems," written by Stanford C. Hooper, 28 October 1931, filing designator A6-3(7), RG80/FCC.

129. Glenn Zorpette, "The Edison of Secret Codes," *American Heritage of Invention & Technology* 10, no. 1 (Summer 1994): 36–40. On appeal, a judge overturned Hebern's conviction.

130. Captain Ridley McClean to Commander Edward J. Foy, 9 October 1926, filing designator A6-3(7), RG80/FSC.

131. CNO to Commandant Twelfth Naval District, 9 October 1926, filing designator A6-3(7), RG80/FSC; Director of Naval Communications (Thomas T. Craven) to Walter D. Reed, 1 December 1927; and Craven to Hannibal C. Ford, 2 December 1927, filing designator A6-3(7), RG80/FCC. The quoted passage is from Craven's letter to Ford.

132. Ridley McLean to Thomas T. Craven, 16 April 1928, filing designator A6-3(7), RG80/FCC.

133. Ibid. McLean stated that he expected the other individual, Lieutenant Commander Laurance F. Safford, to write to OPNAV as well. Safford is likely to have written such a letter, but if he did, it appears not to have survived.

134. CNO to Commandant Twelfth Naval District, 25 April 1928, filing designator A6-3(7), RG80/FCC.

135. C-in-C Battle Fleet to CNO, 5 April 1929; and C-in-C U.S. Fleet to CNO, 29 April 1929, filing designator A6-3(7), RG80/FSC.

136. C-in-C U.S. Fleet to CNO, 28 May 1930; and Memorandum for the CNO, written by Stanford C. Hooper, 28 May 1930, filing designator A6-3(7), RG80/FSC.

137. CNO to C-in-C U.S. Fleet, 18 June 1930, filing designator A6-3(7), RG80/FSC; and Edward Hebern to Laurance Safford (telegram), 3 April 1931, filing designator A6-3(7), RG80/FCC.

138. Director of Naval Communications to BuEng, 31 March 1931; and CNO to Commander Battle Force et al., 5 September 1931, filing designator A6-3(7), RG80/FCC.

139. BuEng to C-in-C U.S. Fleet, 24 December 1931, filing designator A6-3(7), RG80/FCC.

140. Commander Battle Force to CNO, 18 March 1932; and C-in-C U.S. Fleet to CNO, 8 April 1932, filing designator A6-3(7), RG80/FCC.

141. C-in-C U.S. Fleet, "Annual Report, 1 July 1931 to 30 June 1932," 21 July 1932, p. 66, roll 9, M971/FR.

142. Director of Naval Communications to BuEng, 31 March 1931; and Commander Scouting Force to CNO, 11 August 1932, filing designator A6-3(7), RG80/FCC. The quoted passage is from the latter document.

143. Director of Naval Communications to BuEng, 31 March 1931, filing designator A6-3(7), RG80/FCC; and Zorpette, "The Edison of Secret Codes," 41–43.

144. Officer in Charge, Code and Signal Section (Op–20-G) to Director of Naval Communications, 24 March 1932, filing designator A6-3(7), RG80/FCC. Cryptographic historian David Kahn credits U.S. Army officer Parker Hitt with conceiving of and constructing a strip cipher during the First World War, but Kahn does not elaborate on why the army subsequently chose to adopt the cylindrical over the strip cipher. I suspect the army chose the former over the latter mainly because metal is more durable than paper. Whatever the rationale, Navy Department correspondence from the early 1930s indicates that the army was slow to accept the inherent vulnerabilities of the cylindrical cipher, which helps to explain why the navy adopted the strip cipher before the army. David Kahn, *The Codebreakers: The Story of Secret Writing* (New York: Scribner, 1996), 324–25; and Memorandum for Assistant Chief of Naval Operations, "Use of Decryptors in Fleet Problems," written by Stanford C. Hooper, 28 October 1931, filing designator A6-3(7), RG80/FCC.

145. Director of Naval Communications to BuEng, 31 March 1931, filing designator A6-3(7), RG80/FCC.

146. Officer Biography Sheet, 30 June 1954, Morris Smellow folder, NOA/OB.

147. Lieutenant (j.g.) Morris Smellow to CNO, 10 July 1929; and Smellow to CNO, 12 September 1929, filing designator A6-3(7), RG80/FCC.

148. CNO to Lieutenant (j.g.) Morris Smellow, 30 August 1929; and Commander Scouting Fleet to CNO, 14 September 1929, filing designator A6-3(7), RG80/FCC.

149. Officer in Charge, Code and Signal Section (Op–20-G) to Director of Naval Communications, 24 March 1932; and CNO to Commander Special Service Squadron et al., 12 April 1932, filing designator A6-3(7), RG80/FCC.

150. Commanding General, USMC Second Brigade to CNO, 26 July 1932; and Commander Special Service Squadron to CNO, 3 August 1932, filing designator A6-3(7), RG80/FCC.

151. CNO to BuEng, 18 August 1932, filing designator A6-3(7), RG80/FCC; and CNO to C-in-C U.S. Fleet, 1 September 1932, roll 15, M964/FP.

152. Commandant Eleventh Naval District to CNO, 3 March 1933, roll 15, M964/FP.

153. Probably the most helpful suggestions came from Lieutenant (j.g.) Lewis W. Markham. Markham to CNO, 26 May 1933; and CNO to Commanding Officer, USS *Melville*, 11 July 1933, filing designator A6-3(7), RG80/FCC.

154. CNO to Commander John W. McClaren, 5 May 1932, filing designator A6-3(7), RG80/FCC. Along with Smellow, three other naval officers received formal recognition: John W. McClaren, Laurance F. Safford, and Joseph N. Wenger.

155. Commander Scouting Force to CNO, 11 August 1932; and C-in-C U.S. Fleet to CNO, 15 August 1932, filing designator A6-3(7), RG80/FCC.

156. BuEng to CNO, 25 August 1932; and CNO to BuEng, 28 March 1933, filing designator A6-3(7), RG80/FCC.

157. Zorpette, "The Edison of Secret Codes," 43.

158. C-in-C U.S. Fleet, "Annual Report, 1 July 1933 to 15 June 1934," 15 June 1934, p. 29, roll 9, M971/FR. For similar comments, see Commander Battle Force, "Annual Report, 1 July 1933 to 5 May 1934," 5 May 1934, p. 34, roll 9, M971/FR; and Commander Scouting Force, "Annual Report, 1 July 1934 to 31 March 1935," c. April 1935, p. 35, roll 10, M971/FR.

159. CNO to C-in-C U.S. Fleet, 13 August 1934; and CNO to BuEng, 19 November 1934, filing designator A6-3(7), RG80/FSC.

160. C-in-C U.S. Fleet, "Annual Report, 1 July 1935 to 30 June 1936," 24 June 1936, p. 42; and Commander Battle Force, "Annual Report, 1 July 1935 to 24 June 1936," 21 May 1936, p. 26, both on roll 10, M971/FR.

161. Commander Battle Force, "Annual Report, 1 July 1936 to 30 June 1937," c. July 1937, p. 21, roll 10, M971/FR.

162. C-in-C U.S. Fleet, "Annual Report, 1 July 1939 to 30 June 1940," 27 July 1940, p. 24, roll 10, M971/FR; C-in-C U.S. Pacific Fleet, "Annual Report, 1 July 1940 to 30 June 1941, p. 14, roll 3, M971/FR; and Navy Department, "Operating Instructions for ECM Mark 2 and CCM Mark 1," 16 May 1944, p. 17, viewed at http://www.hnsa.org/doc/crypto/ecm/index.htm on 31 May 2011. The quoted phrases are from the latter document.

163. John J. Savard and Richard S. Pekelney, "The ECM Mark II: Design, History, and Cryptology," *Cryptologia* 23, no. 3 (July 1999): 211–28.

164. BuEq to Chief Signal Officer, U.S. Army, 10 April 1905, box 1, RG19/WT.

165. Stone Telegraph and Telephone Co. to BuEq, 13 December 1906, box 59, RG19/BEQ.

166. Typescript of the recollections of Stanford C. Hooper, pp. 200–201, 352–53, boxes 24–25, LOC/HP.

167. Ibid., 355–58.

168. Robert S. Griffin, *History of the Bureau of Engineering, Navy Department, during the World War* (Washington: U.S. Government Printing Office, 1922), 96–97. For reasons of wartime security, the navy changed the name of the device from the *Kolstermeter* to the more innocuous *radio compass*.

169. George H. Clark, "Radio in War and Peace," 200–201, unpublished manuscript, box 289, CC/100; and Griffin, *History of the Bureau of Engineering*, 97–98.

170. Alexander Forbes to R. F. Blake, 24 September 1919, box 291, CC/100. This document is a typed transcription done by George Clark in 1941.

171. Griffin, *History of the Bureau of Engineering*, 97.

172. C-in-C Battle Fleet to SecNav, 12 January 1923, roll 3, M971/FR; C-in-C U.S. Fleet, "Annual Report, 1 July 1922 to 30 June 1923," 1 July 1923, p. 32, roll 4, M971/FR; Ship's Observer, USS *Raleigh* to Fleet Observer, Problem V, 15 March 1926, entry for 10 March 1925, and Blue Fleet Observer, "History of Maneuver—Blue Fleet, Chronological Order of Events," n.d., entry for 5 March 1926, roll 3, M964/FP; Commander Battleship Divisions, Battle Fleet to C-in-C Battle Fleet, 17 June 1924, roll 4, M971/FR; and Commander Battleship Divisions, Battle Fleet to C-in-C Battle Fleet, 30 May 1925, roll 5, M971/FR. The first two quotations are from the penultimate document; the last quote is from the final document.

173. BuEng to SecNav, 1 September 1924, in *ARSN*, 1924, 315; C-in-C Battle Fleet, "Annual Report, 4 October 1925 to 30 June 1926," 1 July 1926, p. 101, roll 5, M971/FR; and Curtis D. Wilbur, "Annual Report of the Secretary of the Navy," 15 November 1925, in *ARSN*, 1925, 20.

174. C-in-C U.S. Fleet to CNO, 4 May 1927, roll 8, M964/FP; and C-in-C U.S. Fleet, "Annual Report, 8 November 1927 to 30 June 1928," 26 September 1928, p. 69, roll 7, M971/FR.

175. C-in-C U.S. Fleet, "Annual Report, 1 July 1929 to 30 June 1930," 1 August 1930, pp. 51, 53; BuEng to CNO, 25 March 1931; and Director of Naval Communications to CNO, 2 October 1931, all on roll 8, M971/FR. The quoted passage is from the last of these documents.

176. BuEng to CNO, 25 March 1931, roll 8, M971/FR; Commander Scouting Force, "Annual Report, 1 July 1931 to 15 June 1932," 10 June 1932, p. 45, and C-in-C U.S. Fleet, "Annual Report, 1 July 1932 to 10 June 1933," 10 June 1933, p. 50, roll 9, M971/FR.

177. C-in-C U.S. Fleet, "Annual Report, Fiscal Year 1935," 11 October 1936, p. 28, roll 10, M971/FR.

178. Remarks of Vice Admiral Henry V. Butler, 15 June 1935, roll 18, M964/FP.

179. Kathleen Broome Williams, *Secret Weapon: U.S. High-Frequency Direction Finding in the Battle of the Atlantic* (Annapolis: Naval Institute Press, 1996), 84.

180. Commander Battle Force, "Annual Report, 1 July 1936 to 30 June 1937," c. July 1937, p. 19; C-in-C U.S. Fleet, "Annual Report, 1 July 1938 to 30 June 1939," 4 August 1939, p. 19; and C-in-C U.S. Fleet, "Annual Report, 1 July 1939 to 30 June 1940," 27 July 1940, p. 21, all on roll 10, M971/FR.

181. Depending on the characteristics of the intercepted signal, operators also sometimes could identify the transmitter from which it originated.

182. Remarks of Vice Admiral Henry V. Butler, 15 June 1935, roll 18, M964/FP.

183. Commander Battle Force, "Annual Report, 1 July 1934 to 30 June 1935," 12 July 1935, p. 14, roll 10, M971/FR. This officer was referring not only to RDF-intercepts but also to shipboard decrypting teams.

184. Navy Biography, 15 February 1962, Arthur Japy Hepburn folder, NOA/OB.

185. Remarks of Vice Admiral Henry V. Butler, 15 June 1935, roll 18, M964/FP; and Commander Scouting Force, "Annual Report, 1 July 1935 to 24 June 1936," c. June 1936, p. 23, roll 10, M971/FR.

186. C-in-C U.S. Fleet, "Annual Report, 1 July 1936 to 30 June 1937," 11 September 1937, p. 11, roll 10, M971/FR.

187. CNO to C-in-C U.S. Fleet et al., 12 January 1938, roll 24, M964/FP.

188. Ibid.

189. Lieutenant L. J. Dow (Air Net Control Officer), "Fleet Problem XIX, Report of Performance of West Coast Air Net, High Frequency Direction Finders," c. May 1938, roll 24, M964/FP.

190. CNO to Commander Scouting Force et al., 16 January 1939, roll 25, M964/FP.

191. Commandant Fourteenth Naval District to CNO, 13 May 1938; and Lieutenant L. J. Dow (Air Net Control Officer), "Fleet Problem XIX, Report of Performance of West Coast Air Net, High Frequency Direction Finders," c. May 1938, both on roll 24, M964/FP.

192. Commandant Twelfth Naval District to CNO, 1 June 1938, roll 24, M964/FP. Claude C. Bloch relieved Hepburn as C-in-C U.S. Fleet in early 1938 so Hepburn participated in Fleet Problem XIX as the Twelfth Naval District Commandant.

193. Ibid.

194. CNO to Commander Scouting Force et al., 16 January 1939, roll 25, M964/FP.

195. Adolphus Andrews, "Report on Radio Intelligence Work, Black Unit, during Fleet Problem XX," n.d., enclosure (a) to C-in-C U.S. Fleet to CNO, 8 May 1939, roll 25, M964/FP.

196. Radioman in Charge, U.S. Naval Radio Station, Winter Harbor, Maine to CNO, 4 March 1939; and Commandant Fifth Naval District to CNO, 4 May 1939, both on roll 25, M964/FP.

197. Andrews, "Report on Radio Intelligence Work, Fleet Problem XX," roll 25, M964/FP.

198. Ibid.; and Commander Submarine Force to C-in-C U.S. Fleet, 29 March 1939, enclosure (b) to C-in-C U.S. Fleet to CNO, 8 May 1939, roll 25, M964/FP.

199. Andrews, "Report on Radio Intelligence Work, Fleet Problem XX," roll 25, M964/FP.

200. Williams, *Secret Weapon*, 75–99.

201. Ronald Spector, ed., *Listening to the Enemy: Key Documents on the Role of Communications Intelligence in the War with Japan* (Wilmington, Del.: Scholarly Resources, 1988), esp. 28–39, 154–70. Although the most critical information gained through the Mid-Pacific Net came from decrypted Japanese messages, RDF intercepts were regularly plotted and reported.

202. Extracts from the reports of commanding officers of USS *Yarborough* and USS *Sloat*, in C-in-C U.S. Fleet, "Report on Fleet Problem One," 99, 102.

203. C-in-C Battle Fleet, "Annual Report, 1 July 1923 to 30 June 1924," 22 July 1924,

p. 61; and C-in-C U.S. Fleet, "Annual Report, 1 July 1922 to 30 June 1923," 1 July 1923, p. 29, both on roll 4, M971/FR. The quoted passage is from the latter document.

204. Commanding Officer, USS *Pennsylvania*, "Annual Report, 1 July 1923 to 30 June 1924," c. July 1924, p. 4, roll 4, M971/FR; and C-in-C U.S. Fleet, "Annual Report, 3 October 1925 to 30 June 1926," 25 August 1926, p. 25, roll 5, M971/FR.

205. C-in-C Battle Fleet, "Annual Report, 4 October 1925 to 30 June 1926," 1 July 1926, p. 60, roll 5, M971/FR.

206. Navy Department, *Searchlights and Signal Lights for the Instruction of Officers and Personnel Connected with Searchlights* (Washington: U.S. Government Printing Office, 1918), 115; C-in-C Scouting Fleet, "Annual Report, Fiscal Year 1925," 16 June 1925, p. 10, roll 5, M971/FR; C-in-C U.S. Fleet, "Annual Report, 1 July 1926 to 30 June 1927," 27 September 1927, p. 45, roll 6, M971/FR; C-in-C Scouting Fleet, "Annual Report, 1 July 1928 to 31 March 1929," c. April 1929, p. 29, roll 7, M971/FR; and C-in-C U.S. Fleet, "Annual Report, 1 July 1930 to 30 June 1931," 28 July 1931, p. 30, roll 8, M971/FR.

207. Commander Submarine Force, quoted in C-in-C U.S. Fleet, "Annual Report, 1 July 1931 to 30 June 1932," 21 July 1932, p. 40; Commander Battle Force, "Annual Report, 1 July 1931 to 30 June 1932," c. July 1932, p. 24; and C-in-C U.S. Fleet, "Annual Report, 1 July 1933 to 15 June 1934," 15 June 1934, p. 28, all on roll 9, M971/FR.

208. Of note, submarine searchlights became much more reliable after the adoption of pressure-proof apparatus in the late 1920s and early 1930s. C-in-C U.S. Fleet, "Annual Report, 1 July 1929 to 30 June 1930," 1 August 1930, p. 47, roll 8, M971/FR.

209. Albert Niblack reported a test in which searchlight beams reflected off clouds were picked up eighty-five miles away. Albert P. Niblack, "Naval Signaling," *United States Naval Institute Proceedings* 18, no. 4 (1892): 479.

210. Commander Battle Force, "Annual Report, 1 July 1932 to 15 May 1933," 15 May 1933, p. 33, roll 9, M971/FR.

211. Commander Battleship Divisions, Battle Fleet to C-in-C Battle Fleet, 17 June 1924, roll 4, M971/FR. Fleet Problem One seems to have spurred interest in underwater sound signaling. See extracts from Report of Commander Outguards, in C-in-C U.S. Fleet, "Report on Fleet Problem One," 98; and Nofi, *To Train the Fleet for War*, 56 n. 3.

212. BuEng to SecNav, 1 September 1924, in *ARSN*, 1924, 303.

213. C-in-C U.S. Fleet, "Annual Report, 8 November 1927 to 30 June 1928," 26 September 1928, p. 57, roll 7, M971/FR. Wiley made similar comments the following year in C-in-C U.S. Fleet, "Annual Report, 1 July 1928 to 21 May 1929," 21 May 1929, p. 83, roll 7, M971/FR.

214. University of Pittsburgh Historical Staff at ONR, *The History of United States Naval Research and Development in World War II* (n.p., 1949), 4:1192–93. This typescript is available in the Navy Department Library, Washington, D.C.

215. Commander Battle Force, "Annual Report, 1 July 1934 to 30 June 1935," 12 July 1935, pp. 15–16, roll 10, M971/FR.

216. BuEq to Bruce Cornwall, 17 May 1905, box 34, RG19/BEQ; and President General Board to SecNav, 19 January 1916, box 49, subject file 419, RG80/GB.

217. C-in-C U.S. Fleet, "Annual Report, 1 July 1923 to 30 June 1924," 28 August 1924, roll 4, M971/FR; and Remarks of Rear Admiral William C. Cole (Chief of Staff, U.S. Fleet), 14 March 1925, roll 3, M964/FP.

218. Walter S. Lemmon, "Recent Development of Radio Telephones," *Proceedings of the Radio Club of America* 1, no. 4 (June 1920): 4.

219. Remarks of Rear Admiral William C. Cole (Chief of Staff, U.S. Fleet), 14 March 1925, roll 3, M964/FP.

220. Because the VHF band (30–300 megahertz) was not defined until a later date, I have adopted the terminology used in the late 1920s and 1930s. To personnel of this era, super-frequencies were simply those above HF, a band then defined as 1.5 to 30 megahertz. University of Pittsburgh Historical Staff, *Naval Research and Development in World War II*, 4:1006o n. 1.

221. C-in-C U.S. Fleet, "Annual Report, 1 July 1924 to 30 June 1925," 9 July 1925, p. 39; and C-in-C Scouting Fleet, "Annual Report, Fiscal Year 1926," 25 June 1926, p. 15, both on roll 5, M971/FR.

222. A. Hoyt Taylor, *Radio Reminiscences: A Half Century* (Washington: Naval Research Laboratory, 1948), 244.

223. Comments by Director of Naval Communications, in "Comments on Report of Commander-in-Chief, U.S. Fleet, by Offices in Naval Operations," 5 November 1928, p. 17, roll 7, M971/FR.

224. C-in-C U.S. Fleet, "Annual Report, 1 July 1929 to 30 June 1930," 1 August 1930, p. 49; C-in-C Battle Fleet, "Annual Report, 1 July 1929 to 30 April 1930," c. May 1930, p. 101, both on roll 8, M971/FR; and Taylor, *Radio Reminiscences*, 244, 266.

225. Taylor, *Radio Reminiscences*, 244–45; and BuEng to CNO, 25 March 1931, roll 8, M971/FR.

226. Director of Naval Communications to CNO, 2 October 1931, roll 8, M971/FR; C-in-C U.S. Fleet, "Annual Report, 1 July 1932 to 10 June 1933," 10 June 1933, p. 51, and Commander Battle Force, "Annual Report, 1 July 1932 to 15 May 1933," 15 May 1933, p. 34, roll 9, M971/FR.

227. Taylor, *Radio Reminiscences*, 285; and University of Pittsburgh Historical Staff, *Naval Research and Development in World War II*, 4:1006s–1006t.

228. C-in-C U.S. Fleet, "Annual Report, 1 July 1931 to 30 June 1932," 21 July 1932, p. 45, roll 9, M971/FR; and C-in-C U.S. Fleet, "Annual Report, Fiscal Year 1935," 11 October 1935, p. 25, roll 10, M971/FR.

229. "Memorandum for Op–20-A," written by Op–20-C, 9 December 1935; and Commander Scouting Force, "Annual Report, 1 July 1935 to 24 June 1936," c. June 1936, p. 24, both on roll 10, M971/FR.

230. Commander Battle Force, "Annual Report, 1 July 1935 to 24 June 1936," 21 May 1936, p. 38, roll 10, M971/FR.

231. C-in-C U.S. Fleet, "Annual Report, 1 July 1935 to 30 June 1936," 24 June 1936, p. 35, roll 10, M971/FR.

232. Memorandum Study, "Analysis of Military Communications of Fleet Problem XIX," compiled by Thomas B. Inglis, 20 March 1939, roll 24, M964/FP; and C-in-C U.S.

Fleet, "Annual Report, 1 July 1937 to 30 June 1938," 3 August 1938, p. 12, roll 10, M971/FR. The quoted phrase is from the latter document.

233. C-in-C Asiatic Fleet to CNO, 11 June 1937, roll 23, M964/FP; and Commander Scouting Force, "Annual Report, Fiscal Year 1937," c. July 1937, p. 24, roll 10, M971/FR.

234. Louis A. Gebhard, *Evolution of Naval Radio-Electronics and Contributions of the Naval Research Laboratory* (Washington: U.S. Government Printing Office, 1979), 100–101; and Memorandum Study, "Analysis of Military Communications of Fleet Problem XIX," compiled by Thomas B. Inglis, 20 March 1939, roll 24, M964/FP.

235. H. G. Lindner and S. A. Greenleaf, "Report of Test on Model TBS Receiver," 25 January 1939; and John M. Coe, "Report on Test of Preliminary Model TBS Transmitting Equipment," 25 January 1939. These reports are available in the Ruth H. Hooker Research Library, United States Naval Research Laboratory, Washington, D.C.

236. C-in-C U.S. Fleet, "Annual Report, 1 July 1938 to 30 June 1939," 4 August 1939, p. 20; and C-in-C U.S. Fleet, "Annual Report, 1 July 1939 to 30 June 1940," 27 July 1940, p. 22, both on roll 10, M971/FR.

237. Navy Department, Bureau of Ships, *Instruction Book for TBS Series, TBS to TBS–8 Inclusive* (NAVSHIPS 900,590) (Camden, N.J.: RCA Victor Division of Radio Corporation of America, 1945). This manual is in the author's personal collection.

238. The four carriers lost in 1942 were *Lexington*, *Yorktown*, *Wasp*, and *Hornet*. The two suffering damage were *Saratoga* and *Enterprise*. That year the U.S. Navy sank six Japanese carriers.

239. The Avenger replaced the Devastator in 1942; the Corsair, along with the Grumman F6F Hellcat, began replacing the Wildcat in 1942–43. The Dauntless actually remained in production through the summer of 1944. Grossnick, *United States Naval Aviation*, 493–99.

CHAPTER 5. Creating the Brain of a Warship: Radar and the CIC

Epigraph. Commander Destroyers, Pacific Fleet, "CIC Handbook for Destroyers Pacific Fleet," 24 June 1943, box 1, NWC/LP.

Epigraph. Pacific Fleet Radar Center, "Shoot that bogey down . . .," *C.I.C.*, 15 September 1944, 49. *C.I.C.* was a classified magazine published by OPNAV. At the time I looked at the World War II–era editions of this magazine, they were filed by name in NOA/WWII. The navy has since transferred most records from its World War II Command File to the National Archives in College Park, Maryland. Personal conversation between the author and James Allen Knechtmann, 16 June 2011.

1. Unless otherwise cited details on Outerbridge's and *Ward*'s activities come from the following sources: Summary of Service, [1955?], William W. Outerbridge folder, NOA/OB; and Joint Committee on the Investigation of the Pearl Harbor Attack, *Hearings Pursuant to S. Con. Res. 27*, 79th Cong., 1st sess., part 1, pp. 41–42, and part 36, pp. 55–60, 245–46 (hereafter cited as "Pearl Harbor Hearings").

2. Lawrence C. Grannis in an interview by Gordon W. Prange, 24 July 1964; and Deck Log of USS *Antares*, 7 December 1941, both quoted in Gordon W. Prange with

Donald M. Goldstein and Katherine V. Dillon, *December 7, 1941: The Day the Japanese Attacked Pearl Harbor* (New York: McGraw-Hill, 1988), 91.

3. Gordon W. Prange with Donald M. Goldstein and Katherine V. Dillon, *At Dawn We Slept: The Untold Story of Pearl Harbor* (New York: Penguin Books, 1981), 495; and Prange, *December 7, 1941*, 91–92.

4. Laura Outerbridge, "Family Hero: Tales of Valor Spun from Seat of an E-Z Boy," *Washington Times*, 3 December 1991, newspaper clipping found in William W. Outerbridge folder, NOA/OB.

5. "Japanese Submarine Sunk at Pearl Harbor Is Found," *New York Times*, 30 August 2002, p. A12; and Burl Burlingame, "Midget Sub Found at Pearl," *USNIP* 128, no. 10 (2002): 94–95.

6. USS *Ward* to COM 14 (Commander Fourteenth Naval District), read into the record by Lieutenant Commander Harold Kaminski, 8 January 1942, in Joint Committee, "Pearl Harbor Hearings," part 23, p. 1036.

7. In early 1941 Secretary of the Navy William Franklin "Frank" Knox issued a general order reorganizing the naval forces of the United States into an Atlantic Fleet, a Pacific Fleet, and an Asiatic Fleet. General Order No. 143, 3 February 1941, roll 3, M984/GO.

8. Testimony of Lieutenant Commander Harold Kaminski and Captain John Bayliss Earle, 8, 9 January 1942, in Joint Committee, "Pearl Harbor Hearings," part 23, pp. 1034–44, 1051–54. The passage quoted is from p. 1039.

9. Testimony of Sergeant George E. Elliott, 17 August 1944, in ibid., part 27, p. 520.

10. Testimony of Commander William E. G. Taylor, 27 April 1944, in ibid., part 26, p. 385.

11. Testimony of Lieutenant Colonel Kermit A. Tyler, 17 August 1944, in ibid., part 27, p. 568.

12. Testimony of Commander William E. G. Taylor, 24 August 1944, in ibid., part 32, p. 464.

13. Testimony of Commander William E. G. Taylor, 27 April 1944, in ibid., part 26, pp. 379–83. In recounting events I have focused on tactical and operational errors. For more on the strategic and intelligence errors leading up to Pearl Harbor, see esp. Roberta Wohlstetter, *Pearl Harbor: Warning and Decision* (Stanford, Calif.: Stanford University Press, 1962); Gordon W. Prange with Donald M. Goldstein and Katherine V. Dillon, *Pearl Harbor: The Verdict of History* (New York: Penguin Books, 1986); and Henry C. Clausen and Bruce Lee, *Pearl Harbor: Final Judgement* (New York: Da Capo Press, 2001).

14. Dulany Terrett, *The Signal Corps: The Emergency* (1956; repr., Washington: U.S. Government Printing Office, 1994), 128–29; and Testimony of Sergeant George E. Elliott and First Lieutenant Joseph L. Lockard, 17 August 1944, in Joint Committee, "Pearl Harbor Hearings," part 27, pp. 521–22, 531–33.

15. Prange criticizes heavily the decision of then Lieutenant General Walter Short (commander of the Army's Hawaiian Department) to support the Signal Corps in this intraservice squabble. Prange, *Verdict of History*, 364–72.

16. In 1941 sonar was the only effective method for detecting a submerged subma-

rine, but radar could be (and was) employed against surfaced submarines. As radar resolution improved, even periscopes became vulnerable to detection.

17. David K. Allison, *New Eye for the Navy: The Origin of Radar at the Naval Research Laboratory* (Washington: U.S. Government Printing Office, 1981); and Henry E. Guerlac, *Radar in World War II*, 2 vols. (New York: Tomash Publishers for the American Institute of Physics, 1987), 59–92, 396–426, 537–50, 909–93. Guerlac's history was published posthumously in 1987, but he actually completed it in 1947 as an official report on the radar program of the National Defense Research Committee (NDRC) for the period from June 1940 to August 1945. Also informative is Louis A. Gebhard, *Evolution of Naval Radio-Electronics and Contributions of the Naval Research Laboratory* (Washington: U.S. Government Printing Office, 1979), 169–89.

18. The CIC, now computerized, remains the brain of the modern warship. Although this chapter covers in some detail the U.S. Navy's development of the CIC during World War II, the subject is one deserving of its own book-length treatment.

19. RADAR is an acronym for "radio detection and ranging." During its developmental years, radar was referred to variously as "radio ranging equipment," "radio detection equipment," "radio echo equipment," and "pulse radio equipment." The navy adopted the term radar in 1940–41. CNO to BuShips et al., 18 November 1940; and CNO to BuShips et al., 9 July 1941, box 50, subject file 419, RG80/GB.

20. A. Hoyt Taylor, *Radio Reminiscences: A Half Century* (Washington: Naval Research Laboratory, 1948), 113–14; and Allison, *New Eye for the Navy*, 39–41.

21. Doppler is the apparent frequency shift that results from relative motion. The archetypal example of this phenomenon is a moving train. As the train approaches a stationary observer, its whistle appears to be blowing at a higher frequency; as the train moves away from the observer it seems to be blowing at a lower frequency. In actuality, of course, the frequency of the sound waves emanating from the whistle is constant.

22. Allison, *New Eye for the Navy*, 61–83, and Robert M. Page, *The Origin of Radar: An Epic of Modern Technology* (Garden City, N.Y.: Doubleday, 1962), 33–36.

23. Allison, *New Eye for the Navy*, 1–3, 85–94; Guerlac, *Radar in World War II*, 70–75; and Page, *Origin of Radar*, 2–4, 64–81.

24. Guerlac, *Radar in World War II*, 75–76.

25. BuEng to Director NRL, 12 June 1936, filing designator C-S67-5, RG19/NRLC.

26. Page, *Origin of Radar*, 80, and Allison, *New Eye for the Navy*, 99.

27. Allison, *New Eye for the Navy*, 100–101, and Page, *Origin of Radar*, 106–25.

28. Gebhard, *Evolution of Naval Radio-Electronics*, 172–76; and Director NRL to C-in-C U.S. Fleet, 7 January 1937, filing designator C-S67-5, RG19/NRLC.

29. Taylor, *Radio Reminiscences*, 299–300, and Allison, *New Eye for the Navy*, 101.

30. Taylor, *Radio Reminiscences*, 295–96.

31. Louis Brown, *A Radar History of World War II: Technical and Military Imperatives* (Philadelphia: Institute of Physics Publishing, 1999), 40–91, 444–55.

32. C-in-C U.S. Fleet to Director NRL, 29 December 1936, filing designator C-S67-5, RG19/NRLC.

33. Page, *Origin of Radar*, 126–28; and interviews of Page by David K. Allison, 26, 27

October 1978, summarized in Allison, *New Eye for the Navy*, 102. The quoted passage is from p. 128 of Page's book.

34. Page, *Origin of Radar*, 80, 129–33, and Guerlac, *Radar in World War II*, 81–82.

35. Robert M. Page, laboratory notebook 346, vol. IV, p. 73, quoted in Allison, *New Eye for the Navy*, 103.

36. Memorandum, "Record of conference in connection with Bureau of Engineering SECRET Problem W5-2S," written by James M. Irish, 28 February 1938, filing designator S-S67-5, RG19/NRLS. At that time, the material bureaus consisted of Aeronautics, Construction & Repair, Engineering, and Ordnance.

37. Taylor, *Radio Reminiscences*, 323–25; and Director NRL to BuEng, 4 March 1938, filing designator S-S67-5, RG19/NRLS. Due to unexpected delays in the manufacture of an antenna support structure, Page and his team would not quite meet the deadline. Director NRL to BuEng, 4 August 1938, filing designator S-S67-5, RG19/NRLS.

38. BuEng to C-in-C U.S. Fleet, 25 April 1938; C-in-C U.S. Fleet to BuEng, 2 May 1938; and C-in-C U.S. Fleet to BuAer, 26 May 1938, filing designator S-S67-5, RG19/NRLS.

39. Taylor, *Radio Reminiscences*, 326–27; BuEng to Director NRL, 5 December 1938, filing designator C-S67-5, RG19/NRLC; and Memorandum for Dr. Taylor, "Radio Ranging Equipment, 200 megacycles," written by Louis A. Gebhard, 6 February 1939, filing designator S-S67-5, RG19/NRLS.

40. BuEng Contract NOs–59870, 26 March 1938, summarized in University of Pittsburgh Programs Research Staff, "The Administration of Naval Development Programs," 19 October 1951, enclosure (c), "Chronological Digest of Documents Relating to the Combat Information Center Program, 1922–1941" (hereafter cited as "CIC Chronology through 1941"), section D–2. I examined this chronology, which consists largely of excerpts from official correspondence, when it was located in NOA/WWII.

41. Allison, *New Eye for the Navy*, 105–7; and Arthur A. Varela to Director NRL, 28 March 1939, filing designator S-S67-5, RG19/NRLS.

42. Commanding Officer, USS *Texas* to Commander Atlantic Squadron, 24 March 1939, filing designator S-S67-5, RG19/NRLS.

43. CXZ operators detected battleships at a maximum range of 15,000 yards and destroyers at a maximum range of 9,000 yards. The corresponding distances for XAF operators were 29,000 and 19,000 yards, respectively. Commanding Officer, USS *Texas* to Commander Atlantic Squadron, 19 March 1939; and Robert M. Page to Commander Atlantic Squadron, 3 April 1939, filing designator S-S67-5, RG19/NRLS.

44. Commanding Officer, USS *Texas* to Commander Atlantic Squadron, 24 March 1939; and Commander Atlantic Squadron to BuEng, 4 April 1939, filing designator S-S67-5, RG19/NRLS.

45. Page to Commander Atlantic Squadron, 3 April 1939, filing designator S-S67-5, RG19/NRLS; and interviews of Page by David K. Allison, 26, 27 October 1978, quoted in Allison, *New Eye for the Navy*, 108–9.

46. Page to Commander Atlantic Squadron, 3 April 1939, filing designator S-S67-5, RG19/NRLS. Regarding submarines, Page wrote: "Submergence of the hull made no

difference in signal as long as the conning tower was above water . . . Signal disappeared about the same time the conning tower went out of sight. No signal could be observed from the periscope alone."

47. Commanding Officer, USS *New York* to Commander Atlantic Squadron, 24 March 1939; and Page to Commander Atlantic Squadron, 3 April 1939, filing designator S-S67-5, RG19/NRLS.

48. Page to Leo C. Young, 3 March 1939; Commanding Officer, USS *New York* to Commander Atlantic Squadron, 24 March 1939; and Commander Atlantic Squadron to BuEng, 4 April 1939, filing designator S-S67-5, RG19/NRLS.

49. Page to Commander Atlantic Squadron, 3 April 1939, filing designator S-S67-5, RG19/NRLS.

50. Memorandum for File, "Conference on Special Project No. 1," 8 May 1939, filing designator S-S67-5, RG19/NRLS.

51. CNO to BuEng, 12 May 1939, summarized in "CIC Chronology through 1941," section D–1.

52. Louis A. Gebhard, Record of Consultative Services, 19, 26 May 1939, filing designator S-S67-5, RG19/NRLS.

53. Louis A. Gebhard, "Record of Consultative Services," 21, 26 June, 31 August, 1 September, and 17 October 1939; and BuEng to Director NRL and Inspector of Naval Material, 25 October 1939, filing designator S-S67-5, RG19/NRLS.

54. Receipt for "one complete set of blueprints of the Model XAF Radio Ranging Equipment," 16 October 1939, signed by J. E. Love of the RCA Manufacturing Company, filing designator S-S67-5, RG19/NRLS.

55. Director NRL to BuEng, 26 February 1940, filing designator S-S67-5, RG19/NRLS.

56. Ibid.

57. Commander Atlantic Squadron to BuEng, 4 April 1939, filing designator S-S67-5, RG19/NRLS; and BuShips to C-in-C U.S. Fleet, 1 August 1940, box 50, subject file 419, RG80/GB.

58. Director NRL to BuEng, 26 February 1940; and BuEng and BuCR to BuSA, 11 June 1940, filing designator S-S67-5, RG19/NRLS.

59. BuEng to CNO, 28 March 1940, filing designator S-S67-5, RG19/NRLS.

60. BuShips to C-in-C U.S. Fleet, 1 August 1940, box 50, subject file 419, RG80/GB.

61. "Service Tests of Model CXAM Equipment," enclosure (a) to Commanding Officer, USS *California* to C-in-C U.S. Fleet, 2 October 1940, filing designator S-S67-5, RG19/NRLS.

62. BuEng to BuSA, 11 June 1940; and Memorandum for Director NRL, "Data on Installations, U.S.S. CHESTER, etc." written by Robert C. Guthrie, 13 February 1941, filing designator S-S67-5, RG19/NRLS. The four cruisers were *Chester*, *Chicago*, *Northampton*, and *Pensacola*.

63. BuEng and BuCR to Commandants of the Navy Yards at Mare Island, Puget Sound, and Pearl Harbor, 6 May 1940, filing designator S-S67-5, RG19/NRLS.

64. Memorandum for Director NRL, "Data on Installations, U.S.S. CHESTER, etc." written by Robert C. Guthrie, 13 February 1941, filing designator S-S67-5, RG19/NRLS.

65. Summary of Service, 20 July 1955, Henry E. Bernstein folder, NOA/OB.

66. "Service Tests of Model CXAM Equipment," enclosure (a) to Commanding Officer, USS *California* to C-in-C U.S. Fleet, 2 October 1940, filing designator S-S67-5, RG19/NRLS; and Irving L. McNally, "Radar Reflections," 1 July 1975, 4. McNally's reminiscences are available at the Navy Department Library in Washington, D.C.

67. Commanding Officer, USS *California* to C-in-C U.S. Fleet, 2 October 1940; and Commanding Officer, USS *California* to BuShips, 22 November 1940, filing designator S-S67-5, RG19/NRLS.

68. Description of events, August to November 1940, in "CIC Chronology through 1941," section E–1.

69. Guerlac, *Radar in World War II*, 930.

70. Memorandum for Director NRL, "Data on Installations, U.S.S. CHESTER, etc." written by Robert C. Guthrie, 13 February 1941, filing designator S-S67-5, RG19/NRLS; and Commanding Officer, USS *Northampton* to CinCPac, 4 June 1941, in "CIC Chronology through 1941," section C–1.

71. Commanding Officer, USS *Chester* to BuShips, 9 August 1940, summarized in "CIC Chronology through 1941," section D–8; and description of events, 3 December 1940, "CIC Chronology through 1941," section C–1.

72. BuShips Contract NOs–59870, 30 November 1940, summarized in "CIC Chronology through 1941," section D–2; BuShips to BuSA, 5 March 1941, filing designator S-S67-5, RG19/NRLS; and description of events, 14 June 1940, in "CIC Chronology through 1941," section D–2.

73. Director NRL to CNO, 1 October 1940; and Technical Aide to SecNav, 7 October 1940, filing designator S-S67-5, RG19/NRLS. The quoted phrase is from the latter document.

74. Director NRL to CNO, 5 November 1940, filing designator S-S67-5, RG19/NRLS; and SecNav to Director NRL, 8 November 1940, summarized in "CIC Chronology through 1941," section B–1. The quoted phrase is from a draft letter written by Bowen as part of his letter to the CNO.

75. Memorandum for Director NRL, "Data on Installations, U.S.S. CHESTER, etc." written by Robert C. Guthrie, 13 February 1941, filing designator S-S67-5, RG19/NRLS.

76. Commanding Officer, USS *Yorktown* to [CinCPac?], 14 June 1941, in "CIC Chronology through 1941," section E–1.

77. Commanding Officer, USS *Yorktown* to CinCPac, 28 March 1941, in "CIC Chronology through 1941," section E–1.

78. Ibid.

79. Commander Cruisers Scouting Force to CinCPac, 24 February 1941; and Commander Aircraft Battle Force to BuShips, 28 April 1941, summarized in "CIC Chronology through 1941," sections E–4 and E–1, respectively.

80. Commanding Officer, USS *Yorktown* to Senior Member, Interior Control Board, 18 July 1941; and CNO to BuShips, 21 August 1941, both summarized in "CIC Chronology through 1941," section E–1.

81. Senior Member, Interior Control Board to SecNav (via CNO), 22 July 1941, quoted in "CIC Chronology through 1941," section E–1.

82. Gebhard, *Evolution of Naval Radio-Electronics*, 251; and Commander Aircraft Battle Force to C-in-C U.S. Fleet, 14 January 1939, filing designator S-S67-5, RG19/NRLS.

83. Page, *Origin of Radar*, 165–66.

84. William C. Bryant and Heith I. Hermans, "History of Naval Fighter Direction," February 1946, box labeled "WWII—Fast Carrier Task Forces," World War II Collection, Naval Aviation History Branch, Naval History and Heritage Command, Washington, D.C.; and Robert M. Page and La Verne R. Philpott, "Identification and Recognition System," U.S. Letters Patent No. 3,296,615, 3 January 1967. Page and Philpott filed for this patent on 19 January 1942. In 2008 the Naval History and Heritage Command reorganized the Naval Aviation History Branch, the Ships History Branch, and the Operational Archives into a single Archives Branch. Because this change was administrative and most documents did not physically move, I have retained the original citation information. Curtis Utz (Director, Archives Branch, Naval History and Heritage Command), e-mail to the author, 20 July 2012.

85. Commander Cruisers Scouting Force to CinCPac, 24 February 1941; and Commanding Officer, USS *Yorktown* to CinCPac, 10 March 1941, both summarized in "CIC Chronology through 1941," section D–5.

86. Commanding Officer, USS *California* to BuShips, 10 August 1941; and CinCPac endorsement to BuShips, 23 August 1941, both summarized in "CIC Chronology through 1941," section D–5.

87. McNally, "Radar Reflections," 3.

88. BuShips to CNO, 24 October 1940; and CNO to BuNav, 14 November 1940, both summarized in "CIC Chronology through 1941," section G–1.

89. David Zimmerman, *Top Secret Exchange: The Tizard Mission and the Scientific War* (Montreal: McGill-Queen's University Press, 1996), 107–29; and CNO to BuNav, 20 November 1940, summarized in "CIC Chronology through 1941," section G–1.

90. BuShips to BuNav, 13 June 1941, summarized in "CIC Chronology through 1941," section G–1.

91. Edwin T. Short to BuNav, 19 September 1941; and BuNav to CinCPac and CinCLant, 10 October 1941, both summarized in ibid.

92. Commander Battleships Battle Force to CinCPac, 17 October 1941; and CinCPac to BuNav, 25 October 1941, both summarized in ibid.

93. BuNav to CNO, 27 November 1941; and CNO to BuShips et al., 19 December 1941, both summarized in ibid.

94. Commander Cruisers Scouting Force to Commanding Officer, USS *Chicago* et al., 25 October 1940; and BuShips to BuNav, 1 May 1941, both summarized in ibid.

95. Director NRL to BuShips, 7 May 1941; BuNav to Commandant First Naval District, 17 May 1941; CNO to BuShips, 28 May 1941; and description of events, 23 June 1941, all summarized in ibid.

96. Fighter directors were the individuals who used radar information to direct friendly aircraft into position to intercept enemy aircraft.

97. BuNav to CNO, 10 June 1941; and BuAer to CNO, 23 June 1941, both summarized in "CIC Chronology through 1941," section G–1.

98. CNO to BuAer, 10 July 1941, summarized in ibid.; and Bryant and Hermans, "History of Naval Fighter Direction," 88.

99. Clipping of a description of the career of Captain John Hook Griffin, March 1996, John Hook Griffin folder, NA/VF; and final draft of an untitled article submitted by Captain Nicholas J. Hammond to the United States Naval Institute, 26 May 1988 (hereafter cited as "Draft Article on World War II Fighter Direction"), p. 1. Hammond, who served as a fighter director officer during World War II, collaborated with two other fighter director officers, Horace Stanwood Foote and Charles D. Ridgway, in writing the article. The U.S. Naval Institute declined to publish it due to excessive length. Paul Stillwell to Hammond, 16 June 1988. I am indebted to David Boslaugh for providing me with copies of these documents.

100. Wing Commander, R.A.F. Station, Biggin Hill to Secretary, Air Ministry, 6 August 1936; and "Agenda for the Meeting to Discuss a Programme for Biggin Hill Experiments [held on 21 April 1937]," both in AIR 2/2625. The quoted passage is from the latter document.

101. David Zimmerman, *Britain's Shield: Radar and the Defeat of the Luftwaffe* (Stroud, U.K.: Sutton Publishing, 2001). See esp. 109–17, 122–31, 156–62, 175–210.

102. Hammond, "Draft Article on World War II Fighter Direction," 1–2.

103. BuNav to Ensign Nicholas J. Hammond, 12 September 1941, quoted in ibid., 1.

104. Hammond, "Draft Article on World War II Fighter Direction," 1–5.

105. R. G. Gray, "History of CIC School, Fleet Training Center, Oahu," 15 September 1945, 1–3; and "Radar Center Trains Thousands for the Pacific War," *C.I.C.*, 15 September 1944, 34–46 (enclosed as appendix IV), in Annex D of Administrative History Appendix 38 (14) (A), NDL/HR.

106. Hammond, "Draft Article on World War II Fighter Direction," 5–11.

107. Jack Griffin to Red Morse [commanding officer of the Fighter Director School in Norfolk, Va.], 6 March 1942, appendix II to Gray, "History of CIC School," NDL/HR.

108. Ibid.

109. Commanding Officer, USS *New York* to Director NRL, 4 May 1939, filing designator S-S67–5, RG19/NRLS.

110. Evidence of Kimmel's intention to seek a decisive fleet engagement is circumstantial; however, surviving war plans suggest this is what he and his staff had in mind. Edward S. Miller, *War Plan Orange: The U.S. Strategy to Defeat Japan* (Annapolis: Naval Institute Press, 1991), 286–322.

111. For more on the role of signals intelligence in the Battle of the Atlantic, see esp. Jürgen Rohwer, *The Critical Convoy Battles of March 1943: The Battle for HX.229/SC122*, trans. Derek Masters (Annapolis: Naval Institute Press, 1977), 229–44; David Syrett, *The Battle of the Atlantic and Signals Intelligence: U-Boat Situations and Trends, 1941–1945* (Aldershot: Scolar Press for the Navy Records Society, 1998); and W. J. R. Gardner, *Decoding History: The Battle of the Atlantic and Ultra* (Annapolis: Naval Institute Press, 1999), 120–45. Syrett's work is particularly informative.

112. Kathleen Broome Williams, *Secret Weapon: U.S. High-Frequency Direction Finding in the Battle of the Atlantic* (Annapolis: Naval Institute Press, 1996), 206–31, and Guerlac, *Radar in World War II*, 708–30.

113. Bryant and Hermans, "History of Naval Fighter Direction," 13–15.

114. John B. Lundstrom, *The First Team: Pacific Naval Air Combat from Pearl Harbor to Midway* (Annapolis: Naval Institute Press, 1984), 87–93.

115. Ibid., 94–107, and Steve Ewing and John B. Lundstrom, *Fateful Rendezvous: The Life of Butch O'Hare* (Annapolis: Naval Institute Press, 1997), 118–40. In the latter book, Lundstrom changes slightly his retelling of events, partly to correct minor errors in the earlier account. Probably because *Fateful Rendezvous* concentrates on O'Hare, *The First Team* provides a more thorough description of Gill's thoughts and actions.

116. "President Honors, Promotes O'Hare," *New York Times*, 22 April 1942, p. 1.

117. Bryant and Hermans, "History of Naval Fighter Direction," 15.

118. Samuel Eliot Morison, *Coral Sea, Midway and Submarine Actions, May 1942–August 1942*, vol. 4 of *History of United States Naval Operations in World War II* (Boston: Little, Brown, 1949), 46–60.

119. Frederick C. Sherman, *Combat Command: The American Aircraft Carriers in the Pacific War* (New York: E. P. Dutton, 1950), 108–16.

120. The attacking aircraft included Aicha D3As (dive bombers), Nakajima B5Ns (torpedo bombers), and Mitsubishi A6Ms (fighters). Later in 1942 Allied forces began calling these types of aircraft Vals, Kates, and Zeroes. The Mitsubishi G4M, which had gun blisters, was known as the Betty, allegedly named after a "well-proportioned nurse." Robert C. Mikesh, *Japanese Aircraft: Code Names and Designations* (Atglen, Pa.: Schiffer Military/Aviation History, 1993).

121. H. P. Willmott, *The Barrier and the Javelin: Japanese and Allied Pacific Strategies February to June 1942* (Annapolis: Naval Institute Press, 1983), 251, 265.

122. Bryant and Hermans, "History of Naval Fighter Direction," 18, 21.

123. Gill thought the Japanese torpedo bombers would approach at a low altitude of about one thousand feet, but they actually came in at ten thousand feet. Lundstrom, *The First Team*, 245–48.

124. Bryant and Hermans, "History of Naval Fighter Direction," 18–19.

125. Commanding Officer, USS *Yorktown*, "Secret Information Bulletin # 1," quoted in Bryant and Hermans, "History of Naval Fighter Direction," 19; and Sherman, *Combat Command*, 108–9.

126. Morison, *Coral Sea, Midway and Submarine Actions*, 53 n. 1.

127. Scholarly literature on the Battle of Midway is voluminous. The most famous accounts of the battle include Morison, *Coral Sea, Midway and Submarine Actions*, 69–159; Mitsuo Fuchida and Masatake Okumiya, *Midway, the Battle that Doomed Japan: The Japanese Navy's Story* (Annapolis: Naval Institute Press, 1955); Walter Lord, *Incredible Victory* (New York: Harper & Row, 1967); and Gordon W. Prange with Donald M. Goldstein and Katherine V. Dillon, *Miracle at Midway* (New York: Penguin Books, 1982). For recent scholarship that reexamines Japanese actions and overturns many accepted canons about the battle, see Jonathan Parshall and Anthony Tully, *Shattered Sword: The*

Untold Story of the Battle of Midway (Washington: Potomac Books, 2005), and Dallas Woodbury Isom, *Midway Inquest: Why the Japanese Lost the Battle of Midway* (Bloomington: Indiana University Press, 2007). Analysis of the aerial combat that took place during the battle is covered best in Lundstrom, *The First Team*, 329–434.

128. Six Mitsubishi A6M fighters escorted each wave. Morison, *Coral Sea, Midway and Submarine Actions*, 132.

129. Bryant and Hermans, "History of Naval Fighter Direction," 21–23.

130. Lundstrom, *The First Team*, 369–411, 425–26. *Yorktown* might have survived, but on 6 June a Japanese submarine put two more torpedoes into the carrier.

131. Bryant and Hermans, "History of Naval Fighter Direction," 21–23; and Lundstrom, *The First Team*, 399, 435–43.

132. John B. Lundstrom, *The First Team and the Guadalcanal Campaign: Naval Fighter Combat from August to November 1942* (Annapolis: Naval Institute Press, 1994), 109–64, 325–459. The performance of the CXAM–1 radars on *Enterprise* and *Hornet* during the Battle of Santa Cruz was extremely poor and contributed significantly to the problems in fighter direction.

133. Commander, Fighting Squadron 62[?], "Action Report," 26 October 1942, quoted in Bryant and Hermans, "History of Naval Fighter Direction," 28.

134. CinCPac, "Action Report, Battle of Santa Cruz," 26 October 1942, quoted in Bryant and Hermans, "History of Naval Fighter Direction," 28–29.

135. Samuel Eliot Morison, *The Struggle for Guadalcanal, August 1942–February 1943*, vol. 5 of *History of United States Naval Operations in World War II* (Boston: Little, Brown, 1949), 17–64; and Denis Warner and Peggy Warner, *Disaster in the Pacific: New Light on the Battle of Savo Island* (Annapolis: Naval Institute Press, 1992), 113–212.

136. University of Pittsburgh Programs Research Staff, "Administration of the Combat Information Center Program to 1947, Final Report," 30 January 1952 (hereafter cited as "CIC Administrative History"), p. 46. I examined this report when it was located in NOA/WWII.

137. Charles Cook, *The Battle of Cape Esperance: Encounter at Guadalcanal* (1968; repr., Annapolis: Naval Institute Press, 1992), 147. Authors who quote Cook tend to overlook the fact that he acknowledged fully the value of radar. To wit, a few pages after giving chance its due, Cook wrote: "It was apparent, too, that American radar was a factor that would take chance progressively out of the picture to the disadvantage of the Japanese." Ibid., 150.

138. David C. Evans and Mark R. Peattie, *Kaigun: Strategy, Tactics, and Technology in the Imperial Japanese Navy, 1887–1941* (Annapolis: Naval Institute Press, 1997), 273–81.

139. Morison, *The Struggle for Guadalcanal*, 235–58; Paul Stillwell, "The Naval Battle of Guadalcanal: The Battleship Night Action," in *Great American Naval Battles*, ed. Jack Sweetman (Annapolis: Naval Institute Press, 1998), 288–90; and "CIC Administrative History," 46–47.

140. Morison, *The Struggle for Guadalcanal*, 270–74.

141. Commanding Officer, USS *South Dakota* to COMINCH, 24 November 1942, "Action report, night engagement 14–15 November, 1942," box 1439, RG38/AR.

142. Ibid.

143. *South Dakota* seems to have been in one of *Washington*'s radar blind spots. Commanding Officer, USS *Washington* to CinCPac, 27 November 1942, "Action Report, Night of November 14–15, 1942," box 1501, RG38/AR.

144. Ibid.

145. Robert Lundgren, "*Kirishima* Damage Analysis," 28 September 2010, viewed at http://www.navweaps.com/index_lundgren/index_lundgren.htm on 30 August 2011.

146. A recent book on the naval actions around Guadalcanal posits that *Washington* accidentally fired and hit the destroyer *Preston*; however, the book's author provides limited evidence in support of this claim. James D. Hornfischer, *Neptune's Inferno: The U.S. Navy at Guadalcanal* (New York: Bantam Books, 2011), 357, 458.

147. Commanding Officer, USS *Washington* to CinCPac, 27 November 1942, "Action Report, Night of November 14–15, 1942," box 1501, RG38/AR.

148. Stillwell, "The Naval Battle of Guadalcanal," 296. Presumably Lee wanted someone familiar with *Washington*'s radar to help him interpret the oral reports he would be receiving.

149. Commanding Officer, USS *Washington* to CinCPac, 27 November 1942, "Action Report, Night of November 14–15, 1942," box 1501, RG38/AR.

150. E. B. Potter, *Nimitz* (Annapolis: Naval Institute Press, 1976), 229.

151. U.S. Pacific Fleet Tactical Bulletin No. 4TB–42, 26 November 1942, filing designator FF12-4, RG313/R179. I am indebted to archivist Nate Patch for pulling this document from a collection that was then undergoing processing.

152. "CIC Administrative History," 137. For most of World War II King served as both COMINCH and CNO. Thomas B. Buell, *Master of Sea Power: A Biography of Fleet Admiral Ernest J. King* (1980; repr., Annapolis: Naval Institute Press, 1995). See esp. chaps. 13 and 14.

153. Summary of correspondence between COMINCH and the Bureau of Ships and the Bureau of Ordnance during the fall of 1942, in "CIC Administrative History," 137–38. The quoted passage is from p. 137.

154. Ibid.

155. "Record of Conference on Combat Operations Center, January 8, 1942 [*sic*]," box 1, NWC/LP.

156. COMINCH and CNO to Vice Chief of Naval Operations, 8 February 1943, summarized and quoted in "CIC Administrative History," 181–85. I have found no direct evidence that King personally approved the name change; however, I am fairly certain no staff officer would have overridden the Pacific Fleet commander-in-chief's proposed designation without first obtaining King's approval. I base this conclusion, in part, from my own experiences serving on the OPNAV and Pacific Fleet staffs from 2002–5 and from 2008–11, respectively.

157. COMINCH and CNO to Vice Chief of Naval Operations, 8 February 1943, quoted in "CIC Administrative History," 184.

158. Senior Member, Interior Control Board to SecNav (via CNO), 22 July 1941, quoted in "CIC Chronology through 1941," section E–1.

159. BuShips internal memorandum from Code 970 to Code 900, 19 January 1943; and CinCPac to COMINCH, 5 February 1943, both summarized in "CIC Administrative History," 233–34. At the time, such facilities were still officially designated as Combat Operations Centers.

160. "CIC Administrative History," 223–70. For more on the debates over the location of CICs on *Essex*-class carriers, see Norman Friedman, *U.S. Aircraft Carriers: An Illustrated Design History* (Annapolis: Naval Institute Press, 1983), 147, 151.

161. Navy Biography, 10 January 1966, Mahlon Street Tisdale folder, Mahlon Street Tisdale Manuscript Collection, Special Collections, Navy Department Library, Washington, D.C.; Tisdale to F. Gordon Barber, 29 November 1949, box 1, NWC/LP; and Russell S. Crenshaw, *The Battle of Tassafaronga* (Baltimore: Nautical & Aviation Publishing Company, 1995), 48–68.

162. Richard B. Frank, *Guadalcanal: The Definitive Account of the Landmark Battle* (New York: Random House, 1990), 515–16. The task force commander was Rear Admiral Carleton H. Wright.

163. Summary of Service, 28 September 1955, Caleb Barrett Laning folder, NOA/OB; and Caleb B. Laning, "Notes on Development of Combat Information Center principles at ComDesPac, 1942–43," September 1987, box 1, NWC/LP.

164. Laning, "Notes on Combat Information Center principles, 1942–43," box 1, NWC/LP. Laning also mentions by name George Phillips [Philip?] and Robert Bookman.

165. Joseph C. Wylie, interview by Paul Stillwell, typescript, Portsmouth, R.I., 21 May 1985, pp. 52–53; and Joseph C. Wylie, interview by Evelyn M. Cherpak, typescript, Newport, R.I., 15 January 1986, pp. 111–12, both in NWC/OH. When recalling matters some forty years later, Wylie accidentally reversed the CIC's fore and aft sections. I have corrected his minor mistake in the text.

166. Commander Destroyers, U.S. Pacific Fleet to COMINCH, 5 April 1943, quoted in "CIC Administrative History," 255.

167. "CIC Administrative History," 256–58.

168. Ibid., 273–75.

169. Joseph C. Wylie, interview by Paul Stillwell, 21 May 1985, pp. 50–51, NWC/OH.

170. Caleb Laning to Ed [?], 23 June 1943; and Commander Destroyers, Pacific Fleet, "CIC Handbook for Destroyers Pacific Fleet," 24 June 1943, box 1, NWC/LP.

171. "CIC Administrative History," 276–81; and signed enclosure to Joseph C. Wylie to Lieutenant Commander Robert M. Lunny, 12 April 1948, box 1, NWC/LP.

172. Joseph C. Wylie, interview by Paul Stillwell, 21 May 1985, p. 51, NWC/OH. See also Joseph C. Wylie, *Military Strategy: A General Theory of Power Control*, with an introduction by John B. Hattendorf (Annapolis: Naval Institute Press, 1989).

173. Commander Destroyers, Pacific Fleet, "CIC Handbook for Destroyers Pacific Fleet," 24 June 1943, box 1, NWC/LP.

174. "CIC Administrative History," 285–87.

175. Commander Destroyers, Pacific Fleet, "CIC Handbook for Destroyers Pacific Fleet," 24 June 1943, box 1, NWC/LP.

176. Russell S. Crenshaw, *South Pacific Destroyer: The Battle for the Solomons from Savo Island to Vella Gulf* (Annapolis: Naval Institute Press, 1998), 206–19; and E. B. Potter, *Admiral Arleigh Burke* (New York: Random House, 1990), 90–107. Potter does not explicitly mention the CIC but his description of Burke's actions make clear that the future CNO relied heavily on information provided by both his own and his subordinates' CICs.

177. Information in this paragraph comes from several sources, including those cited in notes 111 and 112 above. See also Theodore Roscoe, *United States Destroyer Operations in World War II* (Annapolis: Naval Institute Press, 1953); Samuel Eliot Morison, *The Atlantic Battle Won, May 1943–May 1945*, vol. 10 of *History of United States Naval Operations in World War II* (Boston: Little, Brown, 1956); Montgomery C. Meigs, *Slide Rules and Submarines: American Scientists and Subsurface Warfare in World War II* (Washington: National Defense University Press, 1990); and David Syrett, *The Defeat of the German U-boats: The Battle of the Atlantic* (Columbia: University of South Carolina Press, 1994).

178. "CIC Administrative History," 228–29.

179. Roscoe, *United States Destroyer Operations in World War II*, 292–94; Morison, *The Atlantic Battle Won*, 171–77; and Williams, *Secret Weapon*, 221–22.

180. Samuel Eliot Morison, *New Guinea and the Marianas, March 1944–August 1944*, vol. 8 of *History of United States Naval Operations in World War II* (Boston: Little, Brown, 1949), 233.

181. For a battle of such enormous scale and significance, Philippine Sea has attracted surprisingly limited historical analysis. The best-known accounts include Morison, *New Guinea and the Marianas*, 213–340; William T. Y'Blood, *Red Sun Setting: The Battle of the Philippine Sea* (Annapolis: Naval Institute Press, 1981); and Barrett Tillman, *Clash of the Carriers: The True Story of the Marianas Turkey Shoot of World War II* (New York: New American Library, 2005). Of these authors, only Tillman mentions the CIC, at one point making the rather odd claim that the CIC "was seven-tenths science, 1940s style." Tillman, *Clash of Carriers*, 140.

182. I rely here on Y'Blood's figures. Y'Blood, *Red Sun Setting*, 140. These numbers are even more impressive when one considers that Ozawa succeeded in attacking first with four consecutive waves of aircraft.

183. Ibid., 133.

184. Commander Task Force 58, "Operations in Support of the Capture of the Marianas," 11 September 1944, quoted in Bryant and Hermans, "History of Naval Fighter Direction," 81.

185. Gray, "History of CIC School," 5–6, NDL/HR. The Bureau of Navigation became the Bureau of Naval Personnel on 13 May 1942.

186. Memorandum for Captain [Dewitt C.] Ramsey, "Fighter Director School," written by John H. Griffin, 14 September 1942, appendix I to Gray, "History of CIC School," NDL/HR.

187. Lundstrom, *The Guadalcanal Campaign*, 340; and an unsigned letter to John H. Griffin, 29 December 1942, appendix VII to Gray, "History of CIC School," NDL/HR.

188. Gray, "History of CIC School," 9, NDL/HR.

189. Vice Chief of Naval Operations to Commandant USMC and BuAer, [28 March 1943?], quoted in Bryant and Hermans, "History of Naval Fighter Direction," 90; and CinCPac to Commander Service Force Pacific Fleet, 10 April 1943, located in "Radar Operators School History," c. September 1945, in Annex E of Administrative History Appendix 38 (14) (A), NDL/HR.

190. Gray, "History of CIC School," 13, NDL/HR; and CinCPac to Commander Service Force Pacific Fleet, 10 April 1943, enclosure (c) to "Radar Operators School History," NDL/HR.

191. CinCPac to Pacific Fleet (Pacific Fleet Confidential Letter 13CL–43), 25 May 1943, appendix X to Gray, "History of CIC School," NDL/HR. Raymond A. Spruance, Nimitz's chief of staff, signed out this letter.

192. Ibid.

193. "Outline of Policy and Study Course," 10 April 1943, appendix V to Gray, "History of CIC School," NDL/HR.

194. "Selection and Personnel Department, [Pacific Fleet Radar Center], 15 December 1943 to 1 August 1945," c. October 1945, in Annex B of Administrative History Appendix 38 (14) (A), NDL/HR. According to this source, the school later adopted the Cornell Index in an effort to "weed out" individuals who were bright but unsuited for shipboard life.

195. Gray, "History of CIC School," 16, 20; and "Radar Operators School History," part C, NDL/HR.

196. "CIC Administrative History," 435–36.

197. CinCPac to Pacific Fleet (Pacific Fleet Confidential Letter 23CL–43), 25 August 1943, appendix XI to Gray, "History of CIC School," NDL/HR. P. V. Mercer, Nimitz's assistant chief of staff, signed out this letter.

198. Ibid.

199. J. H. Lowe, "History and Development of C.I.C. Team Training Center, NAAS—San Clemente Island, California," 27 September 1945; "Command History, CIC Team Training Center, Astoria, Ore., March 1944 to April 1945," c. September 1945; and "History of CIC Training at USNTS (Tactical Radar), Hollywood, Florida," 30 October 1945, in Administrative History Appendices 21 (18), 21 (16), and 30 (2), NDL/HR.

200. COMINCH to CinCLant and CinCPac, 11 May 1944, 4 August 1944, filing designator P11-1(20), Confidential Correspondence, 1944, Headquarters of the Commander-in-Chief, U.S. Fleet, Records of the Office of the Chief of Naval Operations (Record Group 38), U.S. National Archives, College Park, Md.

201. "CIC Administrative History," 430–31.

202. H. P. Willmott, *The Battle of Leyte Gulf: The Last Fleet Action* (Bloomington: Indiana University Press, 2005), 74–76. According to Willmott, by late 1944 kamikaze tactics offered not the best but the only means of inflicting significant losses on the U.S. fleet.

203. Bryant and Hermans, "History of Naval Fighter Direction," 82–84.

204. John Monsarrat, *Angel on the Yardarm: The Beginnings of Fleet Radar Defense and the Kamikaze Threat* (Newport, R.I.: Naval War College Press, 1985), 151. Monsarrat

attended the Fighter Director and Combat Information School at Camp Catlin in 1943 and then served aboard a carrier through the end of the war. He saw considerable action in 1944–45, including at the Battles of the Philippine Sea, Leyte Gulf, and Okinawa.

205. Samuel Eliot Morison, *Victory in the Pacific, 1945*, vol. 14 of *History of United States Naval Operations in World War II* (Boston: Little, Brown, 1960), 282, 390–92; and Commander Amphibious Forces, U.S. Pacific Fleet (Richmond K. Turner) to COMINCH, 25 July 1945, "General Action Report, Capture of OKINAWA GUNTO, Phases I and II, 17 February 1945 to 17 May 1945" (hereafter cited as "CTF 51, 'Action Report for Okinawa'"), box 535, RG38/AR.

206. CTG 58.1 (Clark) to COMINCH, 5 May 1945, "Report of Operations of Task Group FIFTY-EIGHT POINT ONE (Fast Carrier Group One) in Support of Landings at OKINAWA 14 March to 30 April," box 218, RG38/AR; CTG 58.3 (Sherman) to COMINCH, 18 June 1945, "Operations of Task Group 58.3/38.3 in Support of Occupation of OKINAWA, during the period 14 March–1 June 1945," box 228, RG38/AR; and Commander Task Force (CTF) 54 (Smith) to COMINCH, 4 June 1945, "Action Report—Capture of OKINAWA GUNTO, Phase II, 5 May to 28 May 1945," box 205, RG38/AR.

207. CTF 54 to COMINCH, 4 June 1945, "Action Report—Capture of OKINAWA GUNTO, Phase II, 5 May to 28 May 1945," box 205, RG38/AR.

208. Such criticisms notwithstanding, Mitscher reported that his forces destroyed at least 1,629 airborne enemy aircraft and that only 11 successfully conducted suicide attacks. One official report later estimated that 31 percent of attacking aircraft were suicide planes during the Iwo Jima and Okinawa campaigns (through 30 April 1945). If one extrapolates this figure through the end date of Mitscher's action report, an uncertain but not unreasonable proposition, then Task Force 58 stopped more than 97 percent of the kamikaze attacks launched against it. CTF 58 to COMINCH, 18 June 1945, "Report of Operations of Task Force FIFTY-EIGHT in support of Landings at OKINAWA, 14 March through 28 May" (hereafter cited as "CTF 58, 'Action Report for Okinawa'"), box 216, RG38/AR; and COMINCH, "Anti-Suicide Action Summary," 31 August 1945, pp. 2–4. COMINCH calculated a lower success rate but still praised Task Force 58's performance against kamikaze attacks. This report is available in the Navy Department Library, Washington, D.C.

209. CTF 58, "Action Report for Okinawa," box 216, RG38/AR; and CTF 51, "Action Report for Okinawa," box 535, RG38/AR.

210. CTF 58, "Action Report for Okinawa," box 216, RG38/AR; and Commander Task Group (CTG) 58.4 (Arthur W. Radford) to COMINCH, 25 May 1945, "Action Report—14 March 1945 to 14 May 1945," box 231, RG38/AR.

211. CTF 51, "Action Report for Okinawa," box 535, RG38/AR.

212. John Campbell, *Naval Weapons of World War Two* (Annapolis: Naval Institute Press, 1985), 147–49, 217; Morison, *Victory in the Pacific*, 87–88, 280–81; and Potter, *Nimitz*, 368–69.

213. CTF 54 (Morton L. Deyo) to COMINCH, 5 May 1945, "Action Report—Bombardment and Occupation of OKINAWA," box 205, RG38/AR.

214. Tisdale to F. Gordon Barber, 29 November 1949, box 1, NWC/LP.

215. Monsarrat, *Angel on the Yardarm*, 39–40.

216. Commander Air Force, Pacific Fleet, "Fighter Direction Manual (Tentative)," 22 September 1943, box 1, NWC/LP.

217. Guerlac, *Radar in World War II*, 978.

218. Not every station was put into use, and the location of some shifted over time. Robin L. Rielly, *Kamikazes, Corsairs, and Picket Ships: Okinawa, 1945* (Philadelphia: Casemate, 2008), 5–7.

219. CTG 58.2 (William F. Bogan) to COMINCH, 10 May 1945, "Operations of Task Group FIFTY-EIGHT POINT TWO during the period from 1 April to 17 April, 1945," box 222, RG38/AR; and CTF 54 to COMINCH, 4 June 1945, "Action Report—Capture of OKINAWA GUNTO, Phase II, 5 May to 28 May 1945," box 205, RG38/AR.

220. CTF 58, "Action Report for Okinawa," box 216, RG38/AR; and Lieutenant Commander Carl Ballinger to USS *Langley* CIC, 4 May 1945, reproduced in Monsarrat, *Angel on the Yardarm*, 170.

221. Rielly, *Kamikazes, Corsairs, and Picket Ships*, 351–57.

222. In his after-action report, Turner not only criticized his planners but bemoaned as well the numerous logistical problems that "hindered the rapid establishment of early warning sites" on Okinawa itself. CTF 51, "Action Report for Okinawa," box 535, RG38/AR. Most likely, the planners failed to anticipate that the Japanese would attack picket ships with the same intensity as high-value units.

223. "CIC Yesterday and Today," *C.I.C.*, November 1945, 3.

224. Laning to Leslyn and Robert Heinlein, 15 January 1944. I am indebted to Ed Wysocki for sending me a copy of this letter. For more on the relationship between Laning and Heinlein, see William H. Patterson, *Robert A. Heinlein: In Dialogue with His Century* (New York: Macmillan, 2010), and Edward Wysocki, *The Great Heinlein Mystery: Science Fiction, Innovation, and Naval Technology* (U.S.: CreateSpace, 2012).

225. U.S. Pacific Fleet Tactical Bulletin No. 4TB–42, 26 November 1942, filing designator FF12-4, Confidential Correspondence, 1942, RG313/R179.

Conclusion

Epigraph. Ann M. Blair, *Too Much to Know: Managing Scholarly Information before the Modern Age* (New Haven, Conn.: Yale University Press, 2010), 1.

1. Alfred T. Mahan, *The Influence of Sea Power Upon History, 1660–1783* (Boston: Little, Brown, 1890).

2. See, e.g., Peter Karsten, *The Naval Aristocracy: The Golden Age of Annapolis and the Emergence of Modern American Navalism* (New York: Free Press, 1972); Robert L. O'Connell, *Sacred Vessels: The Cult of the Battleship and the Rise of the U.S. Navy* (New York: Oxford University Press, 1991); William M. McBride, *Technological Change and the United States Navy, 1865–1945* (Baltimore: Johns Hopkins University Press, 2000); and Craig C. Felker, *Testing American Sea Power: U.S. Navy Strategic Exercises, 1923–1940* (College Station: Texas A&M University Press, 2007).

3. According to eminent military historian Jon Sumida, Mahan's strategic arguments

should not be construed as mechanistic principles but rather as scaffolding erected to facilitate expertise in the art and science of command. Jon T. Sumida, *Inventing Grand Strategy and Teaching Command: The Classic Works of Alfred Thayer Mahan Reconsidered* (Washington and Baltimore: Woodrow Wilson Center Press and Johns Hopkins University Press, 1997).

4. Stephen K. Stein, *From Torpedoes to Aviation: Washington Irving Chambers and Technological Innovation in the New Navy, 1876–1913* (Tuscaloosa: University of Alabama Press, 2007).

5. Ann M. Blair, *Too Much to Know: Managing Scholarly Information before the Modern Age* (New Haven, Conn.: Yale University Press, 2010); and Tom Standage, *The Victorian Internet: The Remarkable Story of the Telegraph and the Nineteenth Century's On-Line Pioneers* (New York: Walker, 1998).

Essay on Sources

For the historian, archives can be both a pleasure and a curse. Before each of the many research visits that led to this book, I would eagerly anticipate the chance to mine for hidden gems. Sometimes I would strike pay dirt, finding just the letter, memo, or report I needed to answer one of the vexing questions running through my mind. One such find occurred in the summer of 2008 when I discovered Edward Very's remarkable proposals for improving the U.S. Navy's signaling systems. It made my whole trip. Other times, the archives could be incredibly frustrating. Do official reports from the navy's fleet maneuvers of 1916–17 still exist? Despite the helpful efforts of several archivists, none were found. As Charlie Brown might say, "Aaugh!"

As in most research monographs, archival and manuscript sources are at the heart of this book. A complete list of the archives, record groups, series, and subseries (when applicable), as well as the pertinent manuscript collections, follows this essay. The nature of the topic meant that no single record group or collection dominated my research, although I spent more time in the U.S. National Archives than anywhere else. Nevertheless, this story could not have been told without equally important research at the Imperial War Museum, the Library of Congress, the National Archives of the United Kingdom, Purdue University, the National Museum of American History, the Naval War College, the Navy History and Heritage Command, and the U.S. Naval Academy. Also critical were several printed primary sources, most notably the *Annual Report of the Secretary of the Navy*, the professional journal *United States Naval Institute Proceedings*, and an assortment of documents and reports in the U.S. Congressional Serial Set.

No one publishes in a vacuum, and the work of many fine scholars aided me in situating this book within the extensive literature in American naval history. With respect to shipboard command and control, the most thorough analyses are Michael Palmer's *Command at Sea* (2005) and Norman Friedman's *Network-Centric Warfare* (2009). Palmer's global synthesis reaches back to the sixteenth century and showcases the value of decentralized command and control in battle. Friedman's book, which also adopts an international perspective, provides the best available account of Cold War naval command and control. Complementing Friedman is David Boslaugh's *When Computers Went to Sea* (1999), which concentrates on the U.S. Navy's adoption of shipboard digital computers in the decades following World War II. Also still useful are Arthur Hezlett's *Electronics and Sea Power* (1975) and Linwood Howeth's *History of Communications-Electronics in the United States Navy* (1963).

Numerous books cover aspects of American naval technological development, but only two chronologically parallel this one. Robert O'Connell's *Sacred Vessels* (1991) focuses heavily on the supposed irrationality of American naval officers in their devotion to big guns and big ships. More nuanced is William McBride's *Technological Change and the United States Navy, 1865–1945* (2000), which despite the title focuses mainly on the years after 1890. Although I arrive at different conclusions than McBride, he deserves credit for being the first to examine the complex relationship between technology and the American naval profession over the entire span of the half century from 1890 to 1940.

Previous historians have examined many of the events, people, and technologies covered in this book, and I have benefited immensely from the scholarship of those who came before me. The finest account of the Battle of Mobile Bay is John Friend's *West Wind, Flood Tide* (2004), while the best work on the war scare with Spain in 1873–74 remains William Bradford's *The* Virginius *Affair* (1980). The career of John Grimes Walker is covered well in Daniel Wicks, "New Navy and New Empire: The Life and Times of John Grimes Walker" (Ph.D. diss., University of California at Berkeley, 1979), and one can better understand the naval world Walker inhabited by examining Walter Herrick's *The American Naval Revolution* (1966), Peter Karsten's *The Naval Aristocracy* (1972), William Still's *American Sea Power in the Old World* (1980), and Mark Russell Shulman's *Navalism and the Emergence of American Sea Power, 1882–1893* (1995). Several authors have written about the cruise of the Great White Fleet, but the gold standard remains James Reckner's *Teddy Roosevelt's Great White Fleet* (1988). Susan Douglas's essay "Technological Innovation and Organizational Change: The Navy's Adoption of Radio, 1899–1919," in Merritt Roe Smith's *Military Enterprise and Technological Change* (1985), is essential reading for anyone interested in the history of naval radio in the United States, while Rowland Pocock and Gerald Garratt's *The Origins of Maritime Radio* (1972) and A. J. L. Blond, "Technology and Tradition: Wireless Telegraphy and the Royal Navy, 1895–1920" (Ph.D. diss., University of Lancaster, 1993), cover matters on the other side of the Atlantic.

Extremely useful for understanding the broader contours of radio development in the early twentieth century are Jed Buchwald's *The Creation of Scientific Effects* (1994), Sungook Hong's *Wireless* (2001), and Hugh G. J. Aitken's *Syntony and Spark* (1976) and *The Continuous Wave* (1985). Hong's book in particular was vital in helping me understand the substance and context of scientific and engineering practice during wireless telegraphy's early years. Similarly indispensable was the work of the leading experts in the history of control systems and naval fire control, David Mindell and Jon Sumida, respectively. See especially Mindell's "Opening Black's Box: Rethinking Feedback's Myth of Origin" (*Technology and Culture*, July 2000) and *Between Human and Machine* (2005), and Sumida's *In Defence of Naval Supremacy* (1993) and "The Quest for Reach: The Development of Long-Range Gunnery in the Royal Navy, 1901–1912," in Stephen Chiabotti's *Military Transformation in the Industrial Age* (1996).

There is an abundance of work on the Battle of Jutland, but the three most penetrating analyses are Andrew Gordon's *The Rules of the Game* (1996), Nicholas Lambert, "'Our Bloody Ships' or Our Bloody System?" (*Journal of Military History*, January 1998), and Jon Sumida, "A Matter of Timing" (*Journal of Military History*, January 2003). The best place to start any exploration of the American navy during the first two decades of the twentieth century is William Braisted's two-part *The United States Navy in the Pacific* (1958, 1971), sound scholarship that has certainly stood the test of time. Until recently there existed no comprehensive account of the U.S. Navy's efforts during the First World War, but that changed with publication of William Still's voluminous and comprehensive *Crisis at Sea* (2006). Other useful books covering various aspects of America's wartime naval effort include William S. Sims's *Victory at Sea* (1920), Robert S. Griffin's *History of the Bureau of Engineering, Navy Department, during the World War* (1922), David Trask's *Captains and Cabinets* (1972), Michael Simpson's *Anglo-American Naval Relations, 1917–1919* (1991), Glen Stackhouse, "The Anglo-American Atlantic Convoy System in World War I" (Ph.D. diss., University of South Carolina, 1993), Jerry Jones's *U.S. Battleship Operations in World War I* (1998), Michael Besch's *A Navy Second to None* (2003), Jonathan Winkler's *Nexus* (2008), Joseph Kirschbaum, "The 1916 Naval Expansion Act: Planning for a Navy Second to None" (Ph.D. diss., George Washington University, 2008), and Geoffrey Rossano's *Stalking the U-Boat* (2010).

The extensive literature on the naval arms limitation treaties of the 1920s and 1930s must frame any exploration of interwar naval technology. Especially informative are Raymond O'Connor's *Perilous Equilibrium* (1962), Thomas Buckley's *The United States and the Washington Conference, 1921–1922* (1970), Roger Dingman's *Power in the Pacific* (1976), Emily Goldman's *Sunken Treaties* (1994), Richard Fanning's *Peace and Disarmament* (1995), and John Jordan's *Warships after Washington* (2011). Specific investigations of the treaties' influences on American naval technological development include William McBride, "The Unstable Dynamics of a Strategic Technology" (*Technology and Culture*, April 1997), and John Kuehn's *Agents of Innovation* (2008).

When I began this project there existed no thorough analyses of the interwar fleet problems, but fortunately that changed with the recent publication of Craig Felker's *Testing American Sea Power* (2007) and Albert Nofi's *To Train the Fleet for War* (2010). Both monographs served as welcome crosschecks in my efforts to comprehend the thought processes of interwar American naval officers. An earlier analysis that concentrates on carrier employment during the fleet problems is Mark Campbell, "The Influence of Air Power upon the Evolution of Battle Doctrine in the U.S. Navy, 1922–1941" (M.A. thesis, University of Massachusetts at Boston, 1992), but the two scholars who have done the most to further historians' understandings of the service's interwar tactical doctrine are the father and son team of Thomas and Trent Hone. See especially their *Battle Line* (2006) and the individually authored articles cited in chapters 3 and 4. Thomas Hone also co-wrote, with Norman Friedman and Mark Mandeles, the very informative *American and British Aircraft Carrier Development, 1919–1941* (1999). An equally enlightening book that explores the opposite side of three-dimensional naval warfare is Gary Weir's *Building American Submarines, 1914–1940* (1991).

The first scholar to challenge the commonly accepted view that interwar American naval officers were a technologically conservative lot beholden to the battleship was

Charles Melhorn in *Two-Block Fox* (1974). Melhorn ended his book with a brief recounting of Fleet Problem IX, leaving the details to future historians. One historian who fills in many of those details is Thomas Wildenberg, whose biography of Joseph M. Reeves, *All the Factors of Victory* (2003), paints a telling portrait of the officer who did as much as anyone to turn the American aircraft carrier into an effective instrument of war. Other useful biographies of interwar naval officers include Gerald Wheeler's *Admiral William Veazie Pratt, U.S. Navy* (1974), Clark Reynolds's *Admiral John H. Towers* (1991), and William Trimble's *Admiral William A. Moffett* (1994). World War II flag officers whose biographers shed considerable light on interwar developments include George Dyer's biography of Richmond Kelly Turner, *The Amphibians Came to Conquer* (1972), E. B. Potter's *Nimitz* (1976), and Thomas Buell's *The Quiet Warrior* (1974) and *Master of Sea Power* (1980), biographies of Raymond A. Spruance and Ernest J. King, respectively.

There is some excellent scholarship on the origins and early development of naval radar in the United States, most notably David Allison's *New Eye for the Navy* (1981) and Henry Guerlac's *Radar in World War II* (1987). Also very helpful are several participant histories: Albert Hoyt Taylor's *Radio Reminiscences* (1948), Robert M. Page's *The Origin of Radar* (1962), Irving L. McNally's *Radar Reflections* (1975), and Louis A. Gebhard's *Evolution of Naval Radio-Electronics and Contributions of the Naval Research Laboratory* (1979), which covers not only the history of American naval radar but also radio communications, radio identification, radio direction finding, and electronic countermeasures. Louis Brown's *A Radar History of World War II* (1999) provides a welcome international perspective on radar developments.

The only detailed study of the U.S. Navy's CIC program is an administrative history prepared by the University of Pittsburgh Programs Research Staff in the early 1950s. Also informative is a postwar report written by two U.S. Navy lieutenants, William C. Bryant and Heith I. Hermans, on the history of naval fighter direction. A wonderful complement to both reports is John Monsarrat's *Angel on the Yardarm* (1985), the only published memoir by a World War II fighter director officer. Additional information on all three sources is available in the notes.

The secondary literature on the navy in World War II is too vast to do justice to here, so I will limit myself to the works I drew most heavily upon in conceptualizing and writing chapter 5. Vital for understanding the Battle of the Atlantic are Montgomery Meigs's *Slide Rules and Submarines* (1990), David Syrett's *The Defeat of the German U-boats* (1994), and Kathleen Broome Williams's *Secret Weapon* (1996). With respect to the Pacific War no one is more knowledgeable about naval aviation than John Lundstrom, whose *The First Team* (1984), *The First Team and the Guadalcanal Campaign* (1994), *Fateful Rendezvous* (1997) (coauthored with Steve Ewing), and *Black Shoe Carrier Admiral* (2006) are essential reading. Also very useful are Theodore Roscoe's classics, *United States Submarine Operations in World War II* (1949) and *United States Destroyer Operations in World War II* (1953).

Along with Roscoe, the other usual starting point for anyone interested in American naval operations during the Second World War is Samuel Eliot Morison's fifteen-volume *History of United States Naval Operations in World War II* (1947–62). Although much new archival material has become available since these volumes first appeared,

Morison's service during the war as a combat historian for the navy offers a perspective that can never be replicated. Also very instructive is Ronald Spector's *Eagle against the Sun* (1985). For information about specific battles and campaigns, I relied mainly upon David Zimmerman's *Britain's Shield* (2001); Gordon Prange's Pearl Harbor trilogy *At Dawn We Slept* (1981), *Pearl Harbor* (1986), and *December 7, 1941* (1988); Alan Zimm's *The Attack on Pearl Harbor* (2011); Jonathan Parshall and Anthony Tully's *Shattered Sword* (2005); Dallas Woodbury Isom's *Midway Inquest* (2007); Craig Symond's *The Battle of Midway* (2011); H. P. Willmott's *The Barrier and the Javelin* (1983) and *The Battle of Leyte Gulf* (2005); Richard Frank's *Guadalcanal* (1990); James Hornfischer's *Neptune's Inferno* (2011); Denis and Peggy Warner's *Disaster in the Pacific* (1992); Charles Cook's *The Battle of Cape Esperance* (1968); James Grace's *The Naval Battle of Guadalcanal* (1999); Russell Crenshaw's *The Battle of Tassafaronga* (1995) and *South Pacific Destroyer* (1998); Jürgen Rohwer's *The Critical Convoy Battles of March 1943* (1977); William Y'Blood's *Red Sun Setting* (1981); Barrett Tillman's *Clash of the Carriers* (2005); and Robin Rielly's *Kamikazes, Corsairs, and Picket Ships* (2008).

I would be remiss if I did not mention as well two other groups of secondary sources that aided me greatly in writing this book. The first group consists of what one might label "reference works," although most of these provide more than mere lists of data. Of terrific value are Norman Friedman's Illustrated Design History series on U.S. destroyers (1982), aircraft carriers (1983), cruisers (1984), battleships (1985), and submarines (1995), John Campbell's *Naval Weapons of World War Two* (1985), K. Jack Bauer and Stephen Roberts's *Register of Ships of the U.S. Navy, 1775–1990* (1991), and Roy Grossnick's *United States Naval Aviation, 1910–1995* (1997). Indispensable for understanding the navy's officer personnel system is Donald Chisholm's magnum opus *Waiting for Dead Men's Shoes* (2001). Helpful for establishing general context are three syntheses in particular: Kenneth Hagan's *This Peoples' Navy* (1991), Robert Love's two-volume *History of the U.S. Navy* (1992), and George Baer's *One Hundred Years of Sea Power* (1994).

The second group consists of works that often have little to do with navies but which were instrumental in framing my thoughts and in helping me grapple with larger epistemological issues. These include (but are certainly not limited to) Elting Morison's *Men, Machines, and Modern Times* (1966), John Keegan's *The Face of Battle* (1976), Merritt Roe Smith's *Harpers Ferry Armory and the New Technology* (1977), John Staudenmaier's *Technology's Storytellers* (1985), *The Social Construction of Technological Systems* (1987), edited by Wiebe Bijker et al., Susan Douglas's *Inventing American Broadcasting* (1987), Donald MacKenzie's *Inventing Accuracy* (1990), Daniel Headrick's *The Invisible Weapon* (1991), David Kirsch and Paul Maglio's "On Distinguishing Epistemic from Pragmatic Action" (*Cognitive Science*, October–December 1994), Edwin Hutchins's *Cognition in the Wild* (1995), Jon Sumida's *Inventing Grand Strategy and Teaching Command* (1997), Tom Standage's *The Victorian Internet* (1998), Nicholas Lambert's *Sir John Fisher's Naval Revolution* (1999), Guy Claxton's *Hare Brain, Tortoise Mind* (1999), David Mindell's *War, Technology, and Experience Aboard the USS* Monitor (2000), David Kirsch's *The Electric Vehicle and the Burden of History* (2000), David Edgerton's *The Shock of the Old* (2005), Malcolm Gladwell's *Blink* (2005), Kevin Borg's *Auto Mechanics* (2007), Ann Blair's *Too Much to Know* (2010), and Daniel Kahneman's *Thinking, Fast and Slow* (2011).

Archives and Manuscript Collections

U.S. National Archives and Records Administration, Washington, D.C., and College Park, Md.

Record Group 19, Records of the Bureau of Ships

General Correspondence, 1899–1910, Records of the Bureau of Equipment

Naval Research Laboratory General Files, 1923–1940, Confidential

Naval Research Laboratory General Files, 1923–1940, Secret

Orders and Memoranda, 1906–1930, Records of the Bureau of Equipment

Records Relating to Wireless Stations and Tests of Wireless Equipment, 1904–1910, Records of the Bureau of Equipment

Semiannual Reports of Naval Radio Stations and Ships, 1910–1917, Records of the Radio Division

Record Group 24, Records of the Bureau of Naval Personnel

Fair Copies of Letters Sent, Records of the Signal Office, 1869–1886

Letters Received from the Chief Signal Officer, Letters Received, 1862–1889

Letters Received Relating to Signaling at Sea, Letters Received, 1862–1889

Letters Sent to the Signal Office, Letters Sent to the President, Congressmen, and Executive Departments, 1877–1911

Logs of U.S. Naval Ships, 1801–1915, Logs of Ships and Stations, 1801–1946

Signal-Records Books from Vessels, September 1897–November 1898, Communications "Logs" and Other Records, 1897–1922

Record Group 38, Records of the Office of the Chief of Naval Operations

Confidential Correspondence, 1942–1945, Headquarters of the Commander-in-Chief, U.S. Fleet

Records of the Naval Security Group Repository, Crane, Ind.

Records of the Strategic Plans/War Plans Division (W.P.L. Series)

World War II Action and Operations Reports, Records Relating to Naval Activity during World War II

Record Group 45, Records of the Office of Naval Records and Library

Daily Reports of Movements of Vessels, 1 September 1897 to 31 December 1915, Records of the Office of the Chief of Naval Operations, 1887–1945

Minutes of the Permanent Scientific Commission, Records of Boards and Commissions, 1812–1890

Subject File, 1775–1910

Subject File, 1911–1927

Record Group 80, General Records of the Navy Department

Formerly Confidential Correspondence, 1927–1939

Formerly Secret Correspondence, 1927–1939

General Correspondence, 1897–1915

Subject File, 1900–1947, Records of the General Board

Record Group 125, Records of the Bureau of the Office of the Judge Advocate General (Navy)

Proceedings of Naval and Marine Examining Boards, c. 1890–1941

Record Group 313, Records of Naval Operating Forces

Confidential Correspondence, 1942–1945, General Administrative Files, Commander, South Pacific (Red 179)

Letters Sent to Commanding Officers of Vessels, Squadron of Evolution, 1889–1892

Letters Sent to Commanding Officers of Vessels, Squadrons and Fleets, 1865–1940

Letters Sent to the Navy Department, Squadron of Evolution, 1889–1892

National Archives and Records Administration Microfilm Collections

Microfilm M89, Letters Received by the Secretary of the Navy from Commanding Officers of Squadrons ("Squadron Letters"), 1841–1886

Microfilm M480, Letters Sent by the Secretary of the Navy to Chiefs of Navy Bureaus, 1842–1886

Microfilm M964, Records Relating to United States Navy Fleet Problems I to XXII, 1923–1941

Microfilm M971, Annual Reports of Fleets and Task Forces of the U.S. Navy, 1920–1941

Microfilm M984, Navy Department General Orders, 1863–1948

Microfilm M1140, Secret and Confidential Correspondence of the Office of the Chief of Naval Operations and the Office of the Secretary of the Navy, 1919–1927

Archives and Special Collections, Purdue University, West Lafayette, Ind.

Benjamin F. Miessner Papers

Archives Center, National Museum of American History, Smithsonian Institution, Washington, D.C.

History of Naval Radio (Series 100), George H. Clark Radioana Collection

Department of Documents, Imperial War Museum Collections, London, U.K.

Robert Church Collection

Naval Historical Collection, Naval War College, Newport, R.I.

Caleb B. Laning Papers (Ms. Coll. 116)

Joseph C. Wylie Oral Histories (O.H. 106 and O.H. 185)

Naval Historical Foundation Collection, Manuscripts Division, Library of Congress, Washington, D.C.

Stanford C. Hooper Papers

Public Record Office, National Archives of the United Kingdom, Kew, U.K.

ADM 116/523

ADM 116/567

AIR 2/2625

Special Collections, Navy Department Library, Naval History and Heritage Command, Washington, D.C.

Biographical (ZB) Files

Mahlon S. Tisdale Manuscript Collection

World War II Histories and Historical Reports

Special Collections and Archives, Nimitz Library, United States Naval Academy, Annapolis, Md.

William A. Moffett Papers (Ms. Coll. 198)

Vertical File

U.S. Navy Operational Archives, Naval History and Heritage Command, Washington, D.C.

Officer Biographies

World War II Collection, Naval Aviation History Branch

World War II Command File

Index

Numbers in *italics* indicate images.